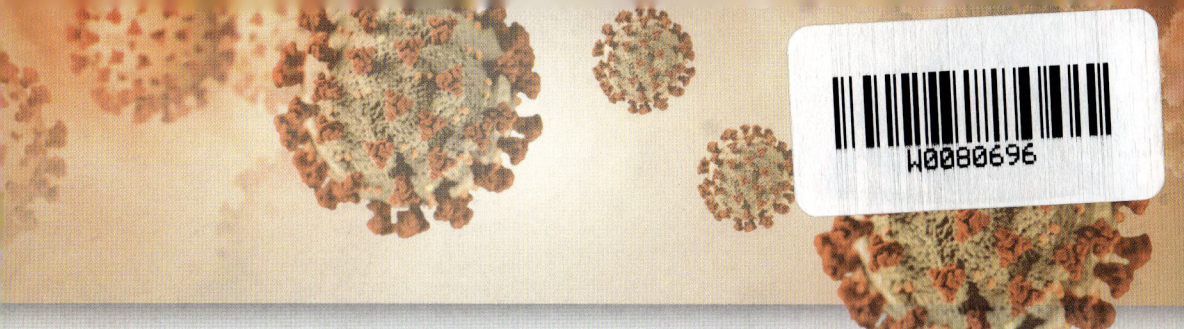

W0080696

Textbook of
Medical Virology

Second Edition

Textbook of
Medical Virology

Second Edition

Baijayantimala Mishra
MD (AIIMS, New Delhi), FIDSA, FRCP (London), FRCP (Glasgow)

Professor and Head
Department of Microbiology
All India Institute of Medical Sciences
Bhubaneswar, Odisha

Former Professor
Post Graduate Institute of Medical Education and Research
Chandigarh

CBSPD

CBS Publishers & Distributors Pvt Ltd

New Delhi • Bengaluru • Chennai • Kochi • Kolkata • Lucknow • Mumbai
Hyderabad • Jharkhand • Nagpur • Patna • Pune • Uttarakhand

Disclaimer
Science and technology are constantly changing fields. New research and experience broaden the scope of information and knowledge. The author has tried her best in giving information available to her while preparing the material for this book. Although all efforts have been made to ensure optimum accuracy of the material, yet it is quite possible some errors might have been left uncorrected. The publisher, the printer and the author will not be held responsible for any inadvertent errors or inaccuracies.

Textbook of
Medical Virology
Second Edition

ISBN: 978-93-90709-98-4

Copyright © Author and Publisher

Second Edition: 2022
Reprint: 2025
First Edition: 2018

All rights reserved. No part of this book may be reproduced or transmitted in any form or by any means, electronic or mechanical, including photocopying, recording, or any information storage and retrieval system without permission, in writing, from the author and the publisher.

Published by Satish Kumar Jain and Produced by Varun Jain for

CBS Publishers & Distributors Pvt Ltd
4819/XI Prahlad Street, 24 Ansari Road, Daryaganj, New Delhi 110 002, India
Ph: 011-23289259, 23266838   Website: www.cbspd.com
                                  e-mail: delhi@cbspd.com

Corporate Office: 204 FIE, Industrial Area, Patparganj, Delhi 110 092
Ph: 011-4934 4934   Fax: 011-4934 4935   e-mail: publishing@cbspd.com; publicity@cbspd.com

Branches

- **Bengaluru:** Seema House 2975, 17th Cross, K.R. Road, Banasankari 2nd Stage, Bengaluru 560 070, Karnataka, India
  Ph: +91-80-26771678/79   Fax: +91-80-26771680   e-mail: bangalore@cbspd.com
- **Chennai:** 7, Subbaraya Street, Shenoy Nagar, Chennai 600 030, Tamil Nadu, India
  Ph: +91-44-26680620, 26681266   Fax: +91-44-42032115   e-mail: chennai@cbspd.com
- **Kochi:** 42/1325, 1326, Power House Road, Opp KSEB, Power House, Ernakulam 682 018, Kerala, India
  Ph: +91-484-4059061-65   Fax: +91-484-4059065   e-mail: kochi@cbspd.com
- **Kolkata:** 147, Hind Ceramics Compound, 1st Floor, Nilgunj Road, Belghoria, Kolkata-700056, West Bengal, India
  Ph: 033-25633055, 033-25633056   e-mail: kolkata@cbspd.com
- **Lucknow:** Basement, Khushnuma Complex, 7-Meerabai Marg (Behind Jawahar Bhawan), Lucknow 226001, India
  Ph: 0522-4000032   e-mail: tiwari.lucknow@cbspd.com
- **Mumbai:** PWD Shed. Gala no. 25/26, Ramchandra Bhatt Marg, Next to JJ Hospital Gate no. 2, Opp. Union Bank of India, Noorbaug, Mumbai-400009, Maharashtra, India
  Ph: 022-66661880/89   e-mail: mumbai@cbspd.com

Representatives

- **Hyderabad** 0-9885175004 • **Jharkhand** 0-9811541605 • **Nagpur** 0-8692091830
- **Patna** 0-9334159340 • **Pune** 0-9664372571 • **Uttarakhand** 0-9716462459

Printed at HT Media Ltd, Greater Noida, UP, India

to

---

my parents
Lt Dr Manisha Mishra and
Lt Shri Sarada Prasad Mishra

Preface to the Second Edition

It gives me immense pleasure to announce the much awaited release of the second edition of *Textbook of Medical Virology*. This edition was scheduled to be released in 2020, but the emergence of COVID-19 put the work on hold.

The emergence of SARS coronavirus 2 seized the entire world in its grip. In the history of mankind, for the first time, the entire world locked itself behind doors. Life became standstill. Hospitals got closed for non-COVID cases, students were sent back to their homes, medical teaching moved from classrooms and clinics to online. The fear of COVID-19 throbbed in every individual's heart. The importance of viruses, viral infections is now realized by every human being, every section of society, every country, and has spread beyond the medical fraternity.

Additional features of second edition

- Seven more chapters have been added to give a holistic picture on all medically important viruses.
- Updated information on epidemiology, pathogenesis, diagnostic guidelines and vaccines in existing chapters.
- Special importance to Indian epidemiology.
- A detailed chapter on COVID-19 (SARS-CoV-2).

The texts have been described in lucid and comprehensive manner with up-to-date information from articles published in reputed journals and highly regarded reference books. It has colored illustrations, photomicrograph, and images to make the chapters interesting. The tables, line diagrams, flowcharts and diagnostic algorithms are given to simplify the explanation, and easy to revise for the students.

A sincere attempt has been made to present the viral infections in a lucid manner. The approach is to build the concept, to provide the insight into the clinical approach and interpretation of viral diagnosis. Like the viruses, virology is often considered as a difficult and dreaded subject by most of the students. This book is a sincere approach to make the virology learning simpler.

I believe this book would fulfill the need of both MBBS and postgraduate students and teachers.

Baijayantimala Mishra

Preface to the First Edition

Since the beginning of the 21st century, world has witnessed the threat of numerous viral infections. The emergence of several novel viruses including some of the deadly viruses such as SARS, bird flu, influenza A (H1N1), and MERS has made the world realized the importance of viral infections. The threats continued with Ebola which devastated Western Africa, crossed the continent for the first time and put the whole world on alert. Outbreaks of dengue, chikungunya and Japanese encephalitis have become serious concern every year worldwide. The problem of arbovirus has now been compounded with emergence of Zika virus and has posed a serious diagnostic challenge for the microbiologists. The increasing number of immunocompromised patients has created a pool of susceptible population for various viral infections posing a challenge for early diagnosis and management. Moreover, the availability of antivirals and better diagnostic techniques for early detection has made the clinical virology more important than ever before. The increased clinical importance of viral infections in past few years has increased its demand tremendously in the medical teaching.

The present book contains all medically important viruses. The approach of the book has been made from a clinical perspective. Each chapter covers all relevant aspects of the viral infection with special emphasis to recent classification, molecular epidemiology, transmission cycle, pathogenesis, diagnostic approach with clinical interpretation, antivirals and vaccine. The texts have been described in clear and comprehensive manner with up-to-date information from articles published in reputed journals and highly regarded reference books. It has self-explanatory color figures and graphs, photomicrograph, images, tables, flowcharts and diagnostic algorithms. Some of the chapters (Chapters 8–10) have been presented according to their clinical manifestations and diagnostic approach.

A sincere attempt has been made to fulfill the wholesome need for concept building, clinical approach and interpretation of viral diagnosis. I believe this book should fulfill the need of both MBBS and postgraduate students.

Baijayantimala Mishra

Acknowledgments

The first edition was prepared with lots of apprehension and speculations. After the release of the first edition, the book was liked by students and received appreciation from my teachers who are renowned virologists of the country. This encouraged me to continue my work, which has culminated in this second edition.

Though the second edition was scheduled to be released in 2020, the emergence of COVID-19 delayed the process and put the things in uncertainty. With God's grace and the blessings of my teachers, I believe it is finally time for the second edition.

My heartfelt thanks to my friend Dr Niveita Karmee for her valuable help in correcting English and grammar in the additional chapters of this edition.

My sincere thanks to Mr YN Arjuna Senior Vice President—Publishing, Editorial and Publicity, CBS Publishers & Distributors Pvt. Ltd., and his publishing team for all their cooperation.

I extend my thanks to all those who had helped me for the first edition of this book.

Baijayantimala Mishra

bmishramicro@gmail.com
bm_mishra@hotmail.com

Contents

Introduction to Medical Virology

Even before the realization of existence of virus, diseases caused by viruses have changed the destiny of the mankind; and after the discovery of virus, the field of virology has given a new shape to science.

The diseases caused by the viruses have been recorded in the history which dates back to more than 3000 BC. But the existence of virus was realized only in late nineteenth century when a series of experiments were carried out in several countries to study the disease affecting the tobacco plants. This happened during the era, when Koch's postulate to associate the microbe as the cause of disease was almost accepted.

DISCOVERY OF VIRUS

Adolf Mayer, an agricultural scientist, was instructed by the Netherlands to investigate the tobacco disease. Mayer named the disease as tobacco mosaic based on the mosaic pattern of appearance on the affected tobacco leaves. He took the sap of the diseased leaves and inoculated onto the healthy tobacco plants and demonstrated the development of disease in the healthy plants. His experiment reveals the infectious nature of the disease agent. But as he could not cultivate or detect the agent under microscope, the causative agent remained unidentified. This also indicated, probably the infectious agent is not cultivable in nature and also small enough to be seen under microscope (submicroscopic particle).

The experiment was continued by a Russian scientist Dmitri Ivanovsky when he was ordered by the Russian Department of Agriculture to find out the cause of the tobacco disease. Ivanovsky reconfirmed Mayer's finding of transmissibility of the causative agent and also showed that the agent is filterable by successfully transmitting the infection from infected sap after passing through Chamberland filter to healthy plant. Ivanovsky's experiment added to the information that the causative agent is small enough and not only submicroscopic but also filterable. As he could not culture this filterable agent, he suggested it as a possible toxin and not a living particle.

However, both the scientists failed to satisfy Koch's postulate as they could not cultivate the infective agent.

The experiment was continued by a Dutch scientist Martinus Beijerinck who collaborated with Adolf Mayer and independently showed that the agent is infectious and filterable. He also showed that the infective fluid is not a toxin but a reproducible particle by showing the diluted sap is regaining the strength in the inoculated healthy plant. This indicated the reproducible capacity of the agent which proved that the agent is a living particle and not a toxin. It was also noticed that the agent was able to reproduce in the living tissue and not in the cell free sap. The later revelation also changed the concept of the then Koch's postulate that, there are some infective agents exists which do not grow outside their host. He named this agent as **contagium vivum fluidum** or contagious living fluid.

The experiments by these three scientists led to the discovery of a novel organism which was smaller than bacteria as it could not be filtered by Chamberland candle filter, could not be visualized by the light microscope and could not be multiplied in artificial media as it could grow only in the living tissue. The agent was then named as **tobacco mosaic virus** as the term **virus (Latin meaning for slimy fluid or poison)** was used for any infective agent.

By end of the experiments that led the discovery of tobacco mosaic virus, the physical characteristics that were discovered for it became part of the definition for "virus"; an infectious filterable agent which is sub-microscopic and grow only within the living tissue. With the further development of technologies, size of the virus was determined to be between 30 and 200 nm and structure, morphology and genomic details were gradually revealed.

MORPHOLOGY OF VIRUSES

The structure of virus is unique and different from other groups of microorganisms like protozoa, bacteria and fungi (Fig. 1.1).

The mature virus particle called "the virion" is comprised of nucleic acid and a protein coat called "capsid" which surrounds the nucleic acid. Capsid is composed of capsomers which are the morphological unit of the capsid. Together both nucleic acid and capsid constitute the nucleocapsid. In addition to this, some of the viruses possess an envelop that surrounds the capsid. Envelop of some viruses possesses surface projections (also called peplomers) which are of glycoprotein in nature. So, from inside outwards, the structure of virus consists of nucleic acid, capsid, and envelop.

SYMMETRY OF VIRUS

The arrangements of the capsomers in human viruses are of two types; isometric with icosahedral symmetry or helical symmetry.

Icosahedral Symmetry

Icosahedral symmetry is formed when the capsomers are arranged in an icosahedrons shape. Icosahedron is a geometrical figure consists of 20 equilateral triangles. So it has **20 faces** (surface of equilateral triangle), **30 edges** (margins of 20 triangles) and **12 vertices** (corners of triangles). It has 5, 3, and 2 rotational symmetry which are formed when the imaginary axis passes through the vertices, faces and edges, respectively (Fig. 1.2a and b).

The icosahedrons shape permits a particular pattern of arrangement of capsomers so that a proper fitting can occur. Along the edges and faces, the capsomers are surrounded by six capsomers and are called "hexamers", whereas the capsomer that is present at the vertices (corner of the triangle) are

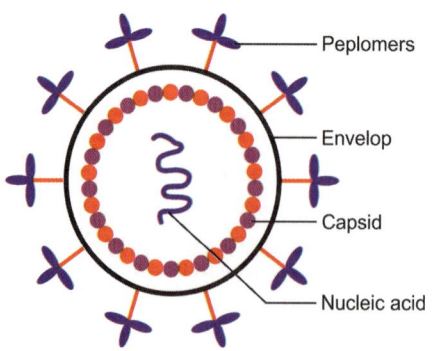

Fig. 1.1: Structure of virus (schematic)

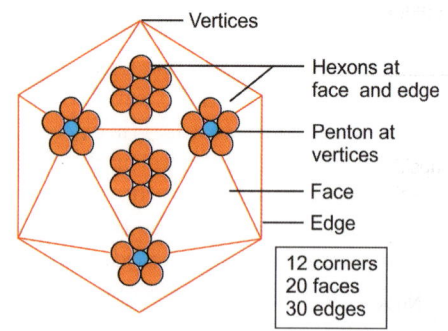

Fig. 1.2a: Icosahedral symmetry showing the location of pentons and hexons

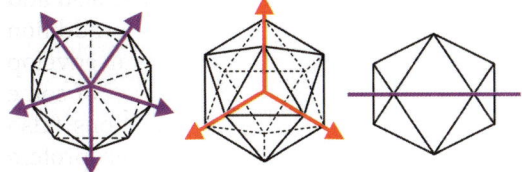

Fig. 1.2b: Purple arrows showing 5-fold symmetry through vertices **red lines** showing 3-fold symmetry through faces

surrounded by five capsomers and are called "pentamers".

The icosahedrons basically a compact shape and thus provides a robust protection to the nucleic acid. Therefore, the presence of envelop is not essential for those viruses. So, viruses with icosahedral symmetry may or may not possess envelops.

Helical Symmetry

Helical symmetry is seen in many of the RNA viruses, but not found in any of the DNA viruses.

The name "helical" is self-explanatory, where the nucleic acid is present in helical manner and surrounded by the protein coat capsid. We can imagine a metal spring or coil with a plastic covering, where the nucleic acid can be compared with the metal spring (present in a coiled fashion) and like the plastic coat, it is covered by the capsid all along it (Fig. 1.3). The entire arrangement, however, appears quite open, so possibly requiring another layer of protection. Therefore, all human viruses with helical symmetry always possess envelop.

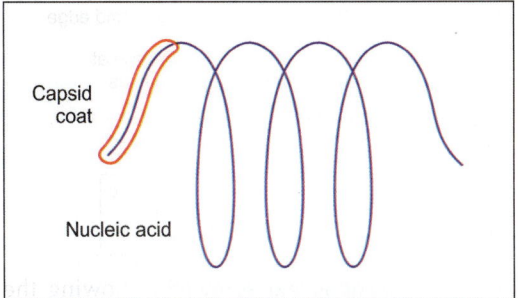

Capsid coat

Nucleic acid

Fig. 1.3: Helical symmetry

Envelop

Viral envelop is the outermost covering of the virus and is of lipoprotein in nature. The lipid component is derived from the host cell membrane, either from the plasma membrane or from the membrane of the cytoplasmic organelle during the process of release of nucleocapsid by budding. The glycoprotein surface peplomers when present on the envelop get incorporated to it by replacing the host cell protein.

As discussed under "helical symmetry", all human viruses with helical symmetry are enveloped and only certain viruses with icosahedral symmetry possess envelops (herpesviruses, togaviruses, retroviruses and flaviviruses).

CHEMICAL COMPOSITION OF VIRUSES

Nucleic Acid

Viruses possess only one type of nucleic acid, i.e. either deoxyribonucleic acid (DNA) or ribonucleic acid (RNA) and never both. This also highlights another unique property of the viruses where genetic information is possessed by RNA (RNA viruses) instead of DNA. Based on the type of nucleic acid the virus contains, they are divided broadly into two types—DNA and RNA viruses. Though DNA is double stranded and RNA is single stranded, DNA and RNA viruses are either double stranded or single stranded. The nucleic acid is present as a single molecule in all the viruses except retroviruses which contain two copies of the nucleic acid.

DNA Viruses

All the DNA viruses are double stranded except parvoviruses which are single stranded DNA viruses and hepadnavirus which has a partially double-stranded genome.

DNA molecule is present either in linear form or in circularized form. The latter form helps the virus in integration of the viral genome with the cellular DNA, also protects the genome against exonucleases. Papilloma-viruses, polyomaviruses and hepadnaviruses possess circular DNA.

The genome of DNA viruses contains around 400 to 4000 nucleotides in the smallest (parvoviruses) and largest group of viruses (poxviruses) (Table 1.1).

RNA Viruses

Based on the different properties, RNA viruses can be divided according to the number of strand they possess, number of molecule (single or multiple as segmented), arrangement of nucleic acid (liner or circular) and polarity (positive, negative or ambi-sense). Table 1.2 gives the list according to each type.

Proteins

The major bulk of the virus is made up of protein. The capsid which is present as a protective coat surrounding the nucleic acid is composed of protein. Thus the major function of protein is to provide the protection. Each capsomer, which is the building blocks of capsid, consists of one to six polypeptide molecules.

The other component of virus that contains protein is the surface projections present on envelop of some of the viruses. Besides the structural component, protein is also present in the form of enzymes which are important for viral replication.

Lipids

The envelop of the virus is of lipoprotein in nature, where the lipid component is derived from the host cell and the protein is of viral origin. The lipid part constitutes of around 30% of the total dry weight of the virus. The major part of the envelop lipid (50–60%) is of phospholipid in nature present in the form of bilayer.

Glycoproteins

The surface projections or peplomers present on some of the enveloped viruses (influenza virus, human immunodeficiency virus) are of glycoprotein in nature which are synthesized by the cellular glycosyltransferase.

CLASSIFICATION OF VIRUSES

Viruses have been classified by several means—based on their epidemiological criteria, organ—system affected, or based on their physicochemi-cal properties. Classification system based on the former criteria though not used as a formal classification system, but useful for clinical practice.

Another type of classification is based on the replication strategy of the viruses called "Baltimore classification". This is based on the observation of Sir David Baltimore that all the viruses must have to generate a positive strand mRNA in order to start their replica-tion. However, this classification is not used for general purpose of virus classification and is discussed under viral replication.

Table 1.1: List of DNA viruses according to different morphological properties	
Property	*Viruses*
Number of NA strand	All DNA viruses are double stranded *except* Parvoviridae (single-stranded DNA) and Hepadnaviridae (partially double-stranded DNA)
Arrangement of NA • Linear • Circular	Adenoviridae, Herpesviridae, Poxviridae, Parvoviridae Papillomaviridae, Polyomaviridae and Hepadnaviridae
Nucleocapsid symmetry	All DNA viruses have icosahedral symmetry *except* Poxviridae which has a complex/ovoid symmetry

NA: Nucleic acid

Table 1.2: List of RNA viruses according to different morphological properties

Property	Viruses
Number of NA strand	All RNA viruses are single stranded except Reoviridae
Number of NA molecule	All RNA viruses contain single molecule of genome *except* retroviruses which contain two molecules of single strand RNA and RNA viruses with segmented genome: Arenaviridae, Reoviridae, Bunyaviridae and Orthomyxoviridae
Arrangement of NA	All RNA viruses are linear *except* arenaviruses and bunyaviruses
Polarity • Positive	The RNA genome that acts as mRNA: Picornaviridae, Caliciviridae, Togaviridae, Coronaviridae, Flaviviridae and Retroviridae
• Negative	When the nucleotide sequence of the genome is complementary to that of mRNA: Orthomyxoviruses, paramyxoviruses, rhabdoviruses, arenaviruses and bunyaviruses
• Ambisense	Arenaviruses and bunyaviruses
Nucleocapsid symmetry • Helical	Orthomyxoviruses, paramyxoviruses, Coronaviridae, Filoviridae, Rhabdoviridae
• Icosahedral	Astroviridae, Caliciviridae, Reoviridae, Togaviridae
Enveloped	All RNA viruses with helical symmetry are enveloped and Retroviridae, Togaviridae and Flaviviridae are enveloped with icosahedral symmetry

NA: Nucleic acid

The International Committee on Taxonomy of Viruses (ICTV) is the virology division of International Union of Microbiological Societies. ICTV has made the formal classification of viruses based on the group of properties of the viruses which constitutes mainly of the type of nucleic acid, capsid symmetry, and possession of envelop, genomic architecture and nucleotide sequence similarities. However, there is no fixed list of properties to describe all virus groups.

The characters that are considered to distinguish between different families and genera of viruses are; morphology of virion, genome organization, size of structural and non-structural viral proteins and the method of viral replication.

The characters that are considered to distinguish between different species within the single genus are—physicochemical properties, antigenic properties of the viral proteins, natural

Table 1.3: Nomenclature of viruses according to hierarchy

Taxon	Suffix	Example
Order	virales	Herpesvirales
Family	viridae	Herpesviridae
Subfamily	virinae	Alphaherpesvirinae
Genus	virus	Simplexvirus

host range, cell and tissue tropism, mode of transmission, pathogenicity, cytopathology and nucleotide sequence relatedness.

Table 1.3 gives the nomenclature of virus as per the hierarchy.

SENSITIVITY TO PHYSICAL AND CHEMICAL AGENTS

It is important to know the sensitivity of viruses to various common physical and chemical agents as the knowledge can be utilized for preservation of viruses and to understand the mode of transmission so that proper preventive measures can be taken.

Temperature Sensitivity

In general, viruses are sensitive to temperature and easily loose the infectivity when exposed to high temperature. Most of the viruses get inactivated or killed when exposed to 60°C. This occurs due to the denaturation of viral surface proteins which is responsible for the attachment to the host cell. Table 1.4 gives the temperature sensitivity of most of the viruses as applicable in various practical situations. (The information is accepted widely in an informal manner.)

As at or above ambient temperature, most of the viruses are destroyed within hours, the storage of viruses is done at lower temperature. In situations where sample for virus diagnosis cannot be sent to the lab immediately or in the lab cannot be processed on the same day, it can be stored in the refrigerator at 4°C for 1–2 days till transportation. This is a common scenario during viral outbreaks when the sample is transported from remote locality to the lab for confirmation.

In the virology testing labs, samples or stock virus when needs to be stored for prolong period, it is kept at or below –70°C or in liquid nitrogen cylinder when possible. It is important to note that virus stock or samples for isolation or molecular diagnosis purpose should never be kept at –20°C/–40°C deep freezer. Storage at this temperature usually leads to the formation of ice vesicles which breaks the virion. This is particularly applicable for respiratory viruses.

In general, the viruses that are transmitted through ingestion of contaminated food and water or infect the gastrointestinal tract such as enteroviruses, hepatitis A virus, rotavirus and other viruses causing diarrhea are relatively resistant to 60°C, usual concentration of chlorination and acidic pH. This indicates how these viruses manage to survive in the environment and are transmitted through food and water.

Sensitivity to Common Chemical Disinfectants

Most of the viruses are susceptible to common chemical agents which act by denaturation of proteins, inactivation of the nucleic acids, amino acids or those which acts as dehydrating agents.

However, enveloped viruses are more susceptible to the lipid solvents like ethyl alcohol, chloroform or chlorine compounds as the lipid component of the envelop gets destroyed by these agents making the virus non-infectious. The use of such agents (hand sanitizers) is important in common practice to prevent infection from respiratory viruses which are transmitted through infected droplets. They also play an important role even for the virus like HIV in case of soiling with blood or body fluids. So, it is important to remember that enveloped viruses can be destroyed more easily by using lipid solvents even if they are highly infectious.

REPLICATION OF VIRUSES

The replication or multiplication of viruses is unique among all the microorganisms as they not only multiply within the host cell but also utilize the host cell machinery for their replication.

Table 1.4: Temperature sensitivity of viruses in various practical situations		
Temperature	*Practical scenario*	*Time of half-life*
60°C	Heating temperature of food and liquid	Seconds
37°C	Ambient temperature in tropical countries/summer season	Minutes
18–22°C	Ambient temperature in air-conditioned room/lab	Hours
+4°C (2–8°C)	Refrigerator temperature	Days
≤ –70°C	Deep freezer	Years

Basic steps of viral replication: To begin the process of virus replication, the virus first has to come into the contact of the host cell (attachment), then the virus or its nucleic acid enters inside the cell (entry or uptake) and the nucleic acid of the virus is released by the process of uncoating. At this stage, the early genes are transcribed to mRNA from which regulatory proteins and enzymes required for viral genome replication are formed along with the proteins which shutdown the host cell nucleic acid and protein synthesis. Replication of viral nucleic acid leads to transcription of late mRNA which is then translated to late proteins. These late proteins are structural proteins of the virus. Finally all the synthesized viral components (nucleic acid, structural proteins) get assembled and released from the host cell in large numbers (Fig. 1.4).

Attachment: The attachment between virus and the host cell depends on the cell and tissue tropism of the virus. For example, certain viruses infect only one type of tissue (liver by hepatitis viruses; hepatitis A virus, hepatitis B virus, hepatitis C virus and hepatitis E virus) and certain viruses infect only one particular type of cells (neurotropism by rabies virus and other encephalitis viruses). This is primarily determined by the presence of specific receptor on those cells and tissues.

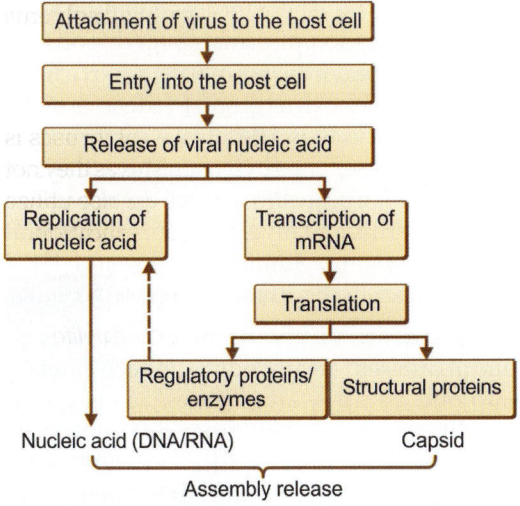

Fig. 1.4: Steps of viral replication

Entry of virus and uncoating: Different types of viruses enter into the host cell by different mechanisms.

Endocytosis: The virus particle after adsorption onto the host cell surface gets invaginated and encircled by endocytic vescicle. The envelop of the virus then fuses with the endocytic vesicle and nucleocapsid is released. It is observed mostly with the enveloped viruses.

Fusion: The viral envelop gets fused with the host cell membrane and nucleic acid is released into the cell. It is observed mostly with the enveloped viruses.

Translocation: The icosahedral capsid of the virus fuses with the host cell membrane. A channel is formed through which the nucleic acid gets translocated to the host cell.

Transcription: After the viral genome is released inside the host cell, the process of transcription is supposed to begin. In molecular biology, the transcription occurs from double-stranded DNA to single-stranded RNA. However, as viruses can be double-stranded DNA, single-stranded DNA, and single-stranded RNA or double-stranded RNA, they have been classified into six types based on their strategy of mRNA transcription. This classification was made by David Baltimore and is known as "Baltimore's classification". According to this principle, mRNA is the beginning point of nucleic acid replication. So, viruses of different types of genome have to first generate a positive mRNA strand. Figure 1.5 shows the transcription strategies of viruses of different types of genome.

Translation: Early proteins are formed from the transcripts of the early genes. Early proteins are viral enzymes, regulatory proteins and the proteins required for nucleic acid replication.

Late proteins are translated from the late mRNA and are structural protein in nature.

Assembly and release: After the replication of nucleic acid and structural proteins, assembly of capsomers occurs to form the

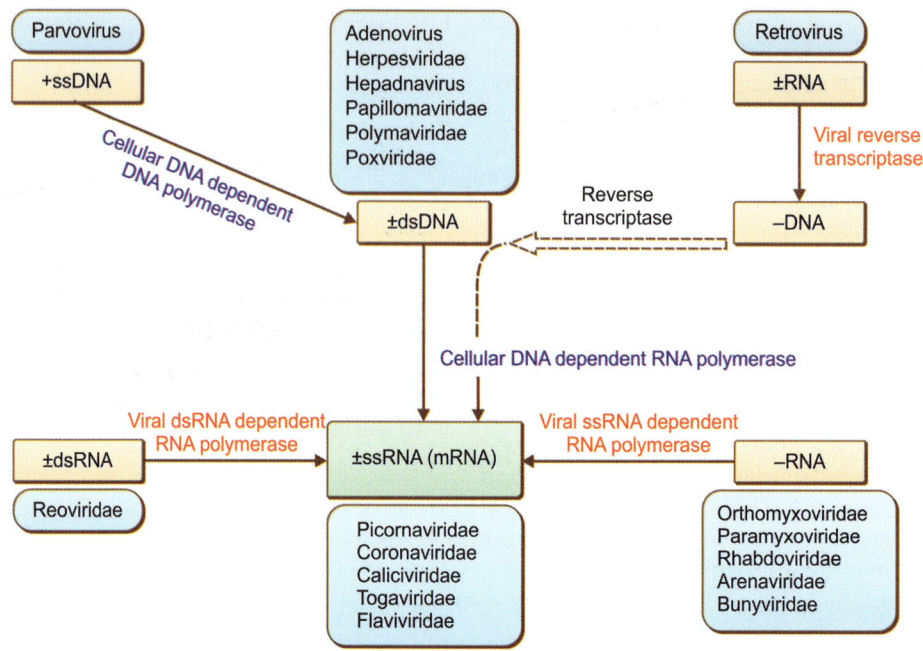

Fig. 1.5: mRNA transcription strategies from different types of viral genomes

procapsid. Packaging of the viral nucleic acid occurs inside the procapsid.

Glycosylation: Glycosylation of the protein occurs that are destined to be incorporated in to the envelop in order to form the peplomers. The process occurs inside the endoplasmic reticulum of the host cell.

Release: After the completion of process of assembly and packaging, the virus particle is released from the host cell either by budding or exocytosis.

VIRAL INFECTIONS

Depending on the cell receptor affinity and tissue tropism, different viruses infect through different routes and involve different organs. According to these properties, viruses can be divided broadly into different groups. This grouping is important from clinical and epidemiological aspects.

Groups of viruses according to their route of infection:

Respiratory viruses: These viruses mainly infect, through infected droplets or aerosol, e.g. influenza virus, parainfluenza viruses, respiratory syncytial virus, human metapneumovirus, rhinoviruses, and coronaviruses. However, some other viruses such as measles virus, mumps virus, parvovirus B19 though infect through respiratory route, manifestations occur through involvement of different organs.

Enteric viruses: These viruses enter the body through ingestion of contaminated food and water, e.g. rotavirus, norovirus, sapoviruses, astrovirus, adenovirus group F (adenovirus 40, 41), enteroviruses and hepatitis A and E viruses. Amongst these, enteroviruses though enter through gastrointestinal tract do not cause gastroenteritis and involve mainly the central nervous system. Similarly HAV and HEV mainly attack the liver tissue and cause hepatitis.

Arboviruses: Viruses transmitted through bite of different arthropods have been clubbed into one group—"arboviruses". This group constitutes of members from different virus families, transmitted by different arthropods. The organ involvement and clinical manifestations are also heterogenous. Table 1.5 gives

Table 1.5: List of important arboviruses	
Vector	*Viral agents according to clinical symptoms*
Mosquito-borne	Fever: Dengue virus (1–4), chikungunya virus, zika virus Arthritis: Chikungunya virus Hemorrhagic fever: Dengue virus (1–4), yellow fever virus Encephalitis: JEV, WNV, EEEV, WEEV, VEEV
Tick-borne	Hemorrhagic fever: CCHFV, KFDV, OHFV Encephalitis: TBEV (Western/European subtype, far eastern subtype, Siberian subtype (previous name Russian spring summer EV), louping ill subtype (previous name louping ill virus)
Sandfly	Encephalitis: Chandipura virus

JEV: Japanese encephalitis virus; WNV: West Nile virus; EEEV: Eastern equine encephalitis viruses; WEEV: Western equine encephalitis viruses; VEEV: Venezuelan equine encephalitis viruses; CCHFV: Crimean-Congo hemorrhagic fever virus; KFDV: Kyasanur forest disease virus; OHFV: Omsk hemorrhagic fever; TBEV: Tick-borne encephalitis viruses.

the list of some of the arboviruses with their vectors and major clinical manifestations.

Transfusion transmitted viruses: Viruses that can be transmitted through blood or blood products have been grouped together, so that appropriate preventive measures can be taken to prevent the transmission, e.g. human immunodeficiency virus (HIV), hepatitis B virus (HBV), hepatitis C virus (HCV), parvovirus B19, cytomegalovirus (CMV). Amongst these, the screening of HIV, HBV and HCV has become mandatory before blood transfusion.

Sexually transmitted viruses: HIV, human papillomavirus (HPV), and herpes simplex virus-2 are the common viruses which are transmitted through sexual route. However, many other viruses which can be transmitted through body fluids can also infect through sexual route.

Congenital viruses: Viruses that can be transmitted either through transplacental route or perinatally through birth canal or breast milk have been largely grouped as congenital virus. It is important from the perspective of prevention when such infections occur during pregnancy. Rubella virus, cytomegalovirus, herpes simplex virus (mostly HSV-2), hepatitis B virus, HIV and parvovirus B19 which causes hydrops fetalis in severe fetal infection are the major causes of congenital viral infections. Recently zika virus which is mainly transmitted through mosquito bite but can be transmitted to fetus through transplacental route and causes fetal microcephaly has also been included in the list.

Viral infections have also been grouped according to the tissue tropism, pathogenesis and clinical manifestations caused by the viruses as described below.

Viral encephalitis: HSV, Japanese encephalitis virus (JEV), West Nile virus (WNV), rabies virus and nipah virus are important viral causes of encephalitis. Several tick-borne viruses are also important cause of encephalitis in certain parts of the globe. Large number of other viruses also can lead to encephalitis in severe form of disease or part of their dissemination which occurs mostly in immunocompromised patients, such as varicella zoster virus, cytomegalovirus, Epstein-Barr virus, influenza virus.

Hepatitis viruses: HAV, HBV, HCV, HEV and hepatitis D virus (HDV) are the major viruses that predominantly target the liver. HAV and HEV are transmitted mainly through enteric route, whereas HBV and HCV enter through parenteral route.

Table 1.6: Human oncogenic viruses

Virus family	Virus	Human cancer
DNA viruses		
Papillomaviridae	Human papillomavirus	Carcinoma cervix (HPV16,18), genital tumors (HPV6&11), oropharyngeal carcinoma
Hepadnaviridae	Hepatitis B virus	Hepatocellular carcinoma (HCC)
Herpesviridae	Epstein-Barr virus	Burkitt's lymphoma, nasopharyngeal carcinoma, Hodgkin's lymphoma
	HHV-8 (human herpes-virus 8)	Kaposi's sarcoma
Polyomaviridae	Merkel cell polyomavirus	Merkel cell carcinoma
RNA viruses		
Retroviridae	Human T cell lymphotropic virus (HTLV)	Adult T cell leukemia
	HIV	AIDS-related malignancies (KSAV, EBV related tumors, HPV)
Flaviviridae	Hepatitis C virus	Hepatocellular carcinoma (HCC)

Hemorrhagic fever viruses: Some of the arboviruses such as dengue virus, Crimean-Congo hemorrhagic fever (CCHF) virus, Kyasanur forest disease (KFD) virus, members of filovirus like Ebola virus and members of Arenaviridae such as Junin and Machupo virus are the predominant cause of hemorrhagic fever. Most of these viruses are restricted to particular geographical location. However, in 2014, the world has experienced the devastating spread of the fatal Ebola virus infection through man-to-man transmission through contact with contaminated blood and body fluids.

Viral myocarditis: A large number of viruses including enteroviruses, adenovirus, parvo-virus, herpesviruses, respiratory viruses and arboviruses like dengue virus and chikungunya virus have been associated with myocarditis. However, coxsackievirus, adenovirus, parvo-virus and human herpes virus-6 are the common viral agents that have been associated with myocarditis and cardiomyopathy.

Oncogenic viruses: Some of the viruses have been associated with tumorogenesis and are known as oncogenic viruses. This is important to know, so that virus associated preneoplas-tic conditions can be screened to prevent the development of neoplasia. Knowledge regarding the association of HBV with hepatocellu-lar carcinoma and HPV with cervical carcinoma has led to the development and implementation of their vaccine. Table 1.6 gives the list of human oncogenic viruses.

Bibliography

1. Knipe DM, Howley PM (eds). Fields Virology, 6th ed. Philadelphia, USA. Wolter Kluwer. Lippincott Williams & Wilkins. 2013.

2. White DO, Fenner FJ (eds). Medical Virology, 4th ed. San Diego, California. Academic Press, Inc. 1994.

3. White DO, Fenner FJ (eds). Medical Virology, 3rd ed. San Diego, California. Academic Press, Inc. 1986.

Laboratory Diagnosis of Viral Infection

LABORATORY DIAGNOSIS

Diagnosis of viral infection is important for appropriate management of the patients. It is also done to confirm any suspected viral outbreak in the community in order to take preventive measures. With the availability of effective antivirals for many viral infections, the necessity of diagnosis has become more important. Monitoring of the viral infection in immunocompromised patients helps to start the pre-emptive therapy to prevent the development of severe fatal disease.

Laboratory diagnosis of viral infection has always remained as a challenge. Viruses being the smallest of the microorganisms are difficult to visualize. As the viruses replicate inside the host cell by utilizing their machinery, they cannot grow in artificial media like bacteria. However, with the development of improved technologies for serological assays, better and rapid isolation and identification system and more importantly advanced molecular techniques have made accurate and rapid diagnosis possible.

The approach of lab diagnosis of viral infections can be divided largely into two types—direct and indirect. **Direct methods** are those by which either virus or its components (viral antigen or viral genome) are detected. **In indirect methods,** the infection is diagnosed by detecting the host response to virus infection, i.e. antibody against the viral antigen is detected.

SAMPLE COLLECTION AND TRANSPORT

The types of sample to be collected for the purpose of diagnosis by direct methods (virus isolation, antigen detection and detection of viral nucleic acid) are same. Samples are collected from the site of virus replication and preferably during the early part of the acute illness when the concentration of virus is expected to be high.

Samples are mostly swabs from the affected part or lesion, aspirate or blood, body fluids or tissue specimens. (a) Blood samples are kept in vial with anticoagulant or plain vial, as the test is mostly done from plasma or serum respectively. (b) Samples that are fluid in nature such as spinal fluid are stored in sterile vial. (c) Swabs and aspirates such as throat/nasal swabs or swabs from skin lesions and nasopharyngeal aspirate or vesicular fluid from the skin lesions are placed in viral transport medium (VTM) (Fig. 2.1).

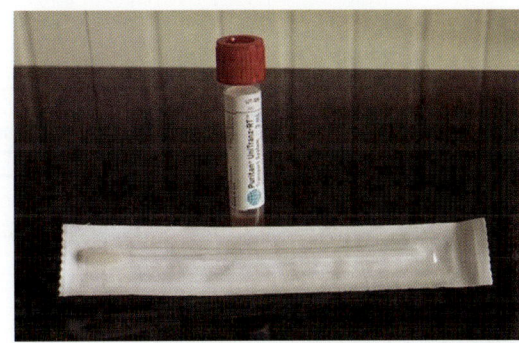

Fig. 2.1: Viral transport medium and swab for collection of sample for viral diagnosis

The samples are transported to the testing lab under cold chain (in a box with ice packs or coolants). In the lab, samples are processed immediately for preparation of smear for immunofluorescence assay and also for isolation. In case of delay in sample processing, the samples can be stored at +4°C for 24 hours and for longer time it should be stored at or below –70°C. Samples collected for virus isolation should not be kept in the deep freezer (–20 to –40°C).

Serum is the most common sample for serological assays. Blood sample is collected in a plain vial without anticoagulant from which serum is separated from the clot by centrifugation. Serum sample can be stored at 4°C, if test is to be done within a few days and at –20°C, if it needs to be kept for longer duration. Flowchart 2.1 shows the algorithm of sample collection, transport and storage for viral diagnosis.

METHODS OF LAB DIAGNOSIS

Traditionally, the lab diagnosis methods for viral infections are direct detection of viral particle or inclusion bodies by microscopy, isolation of virus, serology for detection of virus-specific antibodies and antigen, detection of viral genome and viral load by molecular techniques.

Electron Microscopy

Many of the viruses can be identified by electron microscopy directly from the clinical samples based on their characteristic morphological features. Specific identification of the virus also can be done by immune electron microscopy. However, this method neither offers a practical solution for diagnosis of viral infection, when done for individual patient purpose nor feasible in a common laboratory setting available in most of the places worldwide.

For detection of viral particle in clinical samples, the sample needs to be processed first for clarification followed by staining with negative stain which requires high technical expertise. Secondly, the high limit of detection (a minimum of 10^6–10^7 virions/ml sample is required to be picked up) and high cost of the equipment make it unsuitable as a routine diagnostic tool. However, electron microscopy still plays an important role in discovering novel viruses, searching for unknown

Flowchart 2.1: Algorithm for sample transport and storage for isolation, IF assay and nucleic acid detection for lab diagnosis of viral infection

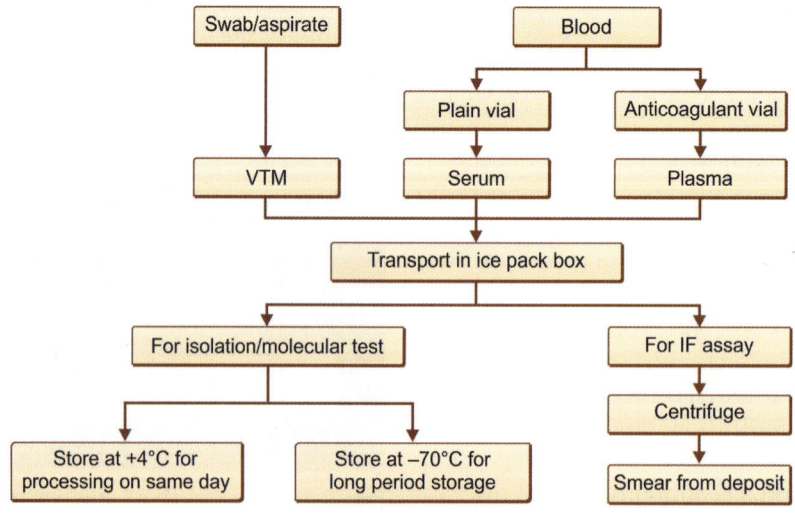

VTM: Viral transport medium, IF: Immunofluorescence

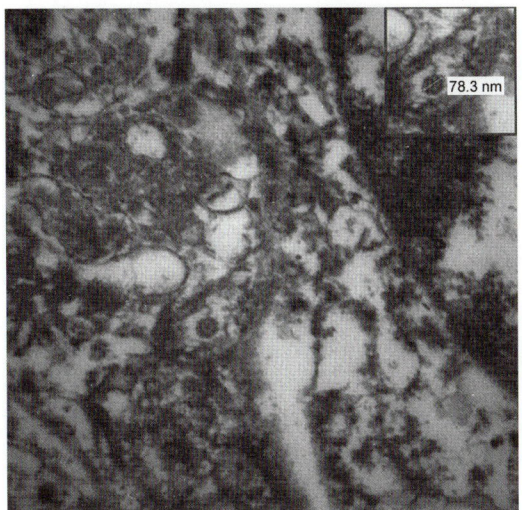

Fig. 2.2: Ultrastructural examination showing viral particles ranging in diameter from 74 to 82 nm, with spikes on the surface (electron microscopy ×18000); Inset highlights one viral particle (*Image courtesy*: Dr Amanjit Bal, Professor, Dept of Histopathology, PGIMER, Chandigarh)

pathogen and study of their morphological characters (Fig. 2.2). The advantages and disadvantages of electron microscopy as a diagnostic tool for viral infections are listed in Table 2.1.

Cytopathology

Certain viruses produce some changes in the infected cells which give a clue to identify the virus. These changes can be detected in the infected tissue or in the exfoliated cells.

Viral **inclusion bodies** are formed by the aggregation of the replicative viral particles inside the host cell. These inclusion bodies (IB) are large enough to be seen under light microscope. Inclusion bodies are characteristic feature for identification of some of the viruses (rabies, CMV) and give a provisional identification for others. In principle, DNA

viruses replicate in the nucleus of the host cell and RNA viruses replicate in the cytoplasm (with a few exceptions). Hence, intranuclear inclusion bodies (INIB) are produced by the DNA viruses (except poxviruses which produce ICIBs as they replicate inside the cytoplasm) and intracytoplasmic inclusion bodies (ICIB) are produced by the RNA viruses (except measles virus which produces both INIBs and ICIBs as it replicates inside cytoplasm as well as nucleus). The image of intranuclear inclusion body is shown in Chapter 3, Fig. 3.2b. Table 2.2 lists the viruses and their inclusion bodies.

Virus Isolation

Viruses require a living system or host for their multiplication as they are obligate intracellular pathogens and require host cell machinery for their multiplication. The isolation of virus is considered as the gold standard for diagnosis of many of the cultivable viruses and newer methods are compared with isolation. Three different host systems are used for isolation of viruses. These are animals, embryonated egg and cell culture.

Animal host: Various susceptible animal hosts have been used for growth of different viruses. Small lab animals, like suckling swiss albino mice, are used commonly for isolation of virus.

The basic steps of virus isolation and identification by animal inoculation are given in Table 2.3.

The requirement of lab animal itself is its main disadvantage. The use of lab animals requires to follow stringent animal ethics guideline which is not possible in most of the diagnostic labs in hospital set up, particularly in developing countries. Moreover, the sensitivity is poor.

Table 2.1: Advantages and disadvantages of electron microscopy in viral diagnosis	
Advantages	*Disadvantages*
• Search for new/unknown pathogen • Study of morphological features and pathogenesis • Research	• High cost of equipment • High detection limit (low sensitivity) • High technical expertise

Table 2.2: Inclusion body produced by different viruses

Virus	Cell/Inclusion body	Description
DNA virus		
Adenovirus	Smudge cell	INIB with degeneration of nuclear membrane
Cytomegalovirus (CMV)	Owl eye appearance	INIB with peri-inclusional halo, granular ICIBs
Herpes simplex virus 1 and 2, Varicella-zoster virus	Multinucleated giant cell	Round INIB, ground glass appearance of nucleoplasm
BK virus	Decoy cell	INIB, enlarged nucleus with ground glass appearance of nucleoplasm
Poxvirus (smallpox virus)	Guarnieri bodies	ICIBs
RNA virus		
Measles virus	Warthin-Finkeldey cells	Multinucleated syncytial giant cell Both INIB and ICIB
Rabies virus	Negri bodies	ICIB in neurons of Ammon's horn of hippocampus
Respiratory syncytial virus	Syncytial giant cell	IC inclusions

IN: Intranuclear; IC: Intracytoplasmic; IB: Inclusion body

Table 2.3: Steps of virus isolation by animal inoculation method

Inoculation of clinical sample into various routes (intracerebral) of animals
↓
Development of symptoms
↓
Animal is sacrificed
↓
Organ (brain) is harvested
↓
Identification of virus antigen/genome

Ex: Suckling Swiss albino mice is used for isolation of arboviruses and rabies virus.

Embryonated egg: Embryonated pathogen free white leg horn hen's eggs are used for the isolation of viruses. Various routes, such as amniotic sac, allantoic cavity, chorioallantoic membrane (CAM) can be inoculated for isolation of different viruses. The route of inoculation is selected depending on the suspected virus in the clinical sample and the purpose of inoculation (Table 2.4).

Because of poor sensitivity, lack of wide availability of pathogen free embryonated egg and high chance of contamination, embryonated egg is no more used for virus isolation from clinical samples. However, the method is still in use for propagation of influenza virus vaccine strain.

Cell culture method is presently the most commonly used method for virus isolation. There are three types of cell lines—primary cell lines, secondary or diploid cell lines and continuous cell lines. Table 2.5 lists the common examples of different types of cell lines.

As the continuous cell lines can be maintained in lab for a long time, these cell lines are most commonly used.

Cell lines are grown on sterile glass tube or more commonly nowadays using the sterile, pyrogen-free tissue culture flask or plates (Fig. 2.3).

Samples are inoculated onto the semi-confluent layer of cell lines which is then incubated at 35°C (temperature may vary depending on the cell line and virus) and observed till development of cytopathic effect. In case of primary isolation, the culture usually is given 2–3 passage before declared as negative. The pattern of cytopathic effect (morphological alteration of infected cells due

Table 2.4: Use of different routes of inoculation in embryonated hen's egg

Route	Age of embryo	Virus	Incubation	Identification	Use
Amniotic	13–14 days	Influenza virus	35°C × 2–3 days	Hemagglutination	Primary isolation
Allantoic	9–11 days	Influenza virus Parainfluenza virus	35°C × 2–3 days	Hemagglutination	Preparation of large quantity virus
CAM	10–12 days	Herpes simplex virus 1 and 2, poxvirus	35°C × 2–3 days	Pock formation on CAM	Isolation, quantitation

Table 2.5: Types of cell lines with their properties

Type of cell lines	Properties	Examples
Primary cell line	Prepared directly from the tissue of origin, not subcultured (after first subculture becomes primary cell line)	Primary human embryonic kidney Primary rabbit kidney Primary monkey kidney
Diploid cell line	Prepared from tissue of origin where 75% of cells contain the original karyotype, can be subcultured but up to 30–50 times *in vitro*	Human fetal diploid kidney (HFDK) Human fetal diploid lung (HFDL or MRC 5 cell line)
Continuous cell line	Can be subcultured for indefinite times *in vitro*, shows absence of contact inhibition, shows the property of aneuploidy or heteroploidy	Human cancer cell lines: HeLa (cervical carcinoma cells) Animal kidney cell lines: Vero (vervet monkey kidney), BGMK (buffalo green monkey kidney), MDCK (Madin Darby Canine kidney)

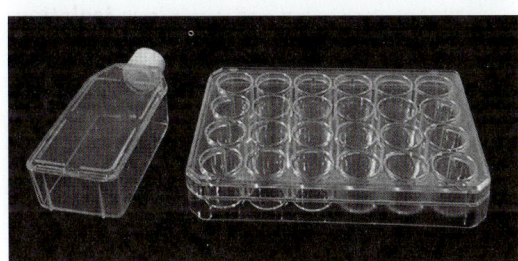

Fig. 2.3: Tissue culture flask and plate

to the pathogenic effect of the virus) may give a clue regarding the virus (Fig. 2.4a and b). Table 2.6 gives the common cell lines used for different viruses with their cytopathic effect.

Besides the demonstration of cytopathic effect (CPE), the presence of virus in the inoculated clinical sample can also be detected by several other methods as follows.

Plaque formation: Plaques are foci of virus infected cells which are formed under solid medium. These foci appear as a clear area when stained with vital stain as it does not take up the vital stain. It is a direct evidence of virus growth and also used as the quantitation method for virus titration (Fig. 2.5).

However, the CPE and plaque formation are not virus specific and the virus has to be identified by detection of viral antigen or viral nucleic acid by immunofluorescence assay or polymerase chain reaction (PCR), respectively.

Metabolic inhibition: The growth of cytopathic virus in the cell culture degenerates the cells hence inhibits the metabolic production of the cells. In absence of virus, cells metabolize in the medium that contains glucose and produces acid which in turn changes the color of the pH indicator phenol red from red to yellow. Whereas, in presence of virus growth, the color of the medium

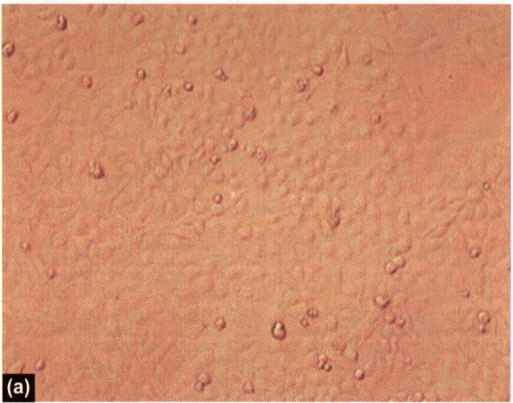
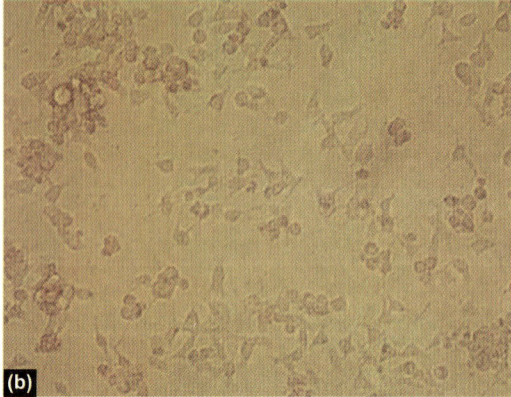

Fig. 2.4a and b: (a) Uninfected Hep 2 cell line; (b) Adenovirus infected Hep 2 cell line showing cytopathic effects of rounding, shrinking, degeneration of cells and surface detachment

Table 2.6: List of common cell lines used for different virus isolation with cytopathic effect		
Virus	*Cell line*	*Cytopathic effect*
HSV	HEK*, Vero	Round, refractile cells with clustering and formation of multinucleated giant cell and intranuclear inclusion body
Measles	Vero/SLAM*, B95a	Multinucleate syncytial giant cell formation
Enteroviruses	L20B, RD, Buffalo green monkey kidney (BGMK) cell	Rounded, refractile cells with detachment from surface

*HEK: Human embryonic kidney; Vero/SLAM: Vero cell transfected with plasmid that encodes the gene for the human signaling lymphocyte activation molecule; RD: Rhabdomyosarcoma.

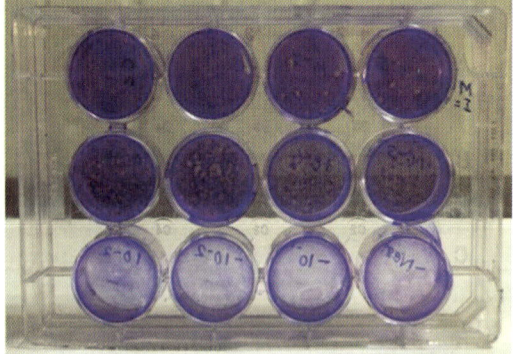

Fig. 2.5: Plaque assay with different dilutions of virus (plaques are seen as unstained foci)

Hemadsorption: The viruses which possess hemagglutinin (members of **Orthomyxovirus, Paramyxoviruses**) have the ability to bind to the erythrocytes of certain species (guinea pig or chick RBC). When the infected cell line is added with those RBCs, after incubation adherence of RBC occur to the cell surface, which can be visualized under microscope.

Hemagglutination: The viruses which possess hemagglutinin have the ability to agglutinate the erythrocytes of certain species. This property of the virus is used for its identification. Hemagglutination is commonly used for identification of **influenza virus** from cell culture supernatant and also from the amniotic and allantoic fluid. Table 2.7 gives the common examples of viruses and their corresponding species of RBC used for hemagglutination.

remains red or reddish orange. However, this is an indirect indicator of growth of virus and liable to give false positive and false negative depending on the rate of growth of virus and cells.

Interference: The growth of cytopathic virus is inhibited in the presence of another non-cytopathic virus. This is based on the principle of interference where the presence of later interferes the growth of the former virus. This is utilized for the detection of non-cytopathic virus, e.g. **rubella virus** (non-cytopathic virus) in presence of a cytopathic enterovirus **(ECHO11)**.

Immunofluorescence (IF): Virus antigen can be detected in the infected cell by IF using specific antibodies (Fig. 2.6a).

Viral nucleic acid detection: This can be done from the infected cells or from the culture supernatant by molecular techniques such as polymerase chain reaction using specific primers (Fig. 2.6b).

Serological Assays

Hemagglutination inhibition (HAI): This test utilizes the property of certain group of viruses that have the ability to hemagglutinate the erythrocyte of certain animals (Table 2.7). This is mostly seen among the members of Orthomyxovirus and Paramyxovirus. The virus-specific antibodies when present in the patient's serum inhibit the hemagglutination. Thus detection of HAI antibody in the serum is an indicator of infection. However, once the HAI antibody appears after the primary infection, it persists for a long time. Thus, (i) mere detection of HAI antibody in the serum may be due to past infection and does not indicate current infection, (ii) rising titer of fourfold or more between acute and convalescent sera indicates ongoing infection. High cutoff titer in the acute sera sample also has been fixed for certain viral infection (Fig. 2.7).

The test is not in common use nowadays, as it requires (i) extensive standardization of each component, (ii) requires the animal RBC which is not feasible in most of the places for routine diagnostic use, and (iii) requirement of paired sera (acute and convalescent sera) for diagnosis of current infection.

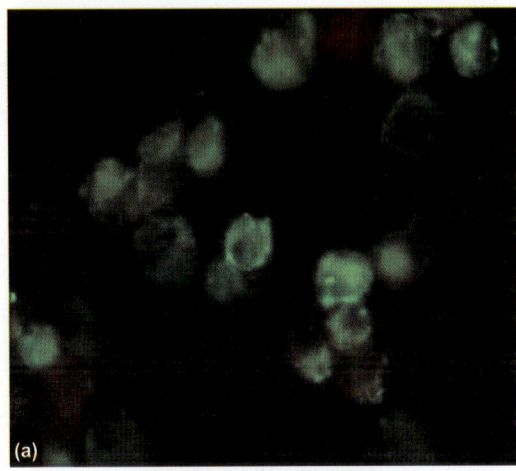

(a)

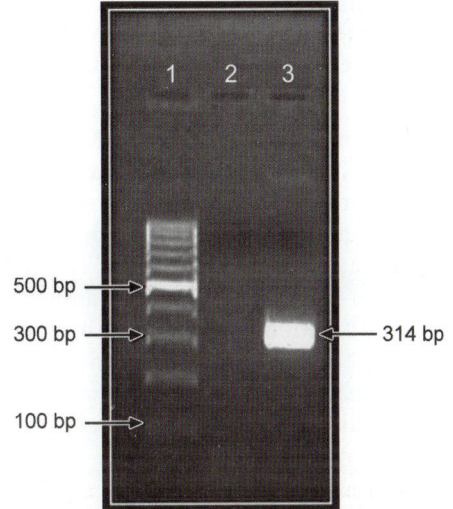

500 bp →
300 bp → ← 314 bp
100 bp →

(b)

Lane-1: 100 bp marker
Lane-2: No template control
Lane-3: Coxsackievirus isolate

Fig. 2.6a and b: (a) Confirmation of coxsackievirus B from cell culture by immunofluorescence showing intracytoplasmic fluorescence; (b) Confirmation of coxsackievirus B from cell culture by PCR

Table 2.7: Common examples of viruses and their corresponding species of RBC

Virus	RBC species
Influenza virus	Guinea pig, fowl, human 'O'
Arboviruses	One day old chick
Measles virus	Monkey
Rubella	Baby chick, trypsinized human O

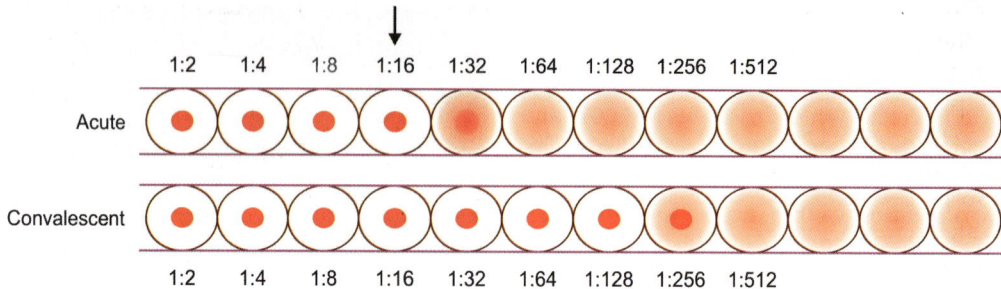

Fig. 2.7: Hemagglutination inhibition test showing >4-fold difference between acute and convalescent sera (schematic)

However, the test is still used for testing of influenza vaccine efficacy. HAI is a good test for confirmation of subacute sclerosing pan-encephalitis (SSPE) which occurs as a late sequel of measles virus infection where ratio of ≤1:64 between CSF and serum titer is considered as significant.

Complement fixation test (CFT): This is based on the principle that complement gets fixed up during antigen–antibody reaction. Two sets of antigen–antibody are used in the test. The first one contains patient sera (in which antibody is going to be detected) and known antigen. If patient's sera contain specific antibody to the antigen given antigen–antibody reaction will occur and complement will be utilized. The second set of antigen–antibody consists of sheep erythrocyte and antibody to sheep RBC (hemolysin) which also acts as an indicator. When complement gets utilized during the first antigen–antibody reaction, it is no more available for the second antigen–antibody reaction. So the second reaction cannot occur and sheep RBC will get settled at the bottom of the U bottom plate. When complement is available for the second antigen–antibody reaction (not utilized during first reaction because of absence of specific antibody in patient's sera), hemolysis will occur.

So, no hemolysis (button formation) indicates positive for CFT antibody in patient's sera and hemolysis indicates sample is negative for CFT antibody (Fig. 2.8).

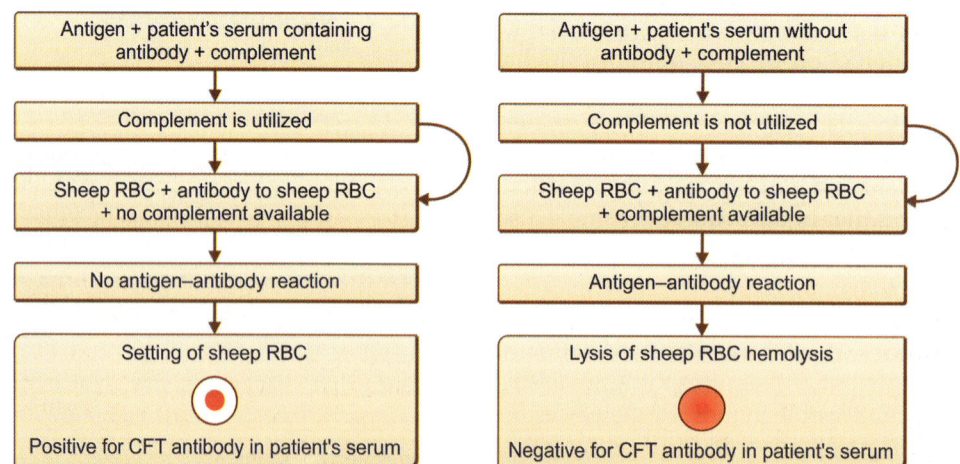

Fig. 2.8: Principle of complement fixation test

CFT antibody is detectable for few months after the subsidence of acute symptoms and unlike HAI antibody does not persist for years. Thus, presence of CFT antibody is an indicator of current or recent past infection. To diagnose current infection, fourfold rise in antibody titer has to be demonstrated between acute and convalescent sera.

However, the test requires preparation and vigorous standardization of each component which makes the test difficult to use for routine diagnosis.

Immunofluorescence (IF) test: This can be of two types: Direct (DIF) and indirect (IIF).

In direct IF, antigen is detected in the infected cell by staining with specific antibody tagged to the fluorescein reagent. The antibody can be either monoclonal or polyclonal in nature. It is a one step procedure, so rapid and easy to perform. It is used for detection of viral antigen from the infected clinical material (nasopharyngeal aspirate) or from the infected cells from cell culture. For example, RSV antigen from nasopharyngeal aspirate.

In indirect IF, infected cells are fixed on the slide, to which primary antibody which is specific to antigen (mouse monoclonal antibody) is added. The unbound antibodies are removed by washing. The secondary antibody (antispecies antibody of the primary antibody; anti-mouse antibody) which is tagged to the fluorescein reagent (FITC) is added followed by counter stain. IIF can be used for detection of either antigen or antibody, e.g. CMV pp 65 antigen (Chapter 3—C: Cytomegalovirus, Fig. 3.6b), HSV antigen (Chapter 3—A: Herpes Simplex Virus, Fig. 3.2a), adenovirus antigen (Chapter 7, Fig. 7.3). In IIF, the secondary antibody is not antigen specific, so it can be used for other IIF assay also. Therefore, IIF is more preferred in diagnostic settings.

Immunohistochemistry: Detection of viral antigen in the tissue by staining it with specific antibody is called immunohistochemistry. The antigen–antibody complex when formed gets

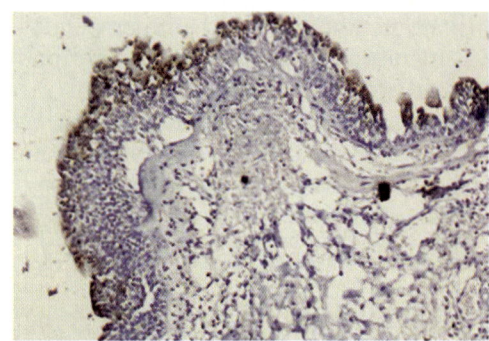

Fig. 2.9: Photomicrograph showing influenza A (H1N1) hemagglutinin (H) antigen localized on the ciliated epithelium of the proximal airways. (*Image courtesy:* Dr Amanjit Bal, Professor, Department of Histopathology, PGIMER, Chandigarh)

deposited at the site of antigen producing a brown color precipitate (Fig. 2.9).

Enzyme immunoassay (EIA): Enzyme immunoassay can be done on a solid phase like wall of a microtiter plate or glass bead, or it can be membrane bound. The test can be employed for the detection of antigen as well as antibody.

In the solid phase based assay, solid phase is coated either with antigen or antibody, to which patient's serum is added and incubated for reaction to occur. The unbound reactants are removed by washing. Secondary antibody that is enzyme labeled is then added followed by the substrate. In case in the first step the reaction between antigen–antibody has occurred, the secondary antibody binds to it. So the enzyme which is tagged to it acts on the substrate in the following step and a color is produced. The change of color is measured by spectrophotometer. Principle of ELISA is explained in schematic format in Fig. 2.10.

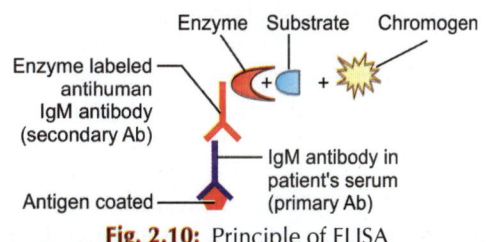

Fig. 2.10: Principle of ELISA

In the membrane bound antigen detection EIA, membrane is bound to viral antibody. On adding the sample (in which viral antigen is to be detected), the viral antigen in the sample binds to the membrane bound viral antibody. In the second step, enzyme labeled viral antibodies are added which binds to the bound antigen. On adding the substrate after removal of unbound antibodies by washing, color is produced on the membrane where primary antibodies were attached.

EIA are also available in different format based on different steps in the procedure, such as indirect ELISA, sandwich ELISA, capture ELISA, avidity ELISA. The choice of ELISA is selected based on the purpose of the test.

For example, for detection of IgM antibody against arboviruses (dengue, Japanese encephalitis, chikungunya), capture ELISA is recommended because of its high specificity.

Avidity ELISA is done to differentiate between primary or secondary infection which is important in case of congenital viral infections (CMV, rubella).

The advantages of ELISA are its high sensitivity and specificity, can be done for large number of samples in one batch and adaptability to automation. The technique does not require sophisticated expensive equipment or high technical expertise, so can be done even at the peripheral lab with minimum facilities.

Table 2.8 lists some common examples of EIA or antigen/antibody detection tests used for viral infection.

Chemiluminescence immunoassay (CLIA): This is based on the principle of generation of electromagnetic radiation in the form of light by release of light from a chemical reaction. The technique is gradually replacing ELISA in high throughput clinical labs. The advantages of CLIA are its high precision, increase sensitivity and specificity, simple to perform, high speed of the test leading to less turnaround time and full automation. CLIA

Table 2.8: Common examples of viral antigen/antibody EIA

Antibody detection tests
IgM/IgG detection:
- Hepatitis viruses (HBV, HAV, HEV)
- Viruses causing congenital infections: CMV, rubella
- Arboviruses: Dengue, JEV, chikungunya, WNV
- Exanthematous viral infections: Measles, VZV

Antigen detection tests
- Respiratory viruses: RSV, influenza virus, parainfluenza virus
- Enteric viruses: Rotavirus
- Hepatitis virus: HBs antigen of HBV
- Dengue virus NS1 antigen

is being increasingly used for many of the serological tests for diagnosis of viral infections particularly for hepatitis B virus surface antigen detection and HCV antibody where large numbers of samples are processed everyday.

Molecular Tests

Target amplification: In these methods, one particular gene is targeted for amplification and the amplified product is detected. Various molecular methods such as polymerase chain reaction (PCR), nucleic acid sequence based amplification (NASBA), strand displacement amplification (SDA), ligase chain reaction (LCR) are used for target amplification.

Amongst these, PCR is the most widely used method worldwide. It involves isolation of the viral nucleic acid from the clinical sample, and then the nucleic acid is subjected to amplification by using specific primers which anneals to the target DNA template. Then amplification occurs with the help of polymerase enzyme and deoxytrinucleotidephosphate. The amplified product is then detected by agarose gel electrophoresis (Fig. 2.11).

Amplification technique amplifies the target gene to the tune of million copies increasing the sensitivity to a great level. The

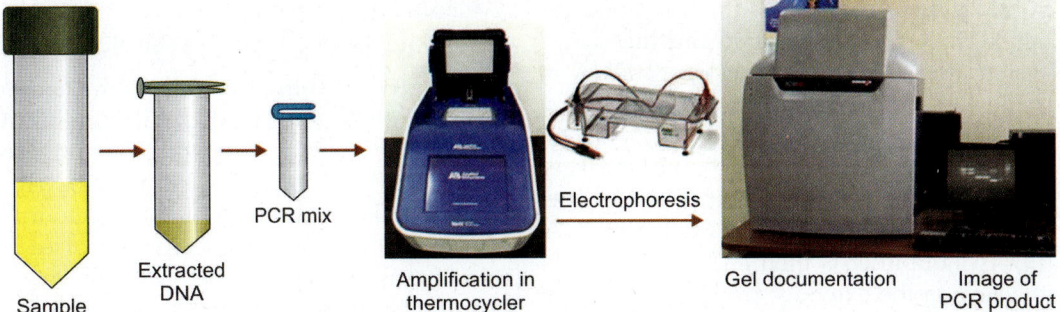

Fig. 2.11: Workflow of PCR

viral genome detection is particularly helpful for those viruses where direct methods of detection play a major role in diagnosis and as the sensitivity of amplification methods is much higher than immunofluorescence assay and virus isolation, it is preferred over them.

However, the requirement for thermocycler, gel documentation system, expensive reagents and technical expertise restricts its wide use. The high sensitivity of the test also makes the test liable for carry over contamination leading to false positive.

Qualitative PCR for detection of viral DNA/RNA can be used for all viral infection. However, considering its significant advantages over other available methods, it is presently used for routine diagnosis of many viral infections (Table 2.9).

Nucleic acid hybridization: This is based on the principle of hybridization of complementary nucleic acid probes to the target nucleic acid of the cell. Target nucleic acid sequences to be detected are pre-identified. Probes with complementary sequences are prepared which is either enzyme labeled or radio-labeled. These probes are allowed to react with the cell for hybridization to occur. When a substrate is added, there is change in color in presence of hybridized probe. It is detected by examining under microscope. The technique is not much in use for lab diagnosis, but confirms the presence of virus inside the cell.

Table 2.9: Common examples of PCR use for diagnosis of viral infections		
Virus	*Clinical condition*	*Sample*
HSV	Encephalitis	CSF
CMV	Congenital infection	Amniotic fluid, blood of newborn
Respiratory viruses	Acute upper/lower respiratory tract infection	Nasopharyngeal aspirate/throat or nasal swab/tracheal aspirate/ bronchoalveolar lavage
Enteric viruses	Viral diarrhea	Stool
EBV, VZV, CMV	CNS symptoms	CSF
Enteroviruses (poliovirus, coxsackievirus)	CNS symptoms	CSF
Japanese encephalitis virus, West Nile virus	Acute encephalitis syndrome	CSF

Signal amplification: By this method, the target DNA is not amplified but the signal system gets amplified. The target DNA is made to single strand, placed on a solid support. Specific oligonucleotide probe with extender probe is added. In presence of target DNA in the sample, hybridization occurs and the target DNA gets captured. Branched DNA (bDNA) amplifier is then added which gets attached to the extender probe followed by enzyme labeled probe which binds to the bDNA. On adding the substrate, enzyme acts on it and the reaction is measured by chemiluminescence.

The method is also used for viral load estimation. For example, viral load estimation of HBV and HCV.

Real-time PCR: This technique can be used for qualitative as well as quantitative viral assay.

Qualitative real-time PCR has an edge over the conventional PCR. It has been observed that the method is more sensitive and rapid as compared to conventional PCR. The reaction can also be checked real-time to assess the result. The chance of cross contamination is also less as it does not involve the handling of the post-amplification product. However, sample-to-sample contamination can occur due to transfer or mixing up of sample from one vial to other. Due to multiple advantages, the technique is almost replacing the conventional PCR in diagnostic labs.

The present uses of qualitative real-time PCR for viral diagnosis are, therefore, same as that of conventional PCR as mentioned in Table 2.9. Besides the use of the technique for single pathogen, more and more pathogen panel detection systems are becoming commercially available for diagnosis using real-time PCR platform. Examples:

- COVID-19 RTPCR
- Respiratory virus panel
- Viral encephalitis panel
- Arbovirus panel
- Influenza viruses
- Acute gastroenteritis panel, etc.

Quantitative real-time PCR: This method is used for estimation of viral load in the clinical sample. The test is performed using multiple standards (3–5 serial log dilutions) with known viral copy number of the target virus in each test run. The copy number of each standard is entered in the software and a standard curve is generated based on the fluorescence (which depends on the amount of the amplified product) and cycle number at which the target is detected. The quantitation of the viral load in the test sample is estimated by comparing the cycle number of the sample with the standard curve of the standards (Fig. 2.12).

Estimation of viral load is indicated in certain clinical conditions where it is important to differentiate between infection and disease. In certain viral infection, the DNA of the virus may be detected in blood or other samples due to persistence of the virus or it is DNA. So, qualitative detection of viral DNA only indicates infection which can occur in asymptomatic cases without causing any symptomatic illness. Whereas the same virus when detected in high amount (high viral load) may lead to clinically overt disease. Thus, it is relevant to determine the viral load in order to assess the progression towards disease and to start the pre-emptive therapy. Following are a few examples:

- Monitoring of CMV load in blood sample during the post-transplant period.
- Parvovirus B19 estimation in blood sample in order to associate the virus with the clinical symptoms.
- BK virus DNA estimation in urine to associate with nephropathy.

Viral load estimation is also important to assess the disease severity and to monitor the response to therapy.

For example, viral load estimation in HBV, HCV, HIV, CMV and EBV.

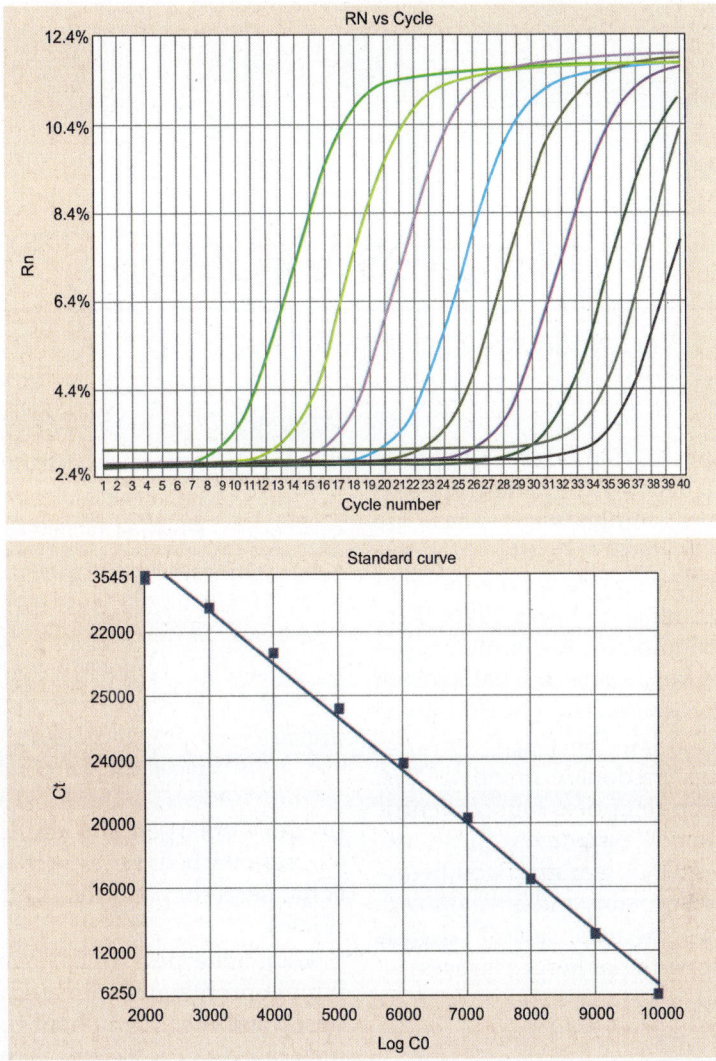

Fig. 2.12: Amplification plot and standard curve of adenoviral standards by TaqMan real-time PCR

Bibliography

1. Cabezas-Fernandez M, Cabeza-Barrera M. Introduction of an automated system for the diagnosis and quantification of hepatitis B and hepatitis C viruses. Open Virol J 2012;6: 122–34.

2. Debiasi RL, Tyler KL. Molecular methods for diagnosis of viral encephalitis. Clin Microbiol Rev 2004;17(4):903–25.

3. Espy MJ, Uhl JR, Sloan LM, et al. Real-time PCR in clinical microbiology: Applications for routine laboratory testing. Clin Microbiol Rev 2006; 19(1):165–256.

4. Gullett JC, Nolte FS. Quantitative nucleic acid amplification methods for viral infections. Clin Chem 2015;61(1):72–78.

5. Lennette EH, Schmidt NJ (eds). Diagnostic procedures for viral rickettsial and chlamydial infections, 5th ed. American Public Health Association, Inc. Washington, D.C. 1979.

6. Lennette EH, Smith TF eds. Laboratory diagnosis of viral infections. 3rd ed. Marcel Dekker AG, Inc. New York. 1999.

3

Herpes Viruses

HERPESVIRIDAE

The word "**herpes**" was used by Hippocrate to describe **creep or crawl**, the pattern of lesion caused by herpes simplex virus. There are hundreds of herpes viruses, they are ubiquitous in nature and have been found to infect almost all animals. Only eight of them naturally infect humans. The ninth herpes virus, herpes B virus, is a natural pathogen of monkeys but can cause fatal encephalitis in humans.

Herpes viruses are double-stranded DNA viruses. The virion comprises four concentric layers. From inside outwards, these are: Double-stranded DNA genome, icosahedral capsid, tegument proteins and envelop with numerous small peplomers. Table 3.1 depicts the common properties of herpes viruses.

CLASSIFICATION

The family Herpesviridae is divided into three subfamilies based on their host range, duration of reproductive cycle, cytopathology and characteristic of latent infection (Table 3.2).

The members of **Alphaherpesvirinae** subfamily show a wide host range in cell culture, can grow rapidly and lyses infected cells and are neurotropic as they have the capacity to establish latent infections primarily in sensory ganglia.

Betaherpesvirinae are characterized by a narrow host range and a slow replication cycle in cell culture with slow virus spread. They

Table 3.1: Common properties of herpes viruses

Virion:	Spherical with ≈150 nm diameter Icosahedral capsid Double-stranded DNA genome
Replication:	Inside nucleus (intranuclear) Acquire envelop through nuclear membrane Release of new virion particles leads to host cell lysis (lytic infection)
Latency:	Capacity to persist in the host for indefinite period in latent form

produce enlarged and multinucleate cells. Viruses may become latent in lymphoreticular cells, secretory glands, kidney and other tissues.

Gammaherpesvirinae are associated with lymphoproliferative diseases in their natural hosts, and infect lymphoid cells *in vitro* and may cause lytic infections in certain epithelial and fibroblastoid cell lines. These viruses are specific for B or T lymphocytes, where infection is frequently arrested and there is no production of infectious progeny. Latent virus may be found in lymphoid tissue. Though these viruses are described as 'lymphotropic', lymphocytes are only semipermissive (Table 3.2).

Viral Latency

Latency is a process in which the virus persists in the host without undergoing active replication. Virus can remain latent in presence of fully developed immune response.

Table 3.2: Classification and biological properties of herpes viruses

Subfamily	Virus	ICTV species name	Site of latency	Biological property
Alpha-herpesvirinae	Herpes simplex virus 1 (HSV-1)	Human herpes virus 1 (HHV-1)	Sensory and cranial nerve ganglia	Fast growing, cytolytic
	Herpes simplex virus 2 (HSV-2)	Human herpes virus 2 (HHV-2)	—do—	
	Varicella-zoster virus (VZV)	Human herpes virus 3 (HHV-3)	—do—	
Beta-herpesvirinae	Cytomegalovirus	Human herpes virus 5	Monocyte, macrophage, CD34+	Slow growing, cytomegalic
	Human herpes virus 6	Human herpes virus 6	CD34+, monocyte, macrophage	
	Human herpes virus 7	Human herpes virus 7	CD34+, monocyte, macrophage	
Gamma-herpesvirinae	Epstein-Barr virus (EBV)	Human herpes virus 4	Memory B cells	Lymphoproliferative
	Kaposi's sarcoma-associated herpes-virus (KSHV)	Human herpes virus 8	B cells	

ICTV: International Committee on Taxonomy of Viruses

The problem with latency is that the chances of reactivation remains throughout the life, which means latent virus can reactivate at any point during the lifetime of the host. On reactivation of the latent virus, normal replication occurs and infectious viruses are produced.

Reactivation of beta and gamma herpes viruses occurs when the immune status of the host is suppressed. Whereas, herpes simplex viruses can get reactivated with variety of non-specific 'triggers' like trauma to tissue innervated by the latently-infected neurons, emotional or physical stress, menstruation, ultraviolet light and hormone imbalance, etc.

During Latency

- Lytic gene expression is repressed.
- No replicating virus or viral antigen can be detected in the sensory ganglia.
- Latency-associated transcript **(LAT)** is expressed.

Latency of HSV: During the primary infection, virus comes into contact with the cutaneous receptors of local sensory nerves. Through these receptors, virus attaches to and penetrates the sensory neuron by fusion at the axonal termini. The nucleocapsid is carried by retrograde axonal transport to the nucleus of the infected neuron. The viral DNA persists in the nucleus in a circular episomal form associated with nucleosomes.

The site of latency depends upon the site of primary infection.

Example:

- Orofacial HSV infection: Trigeminal nerve innervating the face
- Genital HSV infection: Lumbosacral nerves innervating the genitalia.

PATHOGENESIS

Most of the human herpes viruses are shed from the oral mucosa during the asymptomatic phase, whereas VZV is transmitted only from a patient with varicella or zoster.

Symptomatic disease is associated with lytic virus replication, which results in skin lesion in HSV or VZV and visceral lesion in HSV, VZV and CMV. Symptoms of EBV associated infectious mononucleosis occur due to proliferation of T cells in response to infection.

Primary infection due to HSV, CMV, EBV are asymptomatic or leads to mild symptoms, whereas infection with VZV leads to varicella (chickenpox) and HHV6 leads to fever.

A. HERPES SIMPLEX VIRUS

VIRUS

The virion is spherical in shape, of 120–200 nm diameter. The inner core contains the viral genome, which is a linear double stranded DNA consisting of 125–229 kbp. It encodes >90 transcripts and present in a torus or doughnut shape.

The inner core is surrounded by an icosahedral capsid of 100 nm diameters, having 162 capsomers, 150 hexamers and 12 pentamers. Between the capsid and envelop, an amorphous protein material surrounds the capsid known as "tegument". The outermost lipoprotein envelop layer is covered with numerous small peplomers.

EPIDEMIOLOGY

HSV is ubiquitous virus and present all over the world. HSV1 is transmitted mostly through oral secretion and HSV2 through genital secretion. The acquisition of HSV1 occurs during childhood. The seroprevalence of HSV1 reaches almost 100% by adult age in developing countries with low socioeconomic status, whereas the seroprevalence in adults in developed countries is up to 60–70%.

HSV2 is mainly acquired through sexual route. Antibody to HSV2 starts with the sexual exposure and seroprevalence increases with age and is related to a number of sexual partners. The seroprevalence varies from 10 to 70% in different geographic region with high seroprevalence in developing countries. Overall the seroprevalence is two times higher in females than males.

PATHOGENESIS

Herpes simplex virus enters through exposure to mucosal surface or abraded skin sites into the cells of epidermis and dermis and replicates there. Substantial replication occurs both in clinical and subclinical infections after which the virus infects the sensory and autonomic nerve endings. The nucleocapsid of the virus then transported through axon in retrograde manner and reaches the nerve cell bodies. Replication of virus occurs in the ganglion and contiguous neural tissue during this initial phase of infection. Then the virus spreads centrifugally through the peripheral sensory nerves in antegrade manner to the mucocutaneous surface. This spreading fashion of herpes simplex virus explains the site of new lesions at distant places from the initial lesion during the primary infection. Contagious spreads from the initial site and viremia are other modes of extension of lesion. Viremia in near 30–40% of primary HSV2 has been observed.

Neurovirulence of HSV is mainly attributed to its capacity to invade and replicate in the neuronal cells and to persist in latent infection. Amongst several genes that are implicated for the neurovirulence, gene that maps the inverted repeats of the unique long segment of **HSV DNA γ134.5** is the most important one to modulate the neurovirulence. This is evident by the avirulent nature of its deletion mutant.

Acute inflammation, congestions and hemorrhage during the initial weeks and necrosis, perivascular cuffing and liquefaction in the later stage are the common histopathological changes that occur due to HSV encephalitis.

CLINICAL FEATURES

Oropharyngeal lesion: Children and young adults are most commonly affected with primary infection. Gingivostomatitis and pharyngitis are the most common clinical manifestations. Patient presents with fever, malaise, myalgia, irritability and cervical lymphadenopathy. Vesicular lesion appears in the hard and soft palate, gingiva and adjoining facial skin and tissues, which usually ulcerate during the course of the disease. After recovery from the primary oropharyngeal infection, virus remains latent in the trigeminal ganglion. The chance of recurrent infection is at least 50%.

In immunocompromised patients, lesion may extend to deeper tissues and mucosa with pain, necrosis and inability to eat and drink.

Recurrent infection usually precedes with a prodromal period of hyperesthesia. Lesions occur more in the form of cluster and around the mucocutaneous junction of lips called herpes labialis (Fig. 3.1a).

Genital herpes: Majority of primary genital herpes is caused by HSV-2. In primary genital herpes, lesions are multiple, painful, bilateral and punched out shallow ulcers which are present in varying stages, including vesicles, pustules, erythematous ulcers (Fig. 3.1b). Local lesions are mostly accompanied with systemic symptoms such as fever, headache and myalgia. During the primary infection, HSV remains latent in the sacral ganglion, from where periodic reactivation of the virus occurs leading to periodic recurrence of genital lesions. The rate of recurrence is higher in HSV-2 as compared to HSV-1.

Keratoconjunctivitis: Involvement of cornea leading to keratitis is the common manifestation with herpes simplex ocular infection. Keratitis occurs with acute onset of pain with blurring of vision, chemosis and conjunctivitis

leading to characteristic dendritic ulcer (Fig. 3.1c). There may be chorioretinitis in immunocompromised host or in case of disseminated herpes infection.

Herpes encephalitis: HSV encephalitis is the most common cause of acute sporadic encephalitis. More than 95% of HSV encephalitis in immunocompetent adults is caused by HSV-1 and the remainder occurs due to HSV-2.

HSV encephalitis is commonly caused by HSV-1, whereas HSV-2 is mostly associated with meningitis. The mild form of recurrent aseptic meningitis is called **Mollaret's meningitis** which lasts for a few days and may recur after a few months or years. This is mostly seen in young women with or without clinically apparent genital lesions.

HSV enters central nervous system from periphery via olfactory bulb. It causes necrotizing encephalitis commonly involving the medial part of temporal lobe or inferior frontal lobe. The clinical manifestation occurs with an acute onset of fever with focal neurological symptoms and signs. Other viral encephalitides, focal infections and non-infectious process are differential diagnosis.

Presently detection of **HSV DNA in CSF sample is the gold standard** for confirmation of diagnosis. Magnetic resonance imaging (MRI) showing gadolinium enhanced lesions in temporal lobe is characteristic and considered as the choice of neuroimaging technique to diagnose HSV encephalitis. Intravenous acyclovir is used to treat the presumptive HSV encephalitis.

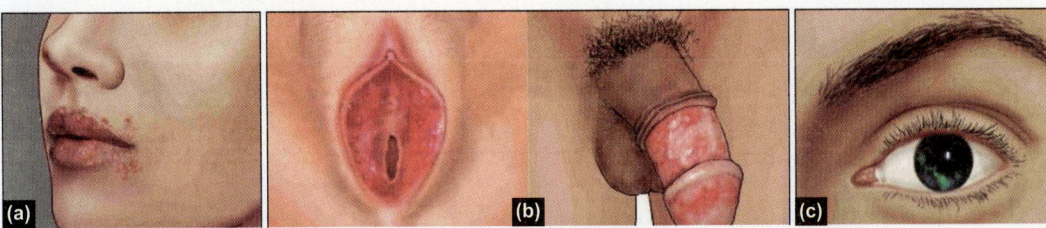

Fig. 3.1a to c: (a) Herpes labialis; (b) Herpetic genital lesions; and (c) Dendritic ulcer

Neonatal herpes: Majority cases of neonatal herpes occur due to **HSV-2 infection** (70%). The infection can be acquired during birth from infection in birth canal, through placenta or from ascending infection from cervix. The risk of neonatal herpes is around 10 times more from mothers with primary herpes virus infection as compared to mothers with recurrent herpes infection. The manifestation in neonate can be localized mucocutaneous disease affecting skin, eye, mucosa or severe form of disease like encephalitis and disseminated disease with high rate of mortality.

Disseminated herpes infection: Disseminated form of manifestation usually occurs in immunocompromised patients due to viremia. Patient usually develops multiorgan involvement or encephalitis.

DIAGNOSIS

Laboratory confirmation is recommended in all suspected herpes simplex virus infections as they can resemble lesion due to other infectious and non-infectious origin.

However, laboratory confirmation should be considered as essential in the following conditions:

- Acute encephalitis syndrome
- Neonatal encephalitis
- Atypical skin and mucosal lesions
- Immunocompromised patients.

Direct detection of virus or viral components such as viral antigens or viral DNA in clinical samples is the mainstay of laboratory diagnosis.

Sample Collection

Herpetic lesion: The affected area should preferably be cleaned with sterile normal saline. Cleaning with alcohol can destroy the virus and should be avoided when the sample is collected for isolation of virus.

In case of skin and mucosal lesion, swab is collected from the base of the lesion. In presence of vesicular lesions, vesicle fluid is collected and swab from base can be collected after deroofing the lesion.

Nylon or cotton swabs are preferred for collection of the sample. Calcium alginate swabs should be avoided.

Swabs meant for virus isolation, antigen or PCR should be put in 1–2 ml of viral transport medium (VTM) and transported to the laboratory at 4°C.

Herpes encephalitis: CSF should be collected for testing of herpes virus DNA in case of suspected herpes simplex encephalitis (HSE) or meningoencephalitis. The sample should be collected in a sterile vial and transported on ice to the testing lab or can be stored at 4°C for a few hours before sending it to the laboratory.

The diagnostic approach of HSV infection is given in Flowchart 3.1.

Flowchart 3.1: HSV infection

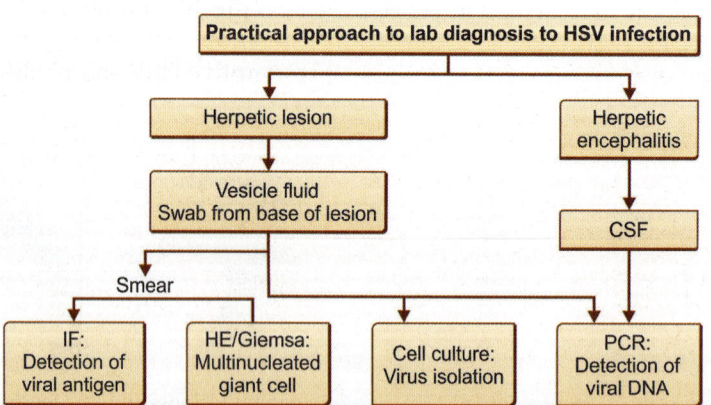

HSV DNA detection: Detection of HSV DNA by PCR from clinical sample has higher sensitivity in all types of clinical samples including both mucosal and visceral involvement and CSF in case of herpes encephalitis. **HSV gB gene and DNA polymerase** genes are commonly employed for detection of HSV DNA. The test is presently the reference test for HSE.

The sensitivity of the DNA PCR from CSF samples in herpes simplex encephalitis (HSE) cases is **>95%** at the time of clinical presentation and specificity is near 100%. In a comparative study with biopsy proven HSE, CSF PCR showed 98% sensitivity and 94% specificity. Considering multiple studies, the overall **sensitivity and specificity** has been estimated as **96% and 99%**, respectively. However, CSF PCR can be negative during the first 1–3 days of the onset of symptoms. Presence of large number of red blood cells in CSF also can lead to false negative PCR, even with high viral load. Antiviral treatment should not be discontinued, if the CSF PCR comes negative within first few days of illness. In neonates, CSF PCR has been reported to be 75% sensitive and 98% specific.

After antiviral therapy with acyclovir, the amount of HSV DNA increases in first 5–6 days and then decline. Majority of the patients become negative after 14 days of therapy and almost all after 30 days. Some authors recommend a second course of acyclovir if CSF PCR persists to be positive after 14 days of therapy.

In case of genital lesions, PCR is presently considered as the preferred method of diagnosis. Though virus isolation by cell culture is considered to be the gold standard, the sensitivity of PCR is four times higher than that of isolation of virus.

HSV antigen detection: Detection of HSV antigen can be done by type specific monoclonal antibody by direct or indirect immunofluorescence test (Fig. 3.2a), ELISA or rapid test (point of care test). This test can be done in swab collected from the suspected herpetic lesion. For example, swab from mucocutaneous lesion, conjunctivitis, ulcerated lesions, crusted vesicles, etc. In general, the sensitivity of these tests is high in symptomatic patients. The sensitivity of IF is around 80% with specificity >90%, whereas >95% sensitivity and specificity has been reported for ELISA. Antigen detection test is considered to be equally sensitive to that of culture but less than PCR.

Cytology: Cytological examination of cells from skin and mucosal lesion can be done for demonstration of **intranuclear inclusion body** or **multinucleated giant cells**. Smears prepared from the cells collected from the base of the lesion are stained with Giemsa **(Tzanck smear)**, Papanicolaou. Demonstration of intranuclear inclusion body or multinucleated giant cells gives a presumptive diagnosis of HSV infection (Fig. 3.2b). However, the test shows poor sensitivity (30%) and specificity.

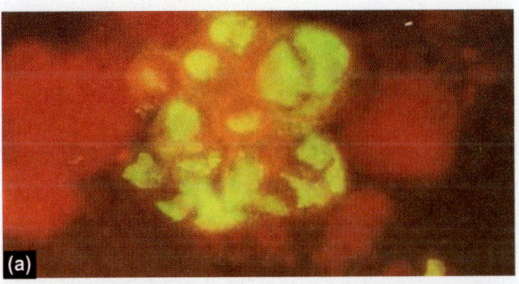
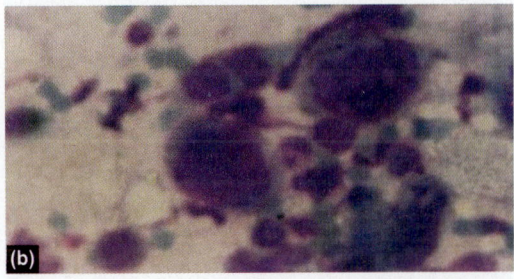

Fig. 3.2a and b: (a) HSV infected cells showing intranuclear fluorescence by indirect IF; (b) HSV infection showing multinucleated giant cells

Virus isolation: Traditionally isolation of virus in cell culture is considered as the gold standard for diagnosis of herpes simplex virus infection. Isolation of HSV can be made by primary **(foreskin fibroblast)**, diploid **(MRC5)** or continuous cell lines **(Vero)**. Higher sensitivity is achieved with primary cell lines. The cytopathic effect (CPE) of HSV is described as clustering of cells with **rounding, ballooning** and **lytic degeneration** or **multinucleated giant cell formation** (Fig. 3.3a and b). However, CPE is not specific and needs to be confirmed by monoclonal antibody or PCR. Isolation followed by direct typing by monoclonal antibody or PCR is considered as preferred method for diagnosis of herpes simplex virus infection. Sensitivity of all these tests is more (i) during primary infection than recurrent infection, (ii) vesicular lesion than ulcerative lesion, and (iii) in immunosuppressed patients than immunocompetent patients due to high level of HSV DNA or antigen. Shell vial method using centrifugation-enhanced technique followed by identification of HSV antigen by IF can also be used for virus isolation. Time required for diagnosis decreases significantly by this method.

Detection of HSV antibody: Role of antibody detection is minimal in diagnosis of acute HSV infection. Presently available IgM detection tests are not reliable for diagnosis of acute HSV infection. Acute and convalescent serum samples can be subjected for demonstration of seroconversion to diagnose primary infection.

Detection of HSV specific IgG antibody in patient's serum indicates the past infection status of the patient. Enzyme-linked immunosorbent assay (ELISA) using type specific glycoprotein G is used to distinguish between HSV1 and 2 infections. Type specific diagnosis is useful particularly in genital herpes as the recurrence is less in HSV-1 infection.

Antibody to HSV can also be detected by conventional serological tests such as complement fixation test (CFT), hemagglutination inhibition test (HAI), immunofluorescence and neutralization tests. However, antibody detected by these tests indicates past infection status of the individual. HSV type specific antibody cannot be detected by these tests. These tests are no more in use for patient diagnosis.

Complement fixation test was used to determine the **significant ratio (≤1:20)** in antibody titer between CSF and serum in herpes encephalitis. The disadvantage of the test is its tedious technical procedure and extensive standardization process.

In case of HSE, intrathecal antibody appears mostly during the second week of illness, hence not useful for diagnosis at the time of presentation. Absence of HSV IgM does not rule out HSV infection as only 30% of HSE are associated with primary HSV infection. The ratio of serum to CSF HSV IgG antibody ≤20 is suggestive of intrathecal antibody production. Previously, conventional serological tests such as hemagglutination inhibition or

 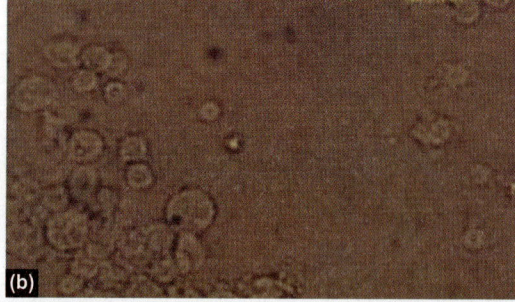

Fig. 3.3a and b: (a) Normal vero cells; (b) HSV infected vero cells showing clustering, rounding, ballooning

complement fixation test were used for determination of serum to CSF antibody ratio. However, these tests are no more preferred now.

THERAPY

Antiviral agents used against HSV are viral DNA polymerase inhibitors. Acyclovir, famciclovir, ganciclovir, and valcyclovir are the approved agents.

All these agents are selectively phosphory-lated from their monophosphate form to the tri-phosphate form by cellular enzymes in the virus infected cells. Triphosphate form of the drug gets incorporated in the DNA chain of the virus and thereby inhibits the replication.

These drugs are available in oral, topical and intravenous formulations and have been shown to reduce the duration of symptoms and lesion in mucocutaneous infection. In seropositive immunocompromised patient, it is given to reduce the risk of reactivation. **Intravenous acyclovir** 10 mg/kg infusion over 1 hour at 8 hours interval is given for 10 days in case of herpes encephalitis. The treatment may be continued, if CSF remains positive for HSV DNA after 10 days.

Acyclovir resistant HSV strains have been reported. Resistance to acyclovir develops due to altered thymidine kinase specificity. Cross-resistance to famciclovir and valcyclovir is usually shown. Resistance is more commonly observed in immunocompromised host and in person with disseminated HSV disease. IV **Foscarnet** is the treatment in case of acyclovir resistance.

Bibliography

1. Arvin A, Campadelli-Fiume G, Mocarski E, et al (eds). Human Herpesviruses: Biology, Therapy, and munoprophylaxis. Cambridge: Cambridge University Press; 2007.

2. Debiasi RL, Tyler KL. Molecular methods for diagnosis of viral encephalitis. Clin Microbiol Rev 2004;17(4):903–25.

3. Johnston C, Corey L. Current concepts for genital herpes simplex virus infection: diagnostics and pathogenesis of genital tract shedding. Clin Microbiol Rev 2016; 29:149–61.

4. Lakeman FD, Whitley RJ, et al. Diagnosis of herpes simplex encephalitis: Application of polymerase chain reaction to cerebrospinal fluid from brain-biopsied patients and correlation with disease. J Infect Dis 1995;171: 857–63.

5. LeGoff, et al. Diagnosis of genital herpes simplex virus infection in the clinical laboratory. Virology Journal 2014, 11:83. http://www.virologyj.com/content/11/1/83.

6. Tang YW, Mitchell PS, Espy MJ, Smith TF, Persing DH. Molecular diagnosis of herpes simplex virus infections in the central nervous system. J Clin Microbiol 1999;37:2127–36.

7. Whitley RJ, Kimberlin DW, Roizman B. Herpes simplex virus. Clin Infect Dis 1998;26:541–55.

B. VARICELLA-ZOSTER VIRUS

Varicella-zoster virus (VZV) causes two separate clinical entities, **varicella and zoster**. The primary infection, known as varicella or commonly called chickenpox, is one of the common contagious diseases of the childhood that is characterized by exanthematous, pruritus and vesicular rash. Zoster, which is otherwise known as **Shingles,** is caused due to reactivation of the latent virus that usually occurs in elderly and immunocompromised individuals and is characterized by painful lesions that are spread over the innervations of a single sensory dermatome.

The **link between varicella and zoster** was established from several evidences:

i. Von Bokay is the first one to observe chickenpox in individuals who had close contact with zoster patients.

ii. In 1925, Kundratiz showed development of chickenpox lesion by inoculating the vesicular fluid from zoster lesion.

iii. Similarities of histopathological lesion like intranuclear inclusion bodies and multinucleated giant cells in both vari-cella and zoster.

iv. Identical appearance of virions isolated from both the lesions.

v. Weller and colleague proved that no difference in biological or immunological properties exists between the virions isolated from varicella and zoster. The same was confirmed further by molecular studies.

VIRUS

Varicella-zoster virus (VZV) is a member of Herpesviridae family and subfamily Alphaherpesvirinae. Structurally it is similar to other members of the family. The diameter of virus is approximately 150–200 nm. The core of the virus consists of linear double-stranded DNA which is surrounded by an icosahedral symmetry and a lipid envelop with glycoprotein spikes. DNA of the virus consists of 125,000 base pairs and encodes for 75 proteins. There are five families of glycoproteins gpI to gpV. These proteins are important for their immunogenicity property and monoclonal antibodies gpI to gpIII have been shown to neutralize the viral infectivity.

VZV being an enveloped virus is sensitive to lipid solvents like detergents, ether, and also to air drying.

EPIDEMIOLOGY

Varicella

Humans are the only known reservoir of VZV. Primary infection **(chickenpox)** occurs due to transmission of virus from patients with chickenpox or herpes zoster to the susceptible or seronegative individual. In temperate countries, majority of primary infections occur during childhood and more than 90% seropositivity is achieved by 10–15 years of age. Whereas the scenario is different in tropical countries where seroconversion occurs at a later age and >80% seropositivity occurs by 30 years of age. The late seroconversion in tropical countries is thought to be due to reduced transmission of virus in hot and humid climate. The transmission of virus is known to be more in cold, temperate climate. The outbreaks

of varicella in tropical countries have been reported more often during winter months.

The severity of disease is considered to be higher in adults as compared to children. Thus, tropical countries are considered to be at higher risk of morbidity and mortality due to seroconversion at a later age.

Incubation period varies from 10 to 21 days, average being 14–21 days. Secondary attack rate is 70–90%.

Herpes Zoster

Herpes zoster occurs due to reactivation of varicella virus from the dorsal root ganglia. The median incidence of zoster is 4–4.5 per 1000 person years as reported by various studies conducted in different populations including USA, Europe, Australia, South America and Asia. It can occur in all age group but the incidence is higher in elderly people with 68% of zoster occurring in individuals above 50 years of age. The mean age at onset of zoster among adults is 59.8 years.

Post-herpetic Neuralgia (PHN)

As with zoster, the risk of PHN also increases with age. 80% of all PHN occur in persons above 50 years of age. The incidence of PHN in zoster patients above 50 years and 80 years are 18% and 33%, respectively.

PATHOGENESIS AND PATHOLOGY

The transmission of VZV mainly occurs when aerosolized virus particle from skin lesion of patient with varicella or zoster is inhaled by the nearby susceptible person or through direct contact with the vesicle fluid of the skin lesion which infects the respiratory tract. It does not spread in absence of the skin lesion and the rate of transmission is high when numbers of vesicles are more.

The entry of virus occurs through respiratory tract or through conjunctiva. The virus then replicate in the nasopharynx, seeded to various reticuloendothelial organs and infects CD4+ and CD8+ T lymphocytes. During the viremic phase, it infects immune cells. Finally through homing of the infected immune cells,

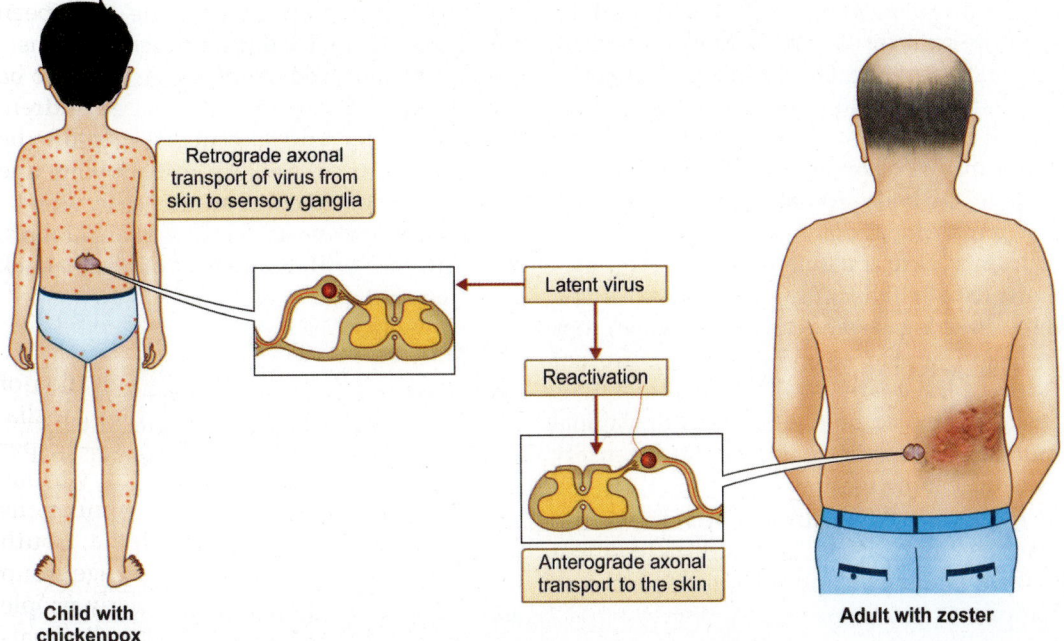

Retrograde axonal transport of virus from skin to sensory ganglia

Latent virus

Reactivation

Anterograde axonal transport to the skin

Child with chickenpox

Adult with zoster

Fig. 3.4: Primary infection and reactivation of VZV

virus reaches to various skin sites, where it replicates in the epidermal cells and produces characteristic varicella rash.

During the primary infection (chickenpox), virus reaches the sensory neuron ganglia either through hematogenous route or through retrograde axonal transport from skin and establishes latency there. When the virus is reactivated it is transported through anterograde axonal transport to the skin where it causes vesicular rash along the dermatome that is innervated by the ganglion from which reactivation has occurred. The decline in VZV specific T cell immunity leads to reactivation of the virus causing zoster (Fig. 3.4).

CLINICAL FEATURES

Varicella-zoster virus is associated with two different but related clinical entities; one is **varicella or chickenpox** which occurs during the primary infection and the other is zoster or shingles which occurs as a manifestation of reactivation of the latent virus.

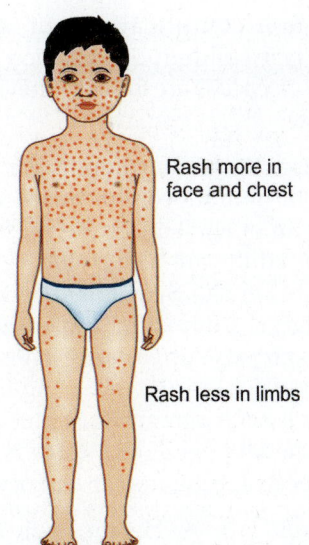

Rash more in face and chest

Rash less in limbs

Fig. 3.5: Child with chickenpox rash

Varicella

Chickenpox often presents with a prodrome of fever, malaise and headache before 24–48 hours of appearance of the rash. Low grade fever, anorexia and malaise usually continue along with the rash for 2–3 days. Skin lesions are the hallmark of infection. Rashes in chickenpox are characteristically **centripetal**, begin in the central part of the body like head, scalp, face and trunk, then **spread centrifugally** to other parts of the body. Initially the nature of the rash is maculopapular, which evolves rapidly to form vesicle, pustule and scab (Fig. 3.2). Crust usually falls off by 10–14 days leaving minor scarring in some cases. Successive crops of lesion appear over 2–4 days. Different stages of lesion are usually seen at one time. Lesions can also be seen on oral or vaginal mucosa. Lesions are **painless but intensely pruritic** leading to scratching.

The severity is usually less in children who may have only a few lesions. In adults, the disease is almost 25 times more severe as compared to children.

Complications

Sepsis: Secondary bacterial infection is one of the common complications of varicella particularly in children. Streptococci group A and Staphylococcus are the common infecting pathogens.

Pneumonia: Varicella pneumonia is more common in immunocompromised children and more so in leukemia patients. Immunocompetent adults are also prone to develop pneumonia (1 in 200). Symptoms of pneumonia usually start 1–3 days after the onset of rash with cough, dyspnea and chest pain. Histology shows the presence of multinucleated giant cells with intranuclear inclusion body. Unlike bacterial pneumonia, there is absence of excess polymorphonuclear leukocytes.

Other complications: Hemorrhagic chickenpox, encephalitis, Guillain-Barré syndrome and Reye's syndrome are rare complications.

Herpes Zoster (HZ)

The disease is characterized by painful unilateral vesicular skin lesion affecting single dermatome. It is also called Shingles. The site of skin lesion depends on the ganglion involved. **T3 to L3 dermatomes** are most commonly involved. Appearance of lesion is usually preceded with abnormal sensation, hyperesthesia and excruciating pain of affected skin area. This is followed by appearance of maculopapular rash which rapidly becomes vesicular. Lesion appears in the form of a **unilateral strap of vesicles** in the affected dermatome.

Complications

Post-herpetic neuralgia: Post-herpetic neuralgia (PHN) is the commonest complication associated with zoster and is defined as the persistence of pain for 90 days or more at the site of herpes zoster skin lesion. In general, pain resolves within a few weeks but can last for months and years too. The risk factors associated with development of PHN are severity of rash, older age and immunocompromised state.

Herpes zoster ophthalmicus (HZO): It occurs in 10 to 20% of HZ cases when the virus reactivation occurs in the first branch of trigeminal nerve.

Other complications: HZ also can lead to several neurologic manifestations such as meningoencephalitis, myelitis, ocular disorders or vasculopathy leading to ischemic or hemorrhagic stroke. Involvement of the geniculate ganglion of the seventh nerve may lead to facial palsy and lesion in the external auditory meatus may lead to hearing loss. Disseminated zoster occurs mostly in the immunocompromised hosts affecting various organs leading to pneumonia, encephalitis or hepatitis.

DIAGNOSIS

The typical clinical manifestation of chickenpox does not require laboratory confirmation in most of the cases. However, disseminated lesion of HSV may be confused with varicella lesions. Sometimes lesions due to enteroviruses, rickettsia and other bacteria like Staphylococcus may also simulate varicella lesions. The rare manifestations of ocular varicella or encephalitis require confirmation. Confirmation by lab diagnosis is also important in immunocompromised individuals due to associated drug resistance and in vaccine related breakthrough varicella cases.

Collection of sample: Vesicular fluid or scrapings from the base of the lesion are collected for demonstration of virus, viral antigen or inclusion body.

Electron microscopy: Herpes virus particle can be demonstrated in the vesicular fluid. The sensitivity is more in the early phase of infection. As the morphology is similar with other herpes simplex viruses, the specific identification of VZV is done by immunoelectron microscopy. This method is not used for routine diagnosis.

Cytology: Smear (Tzanck smear) made up of the vesicular fluid or base of the lesion, stained by Papanicolaou's method or methylene blue. Multinucleated giant cells and intranuclear inclusion bodies can be demonstrated. However, the sensitivity and specificity is poor and cannot be differentiated from HSV induced cytopathology.

Immunofluorescence: Viral antigen can be detected in the infected cells by direct or indirect immunofluorescence method using specific monoclonal antibodies. This technique is more sensitive than culture but less than that of PCR and cannot be used for differentiation of vaccine and wild type virus.

Virus isolation: This was previously considered as the "gold standard". Human embryonic lung fibroblast (HELF) cell line is used for the primary isolation of VZV from clinical samples. Freshly collected vesicular fluid or samples from base of the lesion are inoculated onto the monolayer. The appearance of cytopathic effect takes time and may require blind second passage. The identification in cell culture is done by appearance of enlarged multinucleated giant cell and enlarged cells with intranuclear inclusion body. Presence of

viral antigen or DNA can be detected in the infected cells by monoclonal antibody or PCR, respectively. However, the process is slow and expensive, hence not useful for patient diagnosis. Cell culture technique for VZV is presently useful for determination of drug resistance.

Viral DNA detection: PCR is a highly sensitive and specific method for diagnosis of VZV and presently considered as the test of choice. DNA detection by PCR is required in case of low viral load, old lesion, recurrent varicella and rare manifestations like involvement of CNS or ocular disease. It can be done from various clinical samples, such as vesicle fluid, skin swab, throat swab, blood, CSF, biopsy, saliva, etc. Real-time PCR is available for VZV which can give results in a few hours both for qualitative DNA detection or quantitation of virus. Differentiation of vaccine strain and wild virus strain can be done by restriction fragment length polymorphism (RFLP) of the PCR product. Several multiplex PCR systems are commercially available for diagnosis of viral encephalitis that includes HSV and VZV.

Serology

VZV IgG: Detection of VZV IgG antibodies can be done by various methods like complement fixation test (CFT), immunofluorescence and ELISA. This is useful in determining the immune status of the individual. Fluorescence antibody to membrane antigen assay that detects antibodies to antigens present on the surface of the infected cell, is the most sensitive and specific method to predict the susceptibility.

VZV IgM: Presence of VZV IgM antibodies in serum indicates recent infection but cannot differentiate between primary varicella and recurrent zoster infection as specific, IgM antibodies are also induced during reactivation. During the initial period of infection, IgM antibodies may be negative but present in 100% convalescent sera. IgM antibodies remain positive for around three months from the illness. Several commercial ELISA systems are available for detection of VZV specific IgM antibodies.

TREATMENT

Management of chickenpox patient is mostly directed towards prevention of secondary infections. Maintenance of hygiene with daily bath, avoidance of pruritus by using anti-pruritic agent and prevention of secondary bacterial infections by cropping of nails is the main key.

Oral acyclovir therapy reduces the number and duration of lesion as well as decreases the constitutional symptoms. Acyclovir is recommended in adolescents, adults and high risk children. Recommended oral dose is 20 mg/kg four times a day (up to 800 mg/day) for five days. IV acyclovir is recommended in immunocompromised patients with chickenpox or zoster to prevent severe complications. Patient with acute neuritis or post-herpetic neuralgia are treated with narcotic or non-narcotic analgesics along with acyclovir.

PREVENTION

Passive Immunization

Varicella-zoster immunoglobin (VZIG) is available for prophylaxis. VZIG has been shown to prevent as well as reduce the symptoms of chickenpox. The recommendation of VZIG is given in Table 3.3.

Active Immunization

Varicella vaccine: A live-attenuated vaccine against VZV is licensed and available for human use. The vaccine was first developed in Japan in 1974. The vaccine virus called **Oka strain** is derived from vesicles of a three years old child with typical varicella. The strain was attenuated by passaging in guinea pig embryo fibroblast followed by WI 38 human fibroblast. Approximately 20 amino acids changes unique to vaccine strain have been identified. Eight of these are located in the IE62 protein and responsible for reducing its ability to

Table 3.3: Recommendation of VZIG

Host status	
Immunocompromised	Exposed to chickenpox/zoster patient No history of previous exposure or disease Not vaccinated against chickenpox Household contact with patient
Pregnant woman	Seronegative for VZV and has significant exposure
Newborn	When mother develops chickenpox within the period of 5 days prior delivery to 2 days post-delivery period

transactivate the expression of early VZV proteins. Three mutations in ORF62 are fixed in the vaccine strain which helps in identification of the vaccine strain.

Single dose vaccine containing 1000–3000 pfu of attenuated Oka virus induced seroconversion in >90% children with 85% protection rate to exposure. In adults and immunocompromised children, two doses are required to achieve high seroconversion rate. In around 20% of vaccinated population, there may be mild pain, redness and swelling at the injection site. Vesicular rash may develop in 3–5% cases. The incidence of post-vaccination zoster is significantly less as compared to post-natural infection.

In the US, Oka vaccine is given to all children at age 12–15 months since 1995. Several other countries have also included this vaccine in their universal immunization program. Recently a quadrivalent vaccine of varicella with measles, mumps, rubella (MMRV) has become available. However, the combined vaccine has been associated with higher rate of febrile seizure than monovalent vaccine, when given to 12–13 months old children.

In India, the vaccine is not included in national immunization program. However, Indian Academy of Pediatrics has recommended

Table 3.4: Varicella vaccines

Name	Manufacturer	Available form		Age
Varilrix	GlaxoSmithKline	0.5 ml SC inj		≥12 months
Varivax	Merck	0.5 ml SC inj } 2 doses		≥12 months
Okavax	Sanofi Pasteur India Ltd.	0.5 ml SC inj		≥12 months

two doses of varicella vaccine, first dose at 15 months of age and second dose at least after 3 months of the first dose. For children >13 years, 2 doses of vaccine are given at an interval of minimum 4 weeks. Table 3.4 describes the detail information of available licensed varicella vaccines.

Vaccine for Zoster

Currently two vaccines are licensed by FDA for zoster; Shingrix and Zostavax.

Shingrix: This is a recombinant zoster vaccine, marketed by GlaxoSmithKline (GSK). This recombinant vaccine is a combination of glycoprotein E and envelops protein of VZV with adjuvant AS01B. Glycoprotein E is included as it is the main target of CD4 T cell immune response. The VZV gE subunit is a truncated molecule that lacks the anchor and the carboxy-terminal tail domain. This is the preferred vaccine between the two by CDC and is recommended in immunocompetent persons of 50 years and above. The efficacy of vaccine has been reported to be better than Zostavax. Efficacy to prevent zoster is >90% and to prevent post-herpetic neuralgia is near 90% in persons aged >70 years was found in study involving more than 30,000 participants during a follow-up period of 3.2 years. Efficacy is higher in lesser age people. It is administered intramuscularly. Two doses are given at 2–6 months interval.

Zostavax: This is a live attenuated Oka strain vaccine manufactured by Merck, USA. Each dose of vaccine contains 19,400 pfu of active VZV, and administered single dose subcutaneously. The vaccine has been recommended by CDC for use in immunocompetent persons of more than 60 years age who are allergic to

Shingrix vaccine or prefers Zostavax. It has been shown to decrease the risk of herpes zoster development by 70% in persons of 50–59 years in a follow-up period of 1.3 years. The efficacy for both herpes zoster development and post-herpetic neuralgia decreases with increase in age.

Bibliography

1. Cunningham AL, Lal H, Kovac M, et al; ZOE-70 Study Group. Efficacy of the herpes zoster subunit vaccine in adults 70 years of age or older. N Engl J Med 2016; 375(11):1019–32.

2. Gershon AA, Gershon MD. Pathogenesis and current approaches to control of varicella-zoster virus infections. Clin Microbiol Rev 2013;26(4): 728–43.

3. Johnson RW, Alvarez-Pasquin MJ, Bijl M, Franco E, et al. Herpes zoster epidemiology, management, and disease and economic burden in Europe: a multidisciplinary perspective. Ther Adv Vaccines 2015;3(4):109–20.

4. Lee BW. Review of varicella zoster seroepidemiology in India and Southeast Asia. Trop Med Int Health 1998;3(11):886–90.

5. Marin M, Bialek SR. Varicella (chickenpox). http://wwwnc.cdc.gov/travel/yellowbook/2016/infectious-diseases-related-to-travel/varicella-chickenpox# accessed on 25 September 2016.

6. Wang L, Zhu L, Zhu H. Efficacy of varicella (VZV) vaccination: An update for the clinician. Ther Adv Vaccines 2016 Jan;4(1–2):20–31.

7. Zerboni L, Sen N, Oliver SL, Arvin AM. Molecular mechanisms of varicella zoster virus pathogenesis. Nat Rev Microbiol 2014;12(3):197–210.

8. "https://www.cdc.gov/mmwr/volumes/67/wr/mm6703a5.htm?s_cid=mm6703a5_w" Recommendations of the Advisory Committee on Immunization Practices for Use of Herpes Zoster Vaccines | MMWR (cdc.gov)

C. CYTOMEGALOVIRUS

Cytomegalovirus is one of the important pathogens responsible for causing congenital infection and opportunistic infections in immunocompromised patients.

Human cytomegalovirus (HCMV) is a ubiquitous virus which belongs to family Herpesviridae and subfamily *Betaherpesvirinae*. HCMV is a double-stranded DNA virus having icosahedral nucleocapsid which is surrounded by a membranous protein called tegument and the entire components are enclosed within the lipid bilayer of envelop containing viral glycoproteins. CMV is the largest virus in the family with a genome of ~235 kb encoding ~165 genes. The size of the mature virion is 200–300 nm diameter. The tegument component contains the majority of the viral proteins. Lower matrix phosphoprotein 65 (CMVpp65) also termed unique long 83 (UL83) is the most abundant. Virion transactivator pp71 (upper matrix protein, UL82 gene product), the herpesvirus core virion maturation protein pp150 (large matrix phosphoprotein, UL32 gene product), the largest tegument protein (UL48 gene product), and the UL99-encoded pp28 are the other major tegument proteins.

EPIDEMIOLOGY

CMV is a ubiquitous virus and is distributed all over the world. The prevalence of CMV infection increases with age. In developing countries, infection is acquired in the childhood and majority of the population is seropositive by adult age. The prevalence rate is higher in developing countries in comparison to that of the developed countries which is considered to be due to the lower socioeconomic status of the former.

Epidemiology of vertical infection: Transmission of the virus from mother to fetus occurs transplacentally through vertical transmission. The vertical transmission of CMV from mother to fetus can occur both during primary or recurrent maternal infection. The risk of acquiring primary maternal infection during pregnancy is higher in women of higher socioeconomic background and developed countries as compared to that of the women of lower economic background and developing countries. The rate and severity of fetal infection depends on the immune status of the mother against CMV (Flowchart 3.2).

- The rate of transplacental transfer of CMV infection to fetus is higher during primary maternal infection.

Flowchart 3.2: Outcome of CMV infection in pregnancy

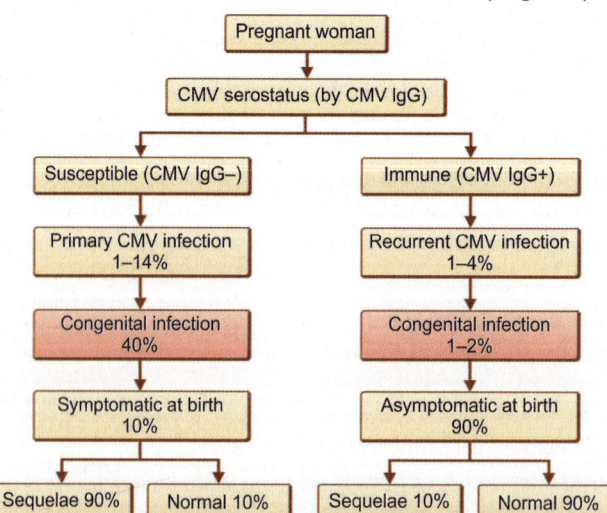

- The rate of symptomatic fetal infection is also more when the infection is transmitted from mother with primary CMV infection.
- The chances of development of sequelae are more in babies having clinical manifestation at birth as compared to the babies who were asymptomatic at birth (90% *vs* 10%).
- The rate of transplacental transmission is higher during the late part of pregnancy but risk of severity of fetal damage is higher when the fetal infection occurs during early part of gestation. The rate of congenital infection has been reported from 0.2 to 2% and higher infection rate is associated with higher maternal seroprevalence. The consequence of CMV infection during pregnancy is depicted in Flowchart 3.2.
- Transmission of virus through breast milk is, however, the most common mode of transmission from mother to baby.

Epidemiology of horizontal infection: Horizontal transmission of CMV infection occurs through exposure to saliva, tears, urine, breast milk, semen, and other infected body fluids. The infection can also be transmitted through blood transfusion and organ transplantation. Transmission of infection among children usually occurs in day care centers mostly through saliva or urine. Mouthing of toys helps in spreading the infection. Adults attending the children in day care centers or nurseries show higher seroprevalence. Between adults, sexual route is one of the commonest modes of infection.

Primary infection of CMV usually occurs during childhood and mostly asymptomatic in nature in immunocompetent individual. Primary infection induces both humoral and cellular immunity. IgM antibody against CMV is raised during the acute phase of infection and acts as an indicator of acute or recent infection. CMV IgG appears within weeks of infection and persists for life, thus is an indicator of previous or past infection.

CMV being a member of herpesvirus family remains inside the host forever in latent form after the primary infection. Reactivation occurs intermittently in immunocompetent persons. In immunocompromised patients, suppression of CMV specific cell-mediated immunity reactivates the virus from latency. Reactivation of CMV can lead to uncontrolled viral replication and CMV disease. This is particularly important in transplant recipients, both stem cell and solid organ transplant and also in AIDS patients.

Depending upon the CMV serostatus of the transplant recipients, CMV infection can be primary or secondary (due to reactivation of latent virus). The post-transplant CMV infection in CMV seronegative recipients (R–ve) is primary in nature which can be acquired from a seropositive donor or due to new infection in the recipient. Whereas in CMV seropositive recipients (R+ve), the infection is secondary mostly due to reactivation of the latent virus because of immunosuppressive drugs and less commonly due to new infection.

CLINICAL MANIFESTATIONS

Infection in immunocompetent host usually does not lead to clinically apparent manifestation. Occasionally it may lead to mononucleosis like features with fever, cervical lymphadenopathy, and sore throat with or without hepatosplenomegaly.

CMV infection in immunocompromised individuals and congenital CMV infection are the two most important clinical entities due to CMV infection.

At this stage, it is important to understand the difference between CMV infection and CMV disease.

CMV infection: Evidence of CMV replication regardless of symptoms (differs from latent CMV).

CMV disease: When CMV infection is accompanied by clinical symptoms and sign, CMV disease is characterized with fever, malaise, leukopenia, and/or thrombocytopenia or as tissue-invasive disease.

Asymptomatic CMV infection is defined as CMV replication without sign and symptoms of CMV disease.

Congenital infection: Around 90% of congenitally infected babies are asymptomatic at birth and 10% present with clinically apparent disease. Amongst the symptomatic neonates, death occurs in around 10% within a few weeks of birth.

Two-thirds of symptomatic neonates show the evidences of neural involvement in the form of microcephaly, lethargy-generalized hypotonia, poor feeding, and seizure. Cranial imaging shows abnormality in about 75% of symptomatic neonates where periventricular calcification is the most common, and other abnormalities can be ventriculomegaly, lissencephaly and cortical atrophy. Sensorineural hearing loss is reported in around 50% symptomatic newborns and chorioretinitis in around 10%. Clinical findings of congenital CMV infection are given in Table 3.5. The risk of developing sequelae is higher in neonates who were symptomatic at birth as compared to those who were asymptomatic. Hearing loss, mental retardation and retinitis are the common sequelae.

Primary maternal infection, infection at early age of gestation, high level of viremia at birth and infection due to CMV gN4 genotype are associated with the severity of congenital disease.

Table 3.5: Clinical manifestations of congenital CMV infection

Common features: Hepatosplenomegaly, petechiae, jaundice , microcephaly

Severe features: Fetal death or fatal cytomegalic inclusion disease in newborn

Lab findings: Thrombocytopenia, hyperbilirubinemia

Other: LBW, intracranial calcification, chorioretinitis

Death in first 6 weeks (DIC)

Long-term sequelae: Sensory neural hearing loss, chorioretinitis

LBW: Low birth weight, DIC: Disseminated intravascular coagulation

Infection in immunocompromised hosts: Incidence of CMV infection or disease in transplant recipients depends on the:
- CMV infection status of the donor (D) and recipient (R).
- Degree of immunosuppressive therapy.
- Viral load.

CMV infection in solid organ transplant (SOT) recipients: According to the CMV immune status, donor (D)/recipient (R) combination has been stratified into four groups. According to the level of risk involved, these are:

1. *D+/R–: Donor seropositive/recipient seronegative:* Donor can transmit the infection to the recipient leading to primary CMV infection in the recipient and as recipient has no previous CMV antibody to fight, the disease severity can also be high. Thus, this group is considered to carry the highest risk for CMV disease in recipient.

2. *D+/R+: Both donor and recipient seropositive:* As recipient is already seropositive, the chance of infection is mostly due to reactivation of the latent virus than due to transmitted infection from donor.

3. *D–/R+: Donor seronegative and recipient seropositive:* In this case, most common cause of infection is reactivation of latent virus in the recipient.

4. *D–/R–:* Both donor and recipients are CMV seronegative. So, the only chance of acquiring infection is from sources other than donor. But if infection will be acquired, it will be primary infection in the recipient which may cause severe disease.

The most common manifestation of CMV disease in SOT recipients is termed **CMV syndrome:**
- This is characterized by general symptoms like fever, malaise, anorexia and arthralgia. Blood count shows, leukopenia, thrombocytopenia and raised hepatic transaminase.
- In severe cases, tissue invasion can occur affecting various organs.

The involvement of gastrointestinal system is the most common amongst them; however, any organ can be affected leading to hepatitis, pneumonitis, nephritis, retinitis, etc.

Indirect effect of HCMV infection can also occur which includes graft rejection, accelerated coronary artery vasculopathy in heart transplant patients and opportunistic infections.

Transplant organ is usually affected in case of primary infection in these patients. This is manifested as renal impairment in kidney transplant, pneumonitis in lung transplant, vasculopathy in heart transplant, etc. In SOT recipients, most of the HCMV infections occur within 4–8 weeks post-transplant. Prophylactic or pre-emptive therapy is given early to prevent the development of CMV disease.

HCMV infection in HSCT recipients: In contrast to SOT recipients, the level of risk according to the donor–recipient CMV serostatus is D–/R+, D+/R+, D+/R– and D–/R–. The risk gets enhanced in case of histocompatibility mismatch.

Gastrointestinal disease and interstitial pneumonitis are the common manifestations in allogenic HSCT recipients. The widespread use of prophylactic or pre-emptive therapy based on pp65 antigenemia or CMV DNA load by real-time PCR has reduced the incidence of CMV disease to a great extent.

The clinical manifestations and diagnosis of CMV disease involving various organs are same in all types of immunocompromised patients.

Gastrointestinal (GI) CMV disease commonly involves the lower GI tract than upper GI tract. It is clinically presented as diarrhea, vomiting, bleeding, anorexia and abdominal pain and often indistinguishable from GI GVHD.

CMV pneumonia is another common clinical entity in transplant recipients. It presents with the features of pneumonia such as hypoxia, tachypnea, dyspnea, appearance of new infiltrates in imaging, etc.

CMV hepatitis, retinitis, encephalitis and disseminated CMV disease are other manifestations.

HCMV infection in HIV patients: Cytomegalovirus disease is one of the AIDS defining illnesses. High HIV viral load and low CD4 counts are the two most important risk factors for HCMV infection in HIV patients. Majority of HCMV infection occurs in HIV positive patients with CD4 count <50/μl. Retinitis, respiratory manifestations and gastrointestinal symptoms are the common presentations. The incidence of HCMV retinitis has decreased up to 80–90% in HIV patients due to use of anti-retroviral therapy.

DIAGNOSIS

Virus isolation: Isolation of CMV from various clinical samples like urine, saliva, blood, amniotic fluid, etc. is done on fibroblast cell lines. The conventional culture is time consuming and also poorly sensitive. This has been replaced with shell vial culture where sample is inoculated with low speed centrifugation onto the cells to enhance the infection. Viral immediate early proteins are expressed within 24 hours in the infected cells, which are then identified by immunofluorescence using monoclonal antibodies. Diploid cell lines like MRC5 or WI 38 are the commonly used cell lines for the growth of HCMV.

CMV inclusion body: CMV produces enlargement of infected cells with production of intranuclear inclusion body with peri-nuclear halo. This cytopathic effect is typically known as **owl's eye** appearance (Fig. 3.6a). Cytomegalic cells bearing intranuclear inclusion body with owl's eye appearance can be demonstrated in the tissue of the affected organ which indicates tissue invasion and thus CMV disease. Presence of these cells in the exfoliated urinary epithelial cells in neonates can be used to diagnose congenital CMV infection.

Detection of viral antigen: Among the various CMV antigens, detection of **pp65 (CMV phosphoprotein 65)** antigen in polymorphonuclear leukocytes is most commonly used. Detection is done by indirect immunofluorescence test using monoclonal antibody to pp65 antigen. The pp65 antigen positive cells show intranuclear apple green fluorescence (Fig. 3.6b).

Presence of CMV pp65 antigen positive cell in blood sample (antigenemia) indicates active infection. Quantitation of antigenemia helps in differentiating CMV infection and disease. This test is of more importance in transplant

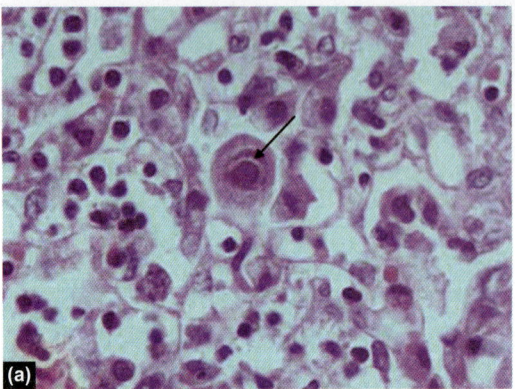
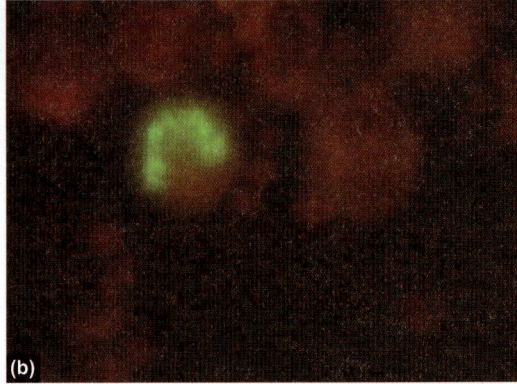

Fig. 3.6a and b: (a) Microphotograph showing intranuclear inclusion with perinuclear halo giving owl's eye appearance of CMV (H&E ×40). (*Photograph courtesy*: Prof Uma N. Saikia, Histopathology, PGIMER, Chandigarh); (b) Microphotograph showing CMV pp65 antigen positivity in polymorphs by IIF using monoclonal antibody (IF ×20).

recipients where antibody response may not be helpful for diagnosis because of immuno-suppression. Number of positive cells over total number of cells counted is determined.

Advantage of the test: Rapidity

Limitations:
a. Subjectivity
b. Sample needs to be processed rapidly
c. Limited value in leukopenic patients

Detection of viral genome: Viral genome can be detected by conventional PCR from various clinical samples. CMV gB gene or immediate early gene are the most commonly targeted genes by PCR.

Qualitative detection of CMV virus DNA in blood of neonate is done to diagnose congenital infection.

In adults, detection of viral genome by conventional PCR can be possible due to active infection or presence of latent virus. Therefore, viral load estimation is important in order to differentiate between CMV infection and CMV disease. High viral load indicates CMV disease.

Serology: Several serological methods are available for detection of antibodies, however, ELISA is most widely employed test.

CMV IgG antibody: CMV IgG antibody once appears during primary infection persists for life. Positive CMV IgG in serum sample in a previously negative individual indicates seroconversion which acts as marker of primary infection.

CMV IgG positivity in a single sample indicates past infection.

CMV IgM antibody: Several types of ELISA are available for detection of CMV IgM anti-body; however, capture ELISA and recombi-nant ELISA methods are preferred because of their high sensitivity and specificity.

Positive IgM antibody in immunocom-petent pregnant women with rapid fall in titer within months is generally considered as primary infection. Whereas, in immunocompro-mised individuals, IgM antibody usually persists for a long time and can give positive result in recurrent infection also. As CMV IgM antibodies can persist from a few months to up to one year after subsidence of the acute phase of infection. The positivity can, therefore, be due to acute infection, convalescent phase of primary infection or due to persistence of antibody. Thus, IgM antibody status in a single sample should be interpreted with caution.

CMV IgG avidity ELISA: It is based on the principle that virus specific IgG antibody of lower avidity is produced during the initial phase of infection which becomes of high avidity with recurrent infection:

- High avidity IgG antibody indicates remote or recurrent infection.
- Low avidity indicates recent infection.

The mean avidity index during the first few months of primary infection has been reported to be of 21% as compared to 78% from patients with remote infection.

Presence of virus specific IgM antibody along with low avidity IgG antibody indicates recent or primary infection, whereas IgM antibody along with high avidity IgG is suggestive of past infection.

Serology has limited role in diagnosis in immunocompromised individuals.

Diagnosis of CMV infection is important in the following clinical settings

- To diagnose the type of infection during pregnancy: Primary or recurrent.
- To diagnose congenital infection.
- To determine the viral load in congenitally infected neonates as a prognostic indicator.
- To diagnose CMV disease in immuno-compromised patients.

Diagnosis of CMV infection during pregnancy: It is now understood that diagnosis of primary infection during pregnancy is important. Serology is the mainstay to distinguish between primary and recurrent infection.

Primary CMV infection in pregnancy is determined by:

- *Seroconversion:* Most authentic marker of primary infection.
- *Positive IgM antibody with low avidity IgG antibody:* More reliable than only IgM positivity.

Diagnosis of congenital infection: Detection of CMV (virus/DNA) within first three weeks of birth is considered as congenital infection. Detection of virus beyond three weeks cannot differentiate between congenital, intranatal and postnatal infections.

Criteria of diagnosis:

- Isolation of virus from urine, saliva or blood sample by shell vial assay.

- Detection of viral genome by conventional or real-time PCR.
- Serum sample positive for IgM antibody.
- Demonstration of four-fold rise of IgG antibody titer.

Prenatal diagnosis is made by detection of virus or viral genome in amniotic fluid at 20 or more weeks of gestation by culture and PCR, respectively.

Diagnosis in immunocompromised host: In immunocompromised patients, reactivation of latent virus is the most common cause of CMV associated disease. However, reactivation and active replication of virus can occur without causing any clinical manifestation. Therefore, mere detection of virus/DNA in blood does not implicate the association of CMV with clinical disease. Hence, it has no clinical relevance.

Presence of virus in the affected organ is the definitive method to associate with the disease. This requires the demonstration of virus/antigen or its inclusion body in the tissue specimen. However, as collection of tissue biopsy is an invasive procedure, it is not feasible in most of the cases. This makes the quantitation or viral load estimation in blood of the patient important which helps (i) to associate with the ongoing clinical disease, (ii) start pre-emptive therapy in post-transplant cases, (iii) to monitor the treatment response.

Methods of CMV quantitation/viral load estimation: This is often done from the blood sample (whole blood or plasma). Quantitation of CMV viral load is done by CMV antigenemia or CMV viral load. There is no consensus regarding the cut of value for viral antigen or DNA load in different types of transplant patients or different groups of immuno-compromised patients. In general, lower viral load (antigenemia/DNA) is associated with disease in HSCT patients as compared to SOT patients. Similarly, the cut-off value for pre-emptive therapy also has been set up at different level in different clinical settings. Between the two methods, viral DNA is better as it can be detected earlier than antigen and also more sensitive.

1. *Quantitation of CMV antigenemia:* Presence of CMV pp65 positive cells is an indicator of active infection. Quantitation is done by counting the number of positive cells over total number of counted cells. Quantitative estimation helps in differentiating between CMV infection and disease, however, as compared to PCR, it is less sensitive and appears later than PCR.

2. *CMV DNA/mRNA load:* This is determined by real-time PCR or NASBA respectively. Detection of CMV mRNA is considered as more specific as this reflects the replicative stage of the virus. However, as it degrades rapidly, the sensitivity of mRNA is less than DNA. CMV DNA in blood is often used to determine the time of pre-emptive therapy initiation and also to monitor the course of the disease. Various PCR platforms are available commercially. One of the problem with CMV PCR is different PCR system is standardized to express the result in different units. Therefore, the result should be normalized as per WHO's reference International standard and should be reported in IU/ml. It is also recommended to use same specimen type, same extraction and PCR assay for serial testing.

Diagnosis of gastrointestinal CMV disease: Detection of antigenemia or PCR from blood sample in gastrointestinal CMV disease is often negative at the time of diagnosis and hence unreliable. This is possibly because, initially it is a local tissue event and thus the CMV load in plasma or whole blood does not reflect properly the CMV replication in gastric mucosa. **Histopathology** and **immunohistochemistry of gastric mucosal biopsy** is the **gold standard** to confirm CMV GI disease. However, some of the recent studies using fresh endoscopic biopsy have shown promising result with quantitative PCR as compared to IHC.

Diagnosis of CMV pneumonia: CMV pneumonia posseses challenge to conform its diagnosis because lung biopsy is not a preferred specimen as it involves invasive procedure. Virus cultivation from bronchoalveolar lavage (BAL) sample does not differentiate between CMV pneumonia and asymptomatic viral shedding. Quantitative PCR in BAL in both HCT and SOT has shown wide variation in viral load cut-off to diagnose CMV pneumonia in various studies (500 IU/ml to 30000 IU/mL in HCT and 3.2×10^5 IU/ml to 1.8×10^7/mL) in SOT respectively.

Though exact cutoff of viral load has not been determined to diagnose any particular type of CMV disease or for initiation of pre-emptive therapy, it is proved beyond doubt that the role of viral load acts as a surrogate marker for disease as evident from below mentioned findings:

1. *Association of high viral load with disease:* In a meta-analysis, 7-fold higher viral load has been shown in CMV disease as compared to asymptomatic infection when studies using one type of PCR platform and >9-fold when all types of studies are included.

2. *Relation of viral load kinetics with development of CMV disease:* More rapid increase in viral load is positively correlated with development of disease.

3. *Rate of viremia during prophylaxis:* The incidence of viremia is significantly less during prophylaxis as compared to the entire period; 3.2% vs 34.3%.

4. *Incidence of CMV disease during prophylaxis:* The pooled incidence of CMV disease during prophylaxis is 0.8% as compared to 13% during the entire period.

5. *Decrease in viral load and symptom resolution:* Correlation between decrease in viral load due to treatment with resolution of symptoms was observed. Patients with viral load **<18000 IU/mL showed faster resolution**. Patients in whom faster decrease in viral load was found, showed faster clinical resolution.

TREATMENT

Antiviral drugs against CMV: Four antiviral agents have been approved for treatment of CMV infections (Table 3.6) and three novel drugs are currently under evaluation.

Ganciclovir (GCV): It is an acyclic guanosine analog. It gets converted to its active form ganciclovir triphosphate by the viral phosphatase coded by gene UL97. Ganciclovir triphosphate selectively inhibits CMV DNA polymerase. It is given either alone or in combination with CMV immunoglobin in severe cases. Majority of the patients treated with ganciclovir show a good response.

Resistance may develop in patients treated for more than 3 months. Mutation in the gene CMV UL97 and less commonly in UL54 is responsible for development of GCV resistance. Myelosuppression leading to leukopenia is the most common manifestation seen with IV ganciclovir.

Valganciclovir: It is a prodrug of ganciclovir. On oral administration, the drug gets hydrolyzed to ganciclovir by the esterase present in the intestine and liver. The oral bioavailability of the drug is 60–70%. The oral valganciclovir of 900 mg once daily is considered to be same as IV ganciclovir of 5 mg/kg once daily. It is considered to be as equally effective of IV ganciclovir therapy for CMV induction (treatment) and maintenance therapy.

Dose regimen:
IV Ganciclovir:
5 mg/kg twice daily
Valganciclovir:
900 mg/twice daily

Foscarnet (phosphonoformic acid): It inhibits DNA polymerase and as it does not require activation by phosphorylation it is active against ganciclovir resistant CMV. However, it is a highly toxic drug that includes renal dysfunction, hypokalemia and hypomagnesemia, hypo- or hyperphosphatemia, hypo- or hypercalcemia. Foscarnet is available only in intravenous formulation.

Dose regimen:
- 90 mg/kg, twice daily × 2–3 weeks for induction therapy.

Cidofovir: This is a cytosine nucleotide analog. It gets phosphorylated to its active diphosphate form which is independent of viral enzyme. It is used mainly in the ganciclovir resistant cases. Renal toxicity is the major adverse effect.

Dose regimen:
- 5 mg/kg/week × 2–3 weeks for induction therapy.

Newer agents: Letermovir, maribavir and **brincidofovir** are the novel antiviral compounds which have been evaluated for CMV prophylaxis in transplant recipients.

Letermovir: This is a 3,4-dihydroquinazoline-4-yl acetic acid derivative. It acts by terminating the CMV viral terminase complex and inhibits the processing and packaging steps in viral replication. Unlike ganciclovir and valganciclovir, it does not have any hematological toxicity, hence better tolerated. In phase 3 randomised control trial in HCT patients, significant reduction in CMV infection was observed with letermovir as compared to standard pre-emptive therapy (37.5% *vs* 60%). Similar result has been reported in SOT patients. It has been suggested that letermovir may decrease the mortality by preventing or delaying the development of CMV disease. The major limitation of this drug is its potential low barrier to develop resistance, no activity against other herpes viruses.

Maribavir and **brincidofovir** are two oral drugs currently under different phases of clinical trial for prophylaxis and treatment of CMV disease in transplant recipients.

Antiviral treatment for HCMV disease is recommended in the following conditions:
a. Prevention and treatment of HCMV disease in transplant recipients (SOT and HSCT).
b. Prevention and treatment of HCMV disease in HIV patients.
c. Treatment of congenital HCMV disease.

Table 3.6: Drugs used for prophylaxis of HCMV disease

Drugs	Property	Adult prophylaxis dose	Route	Major toxicity
Valganciclovir	Prodrug of ganciclovir	900 mg once daily × 2–3 weeks	Oral	Leukopenia
Ganciclovir	Guanosine derivative	1 g three times daily × 2–3 weeks	Oral	Leukopenia
Ganciclovir	Guanosine derivative	5 mg/kg once daily × 2–3 weeks	IV	Leukopenia
Foscarnet	Phosphonoformic acid	90 mg/kg once daily	IV	Renal toxicity
Cidofovir	Nucleotide analog	5 mg/kg 2 weeks	IV	Renal toxicity

Table 3.7: Therapeutic strategies for prevention of HCMV disease in transplant recipients

	Prophylaxis therapy	Pre-emptive therapy
Recommendation	All "at risk" patients early in the post-transplant period for 3–6 months.	Patients who show the evidence of viral replication (CMV pp65 Ag/viral load) before development of symptoms.
Disadvantages	• Exposes the patients to toxicity of drugs • Late onset of CMV disease occurs after discontinuation of therapy and associated with higher rate of mortality	• Risk of toxicity is minimized • Require laboratory logistics for viral load/Ag monitoring • Cutoff threshold to initiate the pre-emptive therapy is not defined for different transplant groups

The approved antiviral drugs used for HCMV disease prevention and treatment are listed in Table 3.6.

Two strategies are commonly employed for prevention of HCMV disease: Prophylaxis and pre-emptive therapy (Table 3.7).

Recommendation of International Consensus group for CMV disease prevention in SOT patients:

• Either prophylaxis or pre-emptive therapy can be employed in:
 – D+/R– kidney/liver transplant recipients.
 – All seropositive recipients (R+) for kidney/liver/heart transplantation.
• Prophylaxis is preferred over pre-emptive strategy in:
 – D+/R– heart and lung transplant recipients.
 – Patients with potent immunosuppressive therapy/antilymphocyte therapy/ and HIV.

Recommendation of expert panel for prevention of CMV infection in HSCT patients:

It is recommended to adapt either prophylaxis or pre-emptive therapy for all CMV seropositive recipients and D+/R– HSCT patients from the day of engraftment till 100 days post-transplantation. Intravenous ganciclovir is recommended for allogenic HSCT recipients.

Antiviral treatment is also recommended to HIV patients with HCMV infection and newborns with signs of congenital CMV infection with ganciclovir.

VACCINE

No licensed vaccine is available so far for prevention of CMV infection or disease. However, vaccine containing CMV gB protein has shown some promising result in various clinical trials involving pregnant women, SOT and HSCT recipients.

Bibliography

1. Buonsenso D, Serranti D, Gargiullo L, Ceccarelli M, Ranno O, Valentini P. Congenital cyto-megalovirus infection: Current strategies and future perspectives. Eur Rev Med Pharmacol Sci 2012;16(7):919–35.

2. Crough T, Khanna R. Immunobiology of human cytomegalovirus: From bench to bedside. Clin Microbiol Rev 2009; 22(1):76–98.

3. Dioverti MV, Razonable RR. Clinical utility of cytomegalovirus viral load in solid organ transplant recipients. Curr Opin Infect Dis 2015;28(4):317–22.

4. Kimberlin DW, Jester PM, Sánchez PJ, et al. National Institute of Allergy and Infectious Diseases Collaborative Antiviral Study Group. Valganciclo-vir for symptomatic congenital cytomegalovirus disease. N Engl J Med 2015; 372(10):933–43.

5. Kotton CN, Kumar D, Caliendo AM, Asberg A, Chou S, Danziger-Isakov L, Humar A. Transplantation Society International CMV Consensus Group. Updated international consensus guidelines on the management of cytomegalovirus in solid-organ transplantation. Transplantation 2013; 96(4):333–60.

6. Lanzieri TM, Dollard SC, Bialek SR, Grosse SD. Systematic review of the birth prevalence of congenital cytomegalovirus infection in develo-ping countries. Int J Infect Dis 2014;22:44–8.

7. Lurain NS, Chou S. Antiviral drug resistance of human cytomegalovirus. Clin Microbiol Rev 2010;23(4):689–712.

8. Razonable RR, Hayden RT. Clinical utility of viral load in management of cytomegalovirus infection after solid organ transplantation. Clin Microbiol Rev 2013;26(4):703–27.

9. Revello MG, Gerna G. Diagnosis and manage-ment of human cytomegalovirus infection in the mother, fetus, and newborn infant. Clinical Microbiology Reviews 2002; 15 (4): 680–715.

10. Jakharia N, Howard D, Riedel DJ. CMV infection in hematopoietic stem cell trans-plantation: Prevention and treatment strategies. Curr Treat Options Infect Dis 2021 Jul 21:1–18.

11. Limaye AP, Babu TM, Boeckh M. Progress and challenges in the prevention, diagnosis, and management of cytomegalovirus infection in transplantation. Clin Microbiol Rev 2020 Oct 28;34(1):e00043–19.

12. Natori Y, Alghamdi A, Tazari M, et al. CMV Consensus Forum. Use of viral load as a surrogate marker in clinical studies of cytomegalovirus in solid organ transplantation: A systematic review and meta-analysis. Clin Infect Dis 2018 Feb 1;66(4):617–31.

13. Ljungman P, de la Camara R, Robin C, et al; 2017 European Conference on Infections in Leukaemia group. Guidelines for the management of cytomegalovirus infection in patients with haematological malignancies and after stem cell transplantation from the 2017 European Conference on Infections in Leukaemia (ECIL 7). Lancet Infect Dis 2019 Aug;19(8):e260–e272.

14. Limaye AP, Babu TM, Boeckh M. Progress and challenges in the prevention, diagnosis, and management of cytomegalovirus infection in transplantation. Clin Microbiol Rev 2020 Oct 28;34(1):e00043–19.

D. EPSTEIN-BARR VIRUS

HISTORY

Sir Dennis Burkitt, an English surgeon, noticed the high incidence of jaw tumor in African children which is now known as Burkitt lymphoma. He also noticed that this tumor is restricted to areas of Africa where malaria is holoendemic. During his lecture on this subject in London, he met Sir Anthony Epstein who was investigating the role of virus in cancer and thereafter Sir Burkitt sent tumor samples to Sir Epstein. Epstein, Bert Achong and Yvonne Barr after several years of study discovered viral-like particles in suspension culture of Burkitt lymphoma tumor material. This particle was finally discovered as a distinct member of herpesvirus group in 1964 and was named as Epstein-Barr virus.

The virus for many years remained as a tumor associated virus till a technician in Henle's lab in Philadelphia developed symptoms resembling infectious mono-nucleosis after accidentally infected with it. She was found to be seroconverted, indi-cating the causal association of EBV with infectious mononucleosis. Later, more studies on college students confirmed the etiological role of EBV with infectious mononucleosis.

Further studies proved the virus to be ubiquitous, with adult seropositivity of >90% worldwide. The virus has also been associated with several types of tumor, mostly various types of lymphoma.

VIRUS

The morphology of EBV is grossly similar to other members of Herpesviridae family. EBV is a double-stranded DNA virus. The nucleic acid is surrounded by icosahedral capsid containing 162 capsomers, which again is surrounded by lipid envelop. The diameter of the mature viral particle is 150–200 nm. Viral envelop is acquired from the infected host cell membrane and like other members of the herpes group of viruses, envelop is thought to be acquired from trans-Golgi network. Envelop contains numerous peplomers.

The genome of EBV contains a linear DNA molecule of 184 kbp lengths. The EBV DNA contains several internal repeats interspersed with unique sequences, flanked by terminal repeat sequences. Two strains of EBV exist: 1 and 2 also called A and B. They differ in the domain of viral genome that code for Epstein-Barr virus nuclear antigen (EBNA) 2, 3A, B and C and also Epstein-Barr virus encoded small RNA (EBER). Type 1 is more common in Western countries, whereas type 2 is less common in Western countries but more common in African countries. Type 1 is more efficient in transformation of B lymphocyte *in vitro* as compared to type 2. However, there is no specific disease association with any particular type. Within types, individual isolates differ in the number of tandem repeats in their internal repeats. This heterogeneity can be used to identify the individual isolates, which can be helpful in analysis of epidemiological studies to monitor transmission of the virus.

Near 100 genes are expressed by EBV during viral replication. Of these, ten are associated with latency. EBV nuclear antigen-1 (EBNA-1) binds to the viral DNA and helps the binding of viral DNA to the host cell chromosome in a form of episome. EBNA-2 causes up-regulation of latent membrane protein 1 and 2 (LMP1, LMP2) and also cellular proteins which are responsible for B cell growth and transformation. LMP1 is the major oncoprotein. This activates the transcription factor nuclear factor kB (NF-kB), activation of c-jun, cytokine production and B cell proliferation.

EBV LIFE CYCLE

EBV infection is primarily transmitted through saliva. Almost all EBV seropositive individuals actively shed the virus in the saliva. The other modes of transmission can be sexual transmission, blood transfusion, solid organ or hematopoietic stem cell transplantation. **Epithelial cells** and **B lymphocytes are the two primary target cells of EBV**.

EBV when infects the epithelial cells, replication of virus occurs and new viruses are produced causing lysis of the infected cells. Infection of B cell by EBV can occur either by these newly released viruses from the infected epithelial cells or through direct entry. Infection of B cell is mediated by the viral envelop glycoprotein gP350 with the B cell surface molecule CD21 or also known as complement receptor 2(CR2). During the primary infection in B cells, some of the infected B cells undergo lytic infection and in others the virus remains in the latent form.

In the lytic cells, viral proteins are expressed but are kept under check by the immune system.

Whereas in the latently infected cells, the linear genome of EBV gets circularizes and remains latent as episome in the nucleus of the host cell. The viral protein EBV nuclear antigen 1 (EBNA1) binds to the viral DNA and maintains it in the episomal form in the B cell.

During latency, EBV expresses only a few proteins (six nuclear proteins, two non-coded RNAs; EBER 1 and 2 and three latent membrane proteins; LMP 1, 2A, 2B) which help the virus to escape the recognition from cytotoxic T cells.

In general, EBV infection of the epithelial cells leads to lysis of the cell, whereas infection of the B cells mostly leads to latent infection with transformation of the B cells.

CLINICAL SYNDROMES

The spectrum of diseases caused by EBV includes both non-malignant and malignant conditions. Table 3.8 enlists the diseases with the respective host immune status and viral markers.

Infectious mononucleosis (IM): Primary EBV infection in adolescents and young adults usually results in infectious mononucleosis (IM). However, it can also be seen in developing countries and also in other age groups. Incubation period is 4–6 weeks. **Fever, sore throat and cervical lymphadenopathy** are the three most common manifestations of infectious mononucleosis and considered as the clinical triad. Tonsilar exudates and posterior cervical lymphadenopathy can be there. The less common manifestations are splenomegaly, hepatomegaly, periorbital edema and rash. Complications, which occur rarely in IM, include upper airway obstruction due to enlarged tonsils, autoimmune hemolytic anemia, granulocytopenia, thrombocytopenia, aplastic anemia, myocarditis, hepatitis, splenic rupture, palatal rash and neurologic complications such as Guillain-Barré syndrome, encephalitis, and meningitis.

Most of the IM patients show leukocytosis with increase T cells and atypical lymphocytes but not increase in B cells. **Atypical lymphocytes** are predominantly activated T lymphocytes with a large amount of cytoplasm with vacuoles. There can be polyclonal B cell activation leading to increase in level of immunoglobulin and **heterophile antibodies**. Serum alkaline phosphatase and serum transaminase level are raised. There occur activation and proliferation of T lymphocytes in response to the EBV infected B cells. Most of the symptoms of IM are attributed to these CD8 T cell expansion and production of cytokines by them and not due to lytic infection of B cell.

Table 3.8: EBV associated diseases

Disease	Host status	
Non-malignant		**Viral markers**
Infectious mononucleosis	Immunocompetent	EBV VCA IgM Heterophil Ab
Chronic active EBV disease: Hemophagocytic lymphohistiocytosis (HLH)	Immunocompetent	EBV VCA IgM
X-linked lymphoproliferative disease		EBV DNA, EBV antigen
Malignant		**EBV gene expression in tissue**
Burkitt lymphoma	Immunocompetent/HIV	EBER, EBNA-1 (latency type I)
Hodgkin's lymphoma	Immunocompetent/HIV	EBER, EBNA-1, LMP-1, LMP-2A, LMP-2B
Nasopharyngeal carcinoma	Immunocompetent	(latency type II)
Lymphoproliferative disorder and PTLD	Immunocompromised	EBER, EBNA-1, EBNA-2, EBNA-3A, EBNA-3B, EBNA-3C, LMP-1, LMP-2A, LMP-2B (latency type III)

Diagnosis of IM: Diagnosis of IM is confirmed by detection of IM specific heterophile antibody or EBV specific IgM antibody.

EBV specific antibody: IgM antibody to EBV virus capsid antigen **(EBV VCA IgM)** appears during acute infection and persists for only 2–3 months, indicating acute or recent infection. During the early part of the illness, EBV VCA IgM may be present in absence of EBV VCA IgG.

EBV VCA IgG antibody, though detectable during the acute infection, persists for years. Therefore, positive EBV VCA IgG positivity in absence of IgM to EBV VCA indicates past infection. Figure 3.7 depicts the schematic representation of kinetics of EBV antibodies in relation to symptoms.

Antibody to **EBNA-1** appears during convalescence period, so positivity of this antibody during acute illness rules out acute infection, whereas seroconversion to EBNA antibody is an indicator of acute infection.

Heterophile antibody: Heterophile antibody against sheep/mammal agglutinin is produced in EBV associated infectious mononucleosis (Table 3.9). These antibodies are IgM in nature and present during acute infection in >95% of IM patients.

Heterophile antibody is detected by presumptive or differential agglutination test.

Presumptive test is known as **Paul-Bunnell test:** This test measures the titer of agglutinin against sheep RBC in patient's serum. **Antibody titer >1:224** is considered as positive for IM. However, because of the non-specificity of the test, the test may be positive in other conditions such as serum sickness, and even in healthy individuals.

Differential agglutination test or Paul-Bunnell-Davidsohn test: This test is employed to remove the Forssman antibodies and differentiates it from IM specific heterophil antibodies.

This test is based on the principle that anti-sheep agglutinins of IM are not absorbed by guinea pig kidney (GPK) powder but absorbed completely by beef RBC. Patient's serum is tested for sheep agglutinin as per the presumptive test and in addition another two rows of serum dilution are tested which have been pre-absorbed separately with guinea pig kidney powder and beef RBC.

In case of true IM specific heterophile antibody:
• Agglutination titer in serum pre-absorbed with GPK should either be same as that of presumptive test or if less, the decrease in titer should not be > 3 tubes than presumptive test.

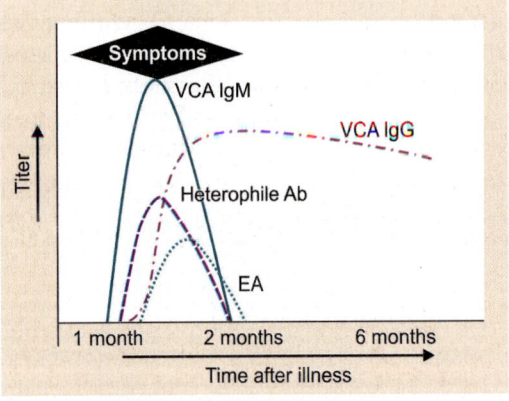

Fig. 3.7: Schematic diagram showing antibody kinetics during primary EBV infection
VCA: Viral capsid antigen, EA: Early antigen

Table 3.9: Salient features of infectious mononucleosis			
Clinical	*Hematological*	*Biochemical*	*Virological*
Fever	Leukocytosis	↑Serum alkaline phosphatase	Heterophil antibody +ve (Monospot test/PBD test)
Sore throat	↑no. T cells	↑Serum transaminase	EBV VCA IgM
Lymphadenopathy	↑no. atypical lymphocytes		EBV DNA (1000–5000 copies/mL of whole blood)
Hepatosplenomegaly			

- Agglutination titer in serum pre-absorbed with beef RBC should decrease and the decrease should be >4 tubes than that of presumptive test.

The main disadvantage of heterophil antibody test is false negative in children as majority of children do not mount heterophil antibody during primary infection.

EBV DNA assay: EBV DNA in blood is detected by PCR or real-time PCR. The detection of EBV DNA is recommended in patients with atypical clinical features or in children with negative heterophil antibody. The major recommendation for EBV DNAemia or quantitation is required for monitoring of therapy in EBV associated malignancies including PTLD.

The average EBV viral load in blood in various conditions:

- Healthy individuals with latent EBV: <1000 copies/mL of whole blood
- Immunocompetent with symptomatic EBV infection: 5000 copies/mL of whole blood
- Transplant recipients: 5000–50,000 copies/mL of whole blood.

Burkitt lymphoma: Burkitt lymphoma (BL) is a high grade malignant B cell tumor. Endemic BL is seen in African children, where malaria is holoendemic. It is manifested as a tumor of jaw. More than 90% of these tumors are associated with EBV. Sporadic BLs are seen in America and Europe. These are mostly present as abdominal tumor and around 20% are associated with EBV. In AIDS patients, 40% BLs are EBV associated.

It is thought that malaria impairs the T cell control, leading to proliferation of EBV infected B cell. This in turn increases the EBV viral load. In BL, there occur chromosomal translocations involving 8:14, 8:22 or 8:2. This leads to dysregulation of c-myc oncogene which gets overexpressed. This in turn inhibits the immunological recognition of EBV proteins by various mechanisms leading to virus proliferation.

Table 3.10: Evidences of EBV as causative agent of BL

- High anti-EBV antibody in serum of BL patients
- Antibody to EBV VCA and ER is high
- Presence of EBV DNA and antigen in BL tumor.
- Children with high EBV antibody titer, high EBV DNA load and EBNA-1: High-risk of developing BL.

Thus BL can be diagnosed by detection of EBV VCA antibody in patients' serum and EBV antigen and EBV DNA in BL tumor tissue (Table 3.10).

Nasopharyngeal carcinoma: Nearly 100% anaplastic or undifferentiated type of nasopharyngeal carcinoma (NPC) are EBV positive. Anaplastic NPC is common in South East Asia more so in Southern China.

EBV associated diagnostic, screening and prognostic markers are given in Table 3.11.

Lymphoproliferative disease: EBV associated lymphoproliferative disease can occur in patients with congenital or acquired immunodeficiency. 1–5% recipients of various transplants; solid organ or bone marrow, develop EBV associated lymphoproliferative disease during the post-transplant period. Recipients of solid organ transplant are at higher risk of developing post transplant lymphoproliferative disorder (PTLD) than hematopoietic stem cell transplant (HSCT) recipients.

Table 3.11: EBV associated markers in nasopharyngeal carcinoma (NPC)

Diagnostic marker:
- High level of EBV DNA in plasma and transformed epithelial cells of the affected tissue
- High level IgA antibody to viral capsid antigen (VCA)/EBNA.
- EBV DNA and EBV VCA IgA antibody: Combined sensitivity and specificity: >95%

Screening marker:
High IgA antibody to VCA in endemic region is used for screening NPC in high endemic region.

Prognostic marker:
High EBV DNA, high EBV VCA IgA antibody at the time of diagnosis is correlated with poor survival and thus acts as a bad prognostic marker.

The risk of developing post-transplant lympho-proliferative disease (PTLD) increases with:

- HLA mismatch bone marrow.
- Primary EBV infection
- Cytomegalovirus infection
- TNF-α promoter polymorphism

Due to the impaired T cell immunity, these patients are unable to control the EBV infected B cell proliferation.

Majority of patients present with infectious mononucleosis like symptoms with fever and local or disseminated lymphoproliferation. Lesions are often extranodal involving liver, lungs, gastrointestinal tract, kidney and central nervous system. Increased level of IL-6, TNF-α and B cell growth factor is usually found in blood of PTLD patients.

PTLD is associated with high EBV viral load, expression of viral proteins (LMP-1, LMP-2, EBNA-1, EBNA-2 and BZLF). Estimation of EBV DNA load is recommended in patients with symptoms suggestive of PTLD or in high-risk patients. Detection of high viral DNA in blood is used as a marker or predictor for diagnosis of PTLD and also for post-treatment follow-up. EBV DNA load along with EBV specific CD8 T cell acts as a better predictive marker than EBV DNA alone.

TREATMENT

Infectious mononucleosis: No specific therapy is available for treatment of IM. Specific antiviral like acyclovir has been shown to be of no use. Corticosteroid is recommended only in life-threatening complications.

Post-transplant lymphoproliferative disease: Reduction of the dose of chemotherapy may resolve the lymphoproliferation. Monoclonal antibody to CD20 positive B cell (rituximab) along with reduced dose of chemotherapy has been reported to be effective in children. In localized lesion, surgery can be helpful. Cytotoxic chemotherapy can be given in case of rituximab failure.

Bibliography

1. Balfour HH Jr, Dunmire SK, Hogquist KA. Clinical and Translational Immunology (2015) 4, e33; doi:10.1038/cti.2015.

2. Cohen JI. Epstein-Barr infection. New Engl J Med 2000;343:481–92.

3. Greena M, and Michaelsa MG. Epstein-Barr Virus infection and post-transplant lympho-proliferative disorder. Am J of Transplantation 2013;13:41–54.

4. Gulley ML, Tang W. Using Epstein-Barr viral load assays to diagnose, monitor, and prevent post-transplant lymphoproliferative disorder. Clin Microbiol Rev 2010 Apr;23(2):350–66.

5. Gulley ML, Tang W. Laboratory assays for Epstein-Barr virus-related disease. J Mol Diagn 2008;10(4):279–92.

6. Ok CY, Li L, Young KH. EBV-driven B-cell lymphoproliferative disorders: From biology, classification and differential diagnosis to clinical management. Exp Mol Med 2015; 47:e132.

7. Oludare A. Odumade,1 Kristin A. Hogquist,1 and Henry H. Balfour, Jr. Progress and problems in understanding and managing primary Epstein-Barr virus infections. Clin Microbiol Rev 2011;24: 193–209.

Human Papillomavirus

Papillomaviruses (PV) are a group of double-stranded DNA viruses that are epitheliotropic in nature. This is known to cause benign skin and mucosal lesions like warts and condyloma, respectively. The virus has gained importance after its causative role was recognized with cervical cancer and several other malignancies.

CLASSIFICATION

Previously, papillomaviruses and polyoma viruses were taxonomically placed in the same family, Papovaviridae. The similar features between two groups of viruses are: (i) Circular double-stranded DNA genome, (ii) non-enveloped icosahedral capsid, (iii) replication inside the nucleus. However, the differences found between them are: (i) Absence of major sequence homology, and (ii) replication strategy, with papillomavirus DNA transcription is unidirectional while polyoma virus DNA replication is bidirectional. Because of these differences, they have been separated into two distinct families, Papillomaviridae and Polyomaviridae by the International Committee on Taxonomy of Viruses (ICTV) in 2000.

The current classification of PV is divided into genus, species, types, subtypes and variants.

Genus: This is the broadest category and there are 29 genera. Each genus is designated by a Greek alphabet. The members sharing more than 60% identity in their L1 ORF DNA sequence are included within one genus.

Species: Members within a genus that share 60 to 70% identity.

Type: Members within a species sharing 71 to 89% identity. They are designated as HPV followed by a number.

Subtype: Members within a type sharing 90 to 98% identity.

Variants: Members within a type with more than 98% identity.

Papillomaviruses (PVs) infect birds, reptiles and mammals including humans. PVs are species specific in nature. Human papillomaviruses (HPVs) are a group of related papillomaviruses that infect humans. Out of 29 genera in the Papillomaviridae family, HPVs constitute five genera. These are alpha, beta, gamma, mu, and nu. HPVs that belong to genus alpha virus are the ones that are associated with genital and mucosal cancer. These HPVs are also known as "genital-mucosal" types as they infect genital and non-genital mucosal surfaces. The remaining four genera infect non-genital skin.

VIRUS STRUCTURE AND ORGANIZATION

Papillomaviruses are small, non-enveloped viruses, with icosahedral symmetry. The diameter of the virion particle is approximately 60 nm which contains the genome of double-stranded DNA of 8000 base pairs. The genome consists of three major regions—early genes, late genes and upstream regulatory region.

The upstream regulatory region is also known as the long control region (LCR) or

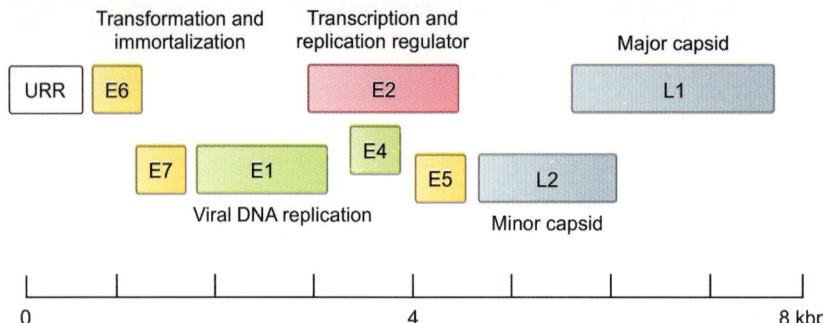

Fig. 4.1: Schematic representation of HPV genome in a linearized form

non-coding region (NCR) and contains the origin of replication, multiple transcription binding sites, and the early and late open reading frames (ORFs).

The early region comprises six ORFs, i.e. E1, E2, E4, E5, E6, and E7. This region encodes for viral regulatory proteins (Fig. 4.1).

- E1 and E2 are involved in initiation of viral DNA replication and regulation of early transcription.
- The E4 protein is expressed during productive infection and facilitates virus assembly and release of virions from infected cells.
- The E5, E6, and E7 genes are associated with virus replication.
- E6 and E7 genes interfere with regulator of host cell cycle and transcription and have the major transforming activity.
- The late genes, L1 and L2, encode the viral capsid proteins.

PATHOGENESIS

The HPV enters through the minor or micro-abrasion in the basal layer of the epithelium (stratum germinativum). This selective infection of basal layer occurs due to its preferential binding with **heparin sulfate proteoglycan (HSPG)** on the basement membrane which is exposed on minor trauma or abrasion. Once the virus enters, the infectious cycle of the virus is dependent upon the differentiation program of the squamous epithelial cells. The replication of virus DNA occurs as the basal cell differentiates and progresses towards the surface of the epithelium. This is evidenced by the presence of viral capsid proteins, viral DNA and viral particles only in the terminally differentiated cells.

Replication of virus is almost non-productive at the basal layer. It establishes itself as a low copy number episome, and with the help of host cell machinery viral DNA is synthesized once per cell cycle.

In the differentiated suprabasal layer, the pattern of viral DNA replication gets switched over to the rolling circle mode which leads to synthesis of high copy number of DNA, capsid proteins and assemble of virions (Fig. 4.2).

When HPV replication starts, transcription of E6 and E7 occurs. The function of E6 and E7 is to destabilize the cell growth regulatory pathway and also to alter the cellular environment to enhance viral replication in a terminally differentiated cell.

The tumor suppressor gene, **p53** and retinoblastoma gene product **pRB** are the two main cellular proteins responsible for cell cycle regulation.

E6 of high-risk HPV types has high affinity to bind p53 protein and causes rapid degradation. It forms complex with p53 with the help of cellular ubiquitin ligase E6AP. This ubiquitylation leads to degradation of p53

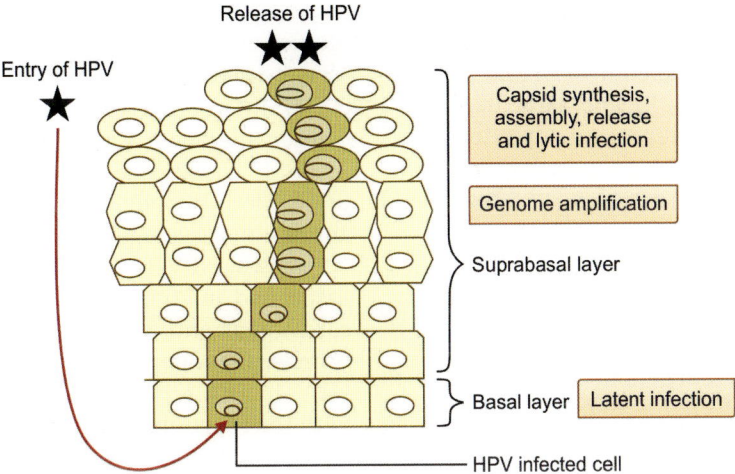

Fig. 4.2: HPV infection of epithelium

protein. It also downregulates p53 by targeting the co-activator of p53, CBP/p300. As a consequence, normal functions of p53 like cell cycle arrest at G1, apoptosis and DNA repair gets arrested, thereby facilitating the process of oncogenesis.

E7 binds to the hypophosphorylated form of Rb. This binding breaks the complex of pRb and cellular transcription factor E2F-1, leading to the release of E2F-1. This leads to transcription of those genes whose products are essential for cell cycle progression and DNA replication, thus leading to cellular proliferation and enhanced cellular DNA synthesis.

E6 and E7 proteins of low-risk HPV types have low affinity to bind with p53 and pRb proteins, respectively. Whereas E6 and E7 of high-risk HPVs like HPV16 and HPV18 are responsible for the immortalization and transformation of the infected cells.

Pathogenesis of Cervical Cancer

HPV infection of the genital tract is self-limiting in most of the cases. Clearance of virus usually occurs within 12 months. Persistence of virus is more commonly seen with the high-risk HPV types; although it has also been observed with some low-risk HPVs. Longer the virus persists chance of clearance becomes less. Amongst the persistent infection, only

some develop CIN3 and amongst the CIN3, only some develop invasive cancer.

The risk of development of invasive cancer depends on several factors:
  i. Persistence of virus,
  ii. infection with high-risk HPV type, and
  iii. variant of HPV within the virus type.

Persistent infection with a high-risk HPV type is the most important risk factor for development of invasive cancer. Early age of sexual debut, multiple partners, chronic inflammation of genital tract, immunosuppression and smoking are the other risk factors for development of invasive cancer.

Molecular Pathogenesis

Integration of Viral DNA

- Integration of HPV DNA occurs mostly with the high-risk HPV types. The rate of integration varies with different HPV types. Most of the lesions produced by HPV18 and HPV45 show viral DNA integration; whereas in HPV 16 infection, this is seen in more than 50% lesions.
- Integration of HPV DNA stabilizes the high expression of E6 and E7; the two oncoproteins responsible for the transformation.
- Integration of viral DNA is associated with more severe lesions (85% in invasive cancer *vs* 16% in CIN3, 5% CIN2 and none in CIN1).

- Integration of viral DNA is associated with deletion of a large part of viral genome which includes deletion of E1 and E2 genes but retaining E6, E7 and LCR. E2 being a transcriptional repressor, deletion of E2 removes the check on transcription of E6 and E7. This in turn leads to high level transcription of E6 and E7.

Activation of telomerase: E6 gene activates telomerase (hTERT). This activation leads to severe dysplasia and cervical cancer.

Aneuploidy and genomic instability due to E6 and E7 in infection with high-risk HPV type also contributes towards development of invasive cancer.

EPIDEMIOLOGY

Transmission of HPV occurs mainly by skin-skin or mucosa-mucosa contact. Near 40% of young women have evidence of HPV infection, with peaks during the teens and early twenties as shown by various cross-sectional and longitudinal studies. Multiple sexual partner increases the risk of acquiring HPV infection and so as the risk of HPV associated malignancy. Monogamous individual may get the virus from the infected partner. The probability of acquiring HPV infection per sexual act has been shown to be high with no difference between HPV types.

Majority of HPV types cause benign warts on the skin or genital region, whereas some HPV types can cause various types of cancer. Alpha-papillomaviruses are categorized as high-risk or low-risk according to the likelihood of development of cancer by that HPV type.

HPV has been attributed as the causative agent for almost all cervical cancer cases, which is the second most common cancer among women. The time taken from HPV infection to development of cervical cancer is more than two decades. Persistent carriers of oncogenic HPV types are at higher risk of developing cervical dysplasia and cervical cancer. Cervical cancer is mostly detected around fifth and sixth decades of life among women in developed country, which occurs almost a decade earlier in developing countries.

The International Agency for Research on Cancer has labelled HPV types 16, 18, 31, 33, 35, 39, 45, 51, 52, 56, 58, and 59 as carcinogenic in the uterine cervix. HPV-16 has been found as the causative agent of cervical cancers in 50% cases. HPV-16 and HPV-18 combinedly associated in 70% of cervical squamous cell carcinomas and 85% of cervical adenocarcinomas. Oncogenic HPV types other than HPV-16 and HPV-18 have been associated with the remaining 30% of cervical cancers.

Table 4.1 depicts the common malignant and non-malignant diseases associated with HPVs.

DIAGNOSIS

Methods of HPV Detection

Isolation of HPV: Method to isolate the virus in *in vitro* system is not possible and also of no importance as infectious virus is not produced in high degree dysplasia and invasive cancer.

Serology: Serology plays limited or no role in clinical diagnosis of invasive cancer or in premalignant screening. Serum antibodies to E6 and E7 proteins are raised in invasive cancer but not in premalignant lesions.

Molecular methods for detection of HPV: Several molecular methods are in use for detection of HPV DNA or RNA. These are polymerase chain reaction (PCR) using consensus primer in conjunction with hybridization, hybrid capture assay using specific RNA probes to capture DNA, multiplex PCR, real-time PCR and microarray system. Majority of these tests target the detection of either L1 gene or E6/E7 genes. Numerous systems are commercially available. Many of them are FDA approved.

Samples: Cervical samples are collected using either spatula or cervical brush or broom. Various such collection devices are commercially

Table 4.1: Malignant and non-malignant diseases associated with HPVs

Diseases	Predominant associated HPV type
Malignancies	
• Anogenital tract cancers	
– Squamous cell carcinoma cervix	16, 18, 31, 33, 35, 39, 45, 51, 52, 56, 58, 59, 68
– Adenocarcinoma cervix	16, 18, 45
– Carcinoma vulva	16, 18
– Carcinoma vagina	16, 18
– Carcinoma penis	16, 18, 6, 11
– Squamous cell carcinoma anus	16, 18
• Oropharyngeal cancer	16
• Non-genital skin	
– Ungual squamous cell carcinoma	16, 18
– Epidermodysplasia verruciformis	5,8
Benign lesion	
• Anogenital diseases	
– Anogenital warts	6, 11, 40, 42, 43, 44, 54, 61, 72, 81, 89
– Condyloma acuminata	6, 11

available with transport medium that can be stored at room temperature or refrigerated temperature. Biopsy samples are collected in case of anogenital or other lesions. Many commercially available tests use specific collection devices along with transport medium.

HPV DNA Detection

Hybrid Capture 2 HPV DNA Test

Hybrid capture 1 (HC1) was the original assay in a tube-based format, was approved in 1995. HC2 is a microtiter format-based test, developed by Digene Corporation, marketed presently by Quigen, got the FDA approval in 1999 and replaced HC1. The test was approved in 2003 for co-testing with routine cytology testing for women over age 30.

Steps and principle of the assay:
- DNA is extracted and denatured from the sample.
- A multigene RNA probe specific for high-risk HPV16, –18, –31, –33, –35, –39, –45, –51, –52, –56, –58, –59, and –68 is added to it.
- If high-risk HPV DNA is present in the sample, DNA–RNA hybrids are formed.

- These are captured onto the well of the microtiter plate coated with monoclonal antibodies to DNA–RNA hybrids.
- A second monoclonal antibody conjugated to alkaline phosphatase is added.
- This binds to the captured hybrids in multiples.
- Light is produced. (The alkaline phosphatase dephosphorylates a chemiluminescent substrate which produces light. As the alkaline phosphatase acts on multiple copies of substrate, it generates an amplified signal for the presence of the target.)

Interpretation of the assay: The emitted light is measured in relative light units (RLU) on a luminometer.

One relative light unit (RLU) is considered as the cut-off value.

An RLU ≥ cut-off value indicates the presence of high-risk HPV DNA but does not distinguish or specify the presence of specific HPV genotype.

An RLU < cut-off value indicates the absence of high-risk HPV DNA or high-risk HPV DNA below the limit of detection of the test.

Sensitivity: The detection limit of the test is 0.2 to 1 pg/ml, equivalent to 1,000 to 5,000 genome copies of HPV. False positivity and false negativity of 5% and 5–12% have been reported, respectively.

Cervista HPV HR test: This test detects 14 high-risk HPV types including 13 that are detected by HC2 test and HPV 66 using sequence-specific probes targeting the L1, E6, and E7 genes. It does not identify individual HPV type. It is based on the signal amplification chemistry. The test is FDA approved.

The detection limit of the test varies according to the HPV type:
- HPV16, –18, –31, –45, –52, –56: 1,250 to 2,500 copies per reaction,
- HPV33, –39, –51, –58, –59, –66, –68: 2,500 to 5,000 copies per reaction,
- HPV35: 5,000 to 7,500 copies per reaction.

Cobas 4800 HPV test: This test is a multiplex real-time PCR and nucleic acid hybridization test. It detects the L1 gene of 14 high-risk HPV types. Four fluorescent reporter probes are used which detect HPV16 and HPV18 individually and HPV –31, –33, –35, –39, –45, –51, –52, –56, –58, –59, and –68 as a pooled result.

The analytical sensitivity of different HPV types at the clinical cut-off is:

HPV45: 150 copies/mL; HPV31, –33, –39, –51, –59: 300 copies/mL, for HPV16, –18, –35, and –58: 600 copies/mL, and HPV56, –66, and –65:1,200 copies/mL, and 2,400 copies/mL for HPV52.

HPV mRNA PCR

This assay targets the detection of HPV E6/E7 oncogene mRNA. During transient infection, the level of E6/E7 expression is less. In persistent infection, there is overexpression of E6/E7 mRNA. Detection of E6/E7 mRNA, therefore, reflects persistent infection and disease progression. Aptima assay is the FDA approved test for detection of HPV E6/E7 mRNA. Complementary oligomers are used for isolation of mRNA which is subjected to transcription mediated isothermal amplification (TMA) followed by hybridization protection assay using chemiluminescence for detection of amplified products.

Clinical implication of HPV DNA testing: The detection of HPV DNA is important in case of cervical cancer screening for premalignant lesions. Cervical cancer screening is done by cytological screening with Pap smear examination.

Most of the major guidelines all over the world recommend the starting age for cervical cancer screening by cytology at 21 years. Screening by HPV DNA detection has been recommended from 25–30 years age as most of the women suffer from transient infection before this age, thereby positivity during that age has no clinical importance. Until 2014, most of the organizations were recommending the testing of HPV DNA along with cytology in women above 30 years of age. In April 2014, modified label of hrHPV test (high-risk HPV) was approved by Food and Drug Administration (FDA) for primary hrHPV testing in women above 25 years age. However, primary screening with hrHPV testing is not yet incorporated in clinical practice guidelines. Flowchart 4.1 depicts the cervical cancer screening guideline as recommended by major organizations along with the interim guidance regarding primary HPV DNA testing.

Advantages of HPV DNA testing: HPV DNA testing has shown higher sensitivity in detecting cervical intraepithelial neoplasia 2 and 3 (CIN 2, CIN3) as compared to that of cytology alone. In addition, negative HPV DNA test has more specificity and provides a higher level of reassurance than cytology.

VACCINE

Presently there are three licensed vaccine available for prevention of HPV infection. Cervarix, manufactured by GlaxoSmithKline, Gardasil and Gardasil 9 are manufactured by Merck. All these vaccines are intended to prevent

Flowchart 4.1: Cervical cancer screening algorithm

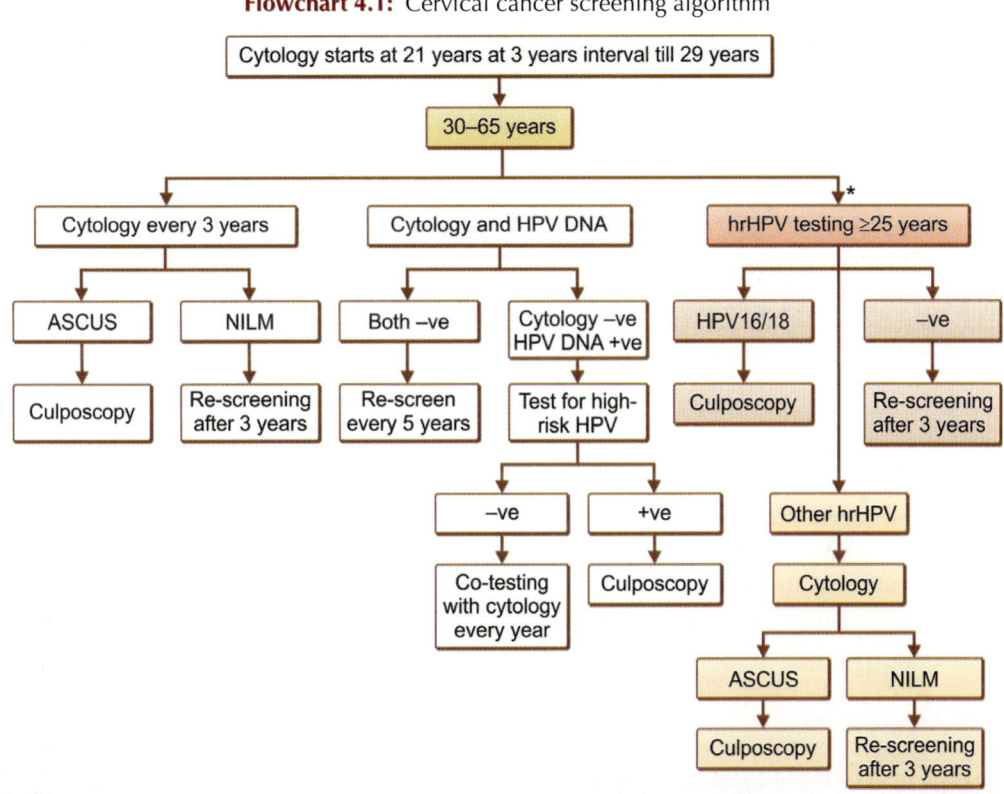

*Interim guidance. ASCUS: Atypical squamous cell of undetermined significance; NILM: Negative for intraepithelial lesion or malignancy.

HPV infection and, therefore, all the HPV associated cancers, particularly cervical cancer.

Cervarix is a **bivalent vaccine**, contains the L1 protein of HPV 16 and HPV 18, the two most common HPV types associated with cancers. **Gardasil** is a **quadrivalent vaccine**, contains the L1 protein of HPV 6 and 11, the two most common HPV types associated with anogenital warts in addition to HPV 16 and 18. **Gardasil 9** is a **9-valent HPV vaccine** which contains all four types that are present in Gardasil quadrivalent vaccine along with five additional types: HPV 31, 33, 45, 52 and 58.

All the three HPV vaccines contain the purified virus like particle of the L1 (major capsid) protein of the respective HPV types. This is based on the evidences that L1 protein can assemble to form virus-like particles (VLP) which resemble the virion both morphologically and immunologically. It induces high titres of neutralizing antibodies of type specific in nature.

All the vaccines are highly effective in preventing from CIN and cervical cancer, and also can prevent other anogenital cancers (vulva, vagina in females and penile in males), anal cancer and cancer of mouth/throat caused by vaccine specific types. In addition, these vaccines also protect against genital warts. Cross-protection to related HPV types has been observed with all the vaccines. Table 4.2 summarises the important components of all three licensed HPV vaccines.

HPV vaccine is recommended both for girls and boys at the age of 11–12 years starting at age 9 through 14. The pre-teenage has been selected as vaccine can prevent the infection before the exposure to virus which occurs

Table 4.2: Licensed HPV vaccines

	Cervarix	Gardasil	Gardasil 9
Manufacturer	GlaxoSmithKline	Merck	Merck
Composition	VLP of L1 of HPV 16 and 18	VLP of L1 of HPV 6, 11, 16 and 18	VLP of L1 of HPV 6, 11, 16, 18, 31, 33, 45, 52 and 58
Adjuvant	Aluminum hydroxide, monophosphoryl lipid A	Aluminum hydroxyphosphate sulfate	Aluminum hydroxyphosphate sulfate
Dose schedule according to age at vaccination			
9–14 years: 2 dose	—	—	0, 6–12 months
9–<15 years: 3 dose	0, 1, 6 months	0, 2, 6 months	0, 2, 6 months
15–45 years	0, 1, 6 months	0, 2, 6 months	0, 2, 6 months

mostly through sexual route and also the immune response is high at this age.

The routine vaccination has been recommended for a 3 dose (intramuscular) schedules at 0, 1/2 and 6 months. In October 2016, US FDA and Advisory Committee on Immunization Practices (ACIP) approved the 2-dose schedule for Gardasil 9 for those who started the vaccination series before 15 years of age except in immunocompromised, whereas 3-dose schedule is recommended for those who start the vaccination series after the age of 15 years.

Overall, all the vaccines are considered as safe and there is no serious side effect. Common side effects are pain at the injection site, fever, dizziness and nausea.

Bibliography

1. Bhatla N, Moda N. The clinical utility of HPV DNA testing in cervical cancer screening strategies. Indian J Med Res 2009; 130(3):261–265.

2. Burd EM. Human Papillomavirus and Cervical Cancer. Clin Microbiol Rev 2003,16:1–17.

3. Burd EM. Human papillomavirus laboratory testing: the changing paradigm. Clin Microbiol Rev 2016;29:291–319.

4. http://www.cdc.gov/vaccines/vpd-vac/hpv/downloads/dis-HPV-color-office.pdf

5. Huh WK, Ault KA, Chelmow D, Davey DD, et al. Use of primary high-risk human papillomavirus testing for cervical cancer screening: interim clinical guidance. J Low Genit Tract Dis. 2015;19(2):91–6.

6. Knipe DM, Howley PM eds. Fields virology. 6th edition. Philadelphia, USA. Wolter Kluwer. Lippincott Williams & Wilkins. 2013.

7. Mark Schiff man, Philip E Castle, Jose Jeronimo, Ana C Rodriguez, Sholom Wacholder. Human papillomavirus and cervical cancer. Lancet 2007; 370: 890–907.

8. Meites E, Kempe A, Markowitz LE. Use of a 2-Dose Schedule for Human Papillomavirus Vaccination: Updated Recommendations of the Advisory Committee on Immunization Practices. MMWR Morb Mortal Wkly Rep 2016;65:1405–1408.

Chapter

5

Polyomavirus

HUMAN POLYOMAVIRUSES

Polyomaviruses are ubiquitously present in avian and mammalian species. In human, their clinical relevance is mainly in immuno-compromised patients. **BK and JC** are the two most important human polyomaviruses which are associated with polyomavirus associated nephropathy **(PVAN)** and progressive multifocal leukoencephalopathy **(PML)**, respectively.

Classification

Human polyomaviruses belong to the family Polyomaviridae. Previously these groups of viruses were classified under Papovaviridae family along with papillomaviruses. In 2000, Polyomaviruses and Papillomaviruses were separated into two different families. The criteria for separation have been discussed in Chapter-4 "Human Papillomavirus".

The proposed classification of Polyomaviridae. The family Polyomaviridae has three genera.

1. Orthopolyomavirus
2. Wukipolyomavirus
3. Avipolyomavirus

Orthopolyomavirus and Wukipolyomavirus contain viruses of mammalian species and Avipolyomavirus contains viruses of avian species.

Polyomaviruses are nonenveloped, 45–50 nm in diameter virion having icosahedral symmetry. It has three capsid proteins—one major (VP1) and two minor proteins (VP2 and VP3). The genome is a double-stranded circular DNA of 5000 base pairs which are covered with cellular histones H2A, H2B, H3 and H4.

Human polyomaviruses: BK and JC viruses were the first two human polyomaviruses (HPyV) discovered and named after the initials of the patients from whom they were detected. After the discovery of these two viruses, 10 more viruses have been added to the list (Table 5.1).

Clinical Features

Human polyomaviruses cause **lytic infection** of the respective target cells, which is the key factor of disease pathogenesis. Primary infection leads to maintenance of virus at a low level because of healthy immune status. Whenever the immune system of the host goes down, either due to immunosuppressive drugs or HIV/AIDS infection or transplantation, replication of virus becomes high. This high viral replication leads to destruction of infected tissues. Thus, it is the reactivation of the viruses which is responsible for the disease process than the primary infection.

BK Virus Associated Diseases

Polyomavirus-associated nephropathy (PVAN): PVAN occurs in 1–10% of renal transplant recipient during the first two years of transplantation. The disease is characterized by high level of BK virus replication in the renal tubular epithelial cells of the trans-planted kidney leading to destruction of cells and epithelial denudation. This leads to tissue

Table 5.1: Human polyomaviruses

Virus	Source of isolation
JCPyV	ICP with progressive multifocal leukoencephalopathy
BKPyV	Urine of renal transplant patient
KIPyV	Respiratory secretion of patient in Karolinska Institute (KI)
WUPyV	Respiratory secretion of patient in Washington University (WU)
Merkel cell PyV (MCPyV)	Skin of Merkel cell carcinoma patient
HPyV 6 and 7	Skin and hair follicule of healthy persons
Trichodysplasia spinulosa associated PyV	Patient with trichodysplasia spinulosa (rare skin disease)
HPyV 9	Serum/plasma of a kidney transplant patient
HPyV 10 (MW and MXPyV)	Condyloma of a patient with WHIM syndrome
Saint Louis PyV	Stool samples of Malawi and US
HPyV 12	Organs of gastrointestinal tract in patient with multiple malignancy

ICP: Immunocompromised patient; WHIM: Warts, hypogammaglobulinemia, infections, myelokathexis.

leakage and infiltration of inflammatory cells into the interstitium, ultimately leading to tubular atrophy and interstitial fibrosis. These changes cause reduction in graft function and graft loss.

Screening: This is recommended in all renal transplant recipients in every 3 months for first 2 years post-transplantation by quantitative PCR of urine or plasma.

- BKPyV >7 log10 genome equivalent/mL in urine or >4 log10 genome equivalent/mL in plasma is considered significant.
- Because of intermittent shedding of BKPyV in urine, presence of virus in high concentration in plasma bears higher clinical significance.
- Presumptive diagnosis is made when the **virus load in plasma is >4 log10 GEq/mL** or **VP1 mRNA >6.5 × 10^5 copies/ng** of total RNA in urine.

Urinary findings: Presence of **decoy cells** and epithelial cells containing viral inclusions and urinary cast of polyomavirus aggregates.

Confirmative evidence: Features of renal histopathology provide the definitive evidence of BKPyVAN which are as follows:

- Intranuclear inclusions in tubular epithelial cells with enlarged nucleus and chromatin damage (Fig. 5.1a and b).
- Atrophy of tubules, interstitial fibrosis.

Summary of different clinical aspects of PVAN is depicted in Fig. 5.2.

Reduction or alteration of immunosuppressive therapy is the usual mode of treatment of PVAN and is effective in majority of cases. Other drugs like cidofovir, leflunomide or fluoroquinolone also have been tried in combination with reduction in immunosuppression, however, with no better result than reduction of immunosuppressive therapy alone.

Other diseases associated with BK virus:

- Polyomavirus associated hemorrhagic cystitis: Mostly in patients with hematopoietic stem cell transplant (HSCT) recipients
- Ureteric stenosis
- Central nervous system involvement
- Interstitial pneumonia.

JCPyV Associated Diseases

JCPyV and progressive multifocal leuko-encephalopathy (PML): JCPyV infection occurs possibly either through respiratory route or ingestion. It infects the stromal cells or immune cells of the respiratory tract. Through lymphocyte, the virus goes to bone marrow and kidney and persists there for life. When the immune system of the host gets suppressed, virus comes out from the bone

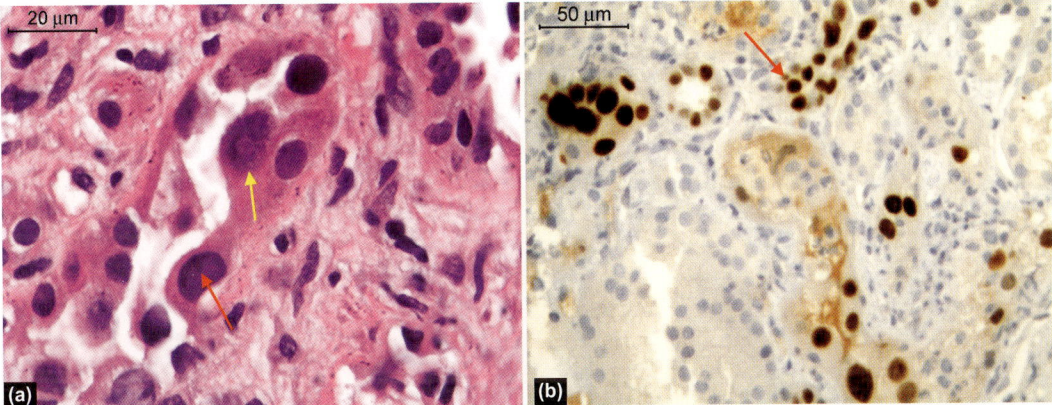

Fig. 5.1a and b: Photomicrograph showing (a) pale eosinophilic (yellow arrow) and smudgy (red arrow) intranuclear inclusions of BK virus in distal tubular epithelial cell in a renal allograft biopsy (H&E, ×100 oil original magnification), (b) intranuclear inclusions (brown color) of BK virus in distal renal tubular epithelial cells by immunohistochemistry (×40 original magnification). (*Photograph courtesy:* Prof Ritambhra Nada, Department of Histopathology, PGIMER, Chandigarh.)

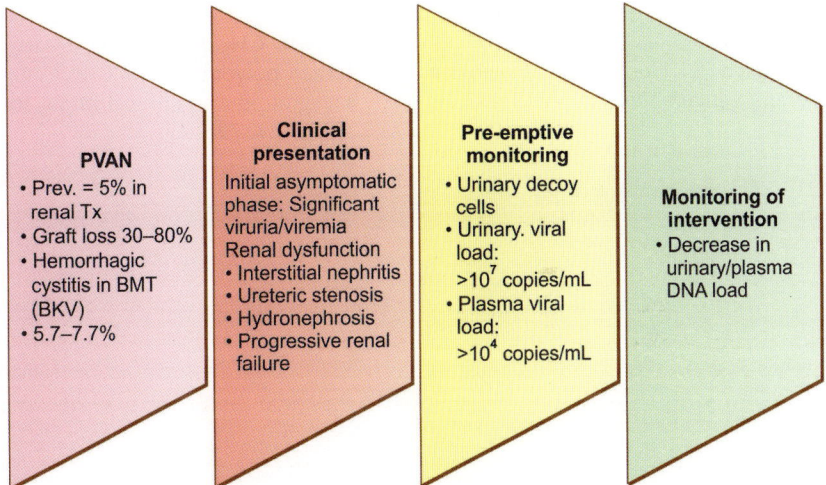

Fig. 5.2: Polyomavirus-associated nephropathy (PVAN)

marrow, crosses the blood–brain barrier and enters brain. Inside the brain, it infects the **oligodendrocytes**, the cell that is responsible for myelin production in the white matter of the human brain. The virus causes lytic infection of the oligodendrocytes and causes progressive multifocal encephalopathy in immunosuppressed hosts.

Almost all progressive multifocal leuko-encephalopathy (PML) cases are positive for JCPyV antibody. PML occurs in various types of immunosuppressive conditions, but is the most common amongst AIDS patients. Near **80% of PML cases are AIDS patients**, making PML as one of the AIDS defining illnesses in 1–3% HIV infected individuals. The disease is also seen in HIV patients with immune reconstitution inflammatory syndrome (IRIS) due to influx of latently infected immune cells into the brain.

The use of therapeutic monoclonal antibodies, natalizumab, efalizumab and rituximab can induce the immune defects which can trigger the development of PML in those patients.

Clinically PML manifests with progressive focal neurological deficit of motor, cognitive and visual functions. Motor deficit usually manifests as ataxia with difficulties to walk, and talk. Cognitive dysfunction is as of the subcortical dementia.

Laboratory confirmation of PML is done by detection of JCPyV in CSF samples. Estimation of viral load helps as prognostic indicator. Low viral load has been shown with long-term survival of PML patients.

Nucleoside analog (cytarabine arabinoside) has been used in JCPyV diseases which acts by blocking the replication of JCPyV. CMX001, an analog of cidofovir, has been reported to reduce the JCPyV viral load in CSF.

Bibliography

1. Dalianis T, Hirsch HH. Human polyomaviruses in disease and cancer. Virology 2013;437(2): 63–72.

2. Ehlers B, Wieland U. The novel human polyomaviruses HPyV6, 7, 9 and beyond. APMIS 2013; 121(8):783–95.

3. Ferenczy MW, Marshall LJ, Nelson CD, Atwood WJ, Nath A, Khalili K, Major EO. Molecular biology, epidemiology, and pathogenesis of progressive multifocal leukoencephalopathy, the JC virus-induced demyelinating disease of the human brain. Clin Microbiol Rev 2012; 25(3):471–506.

4. Gullett JC, Nolte FS. Quantitative nucleic acid amplification methods for viral infections. Clin Chem 2015;61(1):72–78.

5. Hirsch HH. BK virus: opportunity makes a pathogen. Clin Infect Dis 2005;41(3):354–60.

6. Hirsch HH, Kardas P, Kranz D, Leboeuf C. The human JC polyomavirus (JCPyV): virological background and clinical implications. APMIS 2013; 121: 685–727.

7. Rinaldo CH, Tylden GD, Sharma BN. The human polyomavirus BK (BKPyV): virological background and clinical implications. APMIS 2013; 121: 728–45.

Parvovirus

Parvoviruses are small non-enveloped viruses of 25–30 nm in diameter with icosahedral symmetry. Before the discovery of circoviruses, parvoviruses were amongst the smallest DNA viruses to infect mammals (Parvum in Latin means "small"). The genome consists of single-stranded DNA of 5000 base pair in length.

The currently proposed taxonomy of Parvoviridae by International Committee on Taxonomy on Viruses (ICTV):

Family Parvoviridae is divided into two sub-families—Parvovirinae and Densovirinae, based on their ability to infect vertebrates and non-vertebrates, respectively.

The subfamily Parvovirinae contains eight genera:

1. Amdoparvovirus
2. Aviparvovirus
3. Bocaparvovirus
4. Copiparvovirus
5. Dependoparvovirus
6. Erythroparvovirus
7. Protoparvovirus
8. Tetraparvovirus

Amongst these, members of four genera are known to cause infection in humans as shown in Table 6.1.

As the newly proposed names are not yet in common use, the present chapter will use the old names. Human bocavirus has been described in Chapter 10 (Respiratory Viruses).

HUMAN PARVOVIRUS B19 (PVB19)

Parvoviruses are ubiquitous viruses and present worldwide. Infection is commonly acquired during childhood. The seroprevalence of PVB19 increases with age. In children up to 5 years of age, the seroprevalence is 2–5%, in adults it goes up to 40–60%, and in elderly near 85%.

Table 6.1: Parvoviruses causing human infections			
Previous name	*Proposed name*	*Proposed genus*	*Disease in humans*
Human parvovirus B19	Primate erythroparvovirus 1	Erythroparvovirus	TAC, hydrops fetalis, erythema infectiosum
Human bocavirus	Primate bocavirus 1 and 2	Bocaparvovirus	LRTI in children
Adeno associated virus 1	Adeno associated virus A	Dependovirus	No known disease
Human parvovirus 4	Primate tetraparvovirus 1	Tetraparvovirus	Possible association: • VIS in IDU • Resp. infection? • Co-infection with HIV, HBV and HCV in IDU

TAC: Transient aplastic crisis; LRTI: Lower respiratory tract infection; VIS: Virus infection syndrome; IDU: Intravenous drug user

Modes of Transmission

- *Respiratory route:* Through inhalation of virus present in the aerosol.
- *Mother to fetus:* Vertical transmission from mother to fetus has been reported in one-third of mothers with primary PVB19 infection.
- *Other modes:*
  - Through blood and blood-derived products
  - Organ transplantation: Bone marrow and solid organ transplantation.

In temperate climate, PVB19 infection is more commonly seen during late winter or early spring months.

After HPVB19 enters the human host, it targets the erythroid precursor cells by binding to **erythrocyte P antigen or glycospingolipid globoside**. Blood group P antigen is present in hematopoietic precursors, erythroblasts and megakaryocytes. These cells are permissive for the PVB19 as the virus can enter the cell as well as can also replicate efficiently. In endothelial cells, fetal myocytes and placental trophoblasts HPVB19 can enter but complete virions are not produced inside these cells.

Two co-receptors have been identified:

- **α5β1 integrin:** Responsible for cell adhesion and present in erythroid progenitors.
- **Ku80 molecules:** An autoantigen, permits the entry of HPVB19 into the cells.

Expression of both P antigen and co-receptors makes the erythroid progenitor cells more permissive for HPVB19.

Clinical Features

The clinical manifestations due to parvovirus B19 is different in immunocompetent individuals and immunocompromised hosts.

Diseases in Immunocompetent Host

Erythema infectiosum: This is the most common manifestation of HPVB19 in immunocompetent individuals. Also known as **Fifth disease** according to its position in the list of common childhood infections.

The disease starts with prodromal symptoms of fever, malaise and coryza. The characteristic manifestation is the erythematous rash on the cheek, known as **slapped-cheek appearance**. The second stage is followed with extension of rash on the trunk, back and extremities. There may be central clearing of the rash giving rise to lacy reticular pattern. Symptoms are normally self-limiting and resolved within 1–3 weeks. In some cases, rash may recur on exposure to heat, sunlight or exercise. In comparison to children, rash is less pronounced, and arthralgia is more common in adults.

Rash and joint symptoms presumably occur due to antigen–antibody immune complex deposition in skin, blood vessels and synovial tissue as it coincides with the appearance of antibody.

Arthropathy: It may occur in association with erythema infectiosum or as a primary infection. It is more common in adults affecting almost 50–60% of individuals and more commonly seen in women than men.

In adults, arthropathy is **polyarticular, symmetric** in nature. Proximal metacarpophalyngeal and interphalyngeal joints are commonly involved. Knee, ankle and wrist joints are less commonly involved. Symptoms last for a few weeks. In contrast to adults, involvement of joints may be symmetric or asymmetric in children and knee and ankle joints are affected more commonly in them. Absence of erosion in the joints helps it differentiating from rheumatoid arthritis.

Transient aplastic crisis: In disease conditions with increased turnover of erythrocytes (hemolytic anemia, hereditary spherocytosis, sickle cell disease, etc.), HPVB19 infection can lead to **transient aplastic crisis**. This is manifested by cessation of erythrocyte production and drop in hemoglobin level requiring multiple transfusions. Congestive heart failure, cerebrovascular accident and splenic sequestration can occur as complications. In addition to the involvement of erythrocyte lineage, it can also cause thrombocytopenia, neutropenia and pancytopenia.

Hydrops fetalis: HPVB19 infection during pregnancy poses risk of around 30%

transplacental transmission to fetus. The overall risk of adverse outcome in fetus is around 5–10%. The infant may develop severe anemia, hypoalbuminemia, hepatitis and myocarditis. Combination of all these features leads to congestive heart failure and hydrops fetalis. It has been estimated that 15–20% of non-immune hydrops fetalis are due to HPVB19 infection. The rarity of the event does not indicate routine screening for parvovirus infection during pregnancy. However, on confirmation of acute infection during pregnancy, weekly/bi-weekly ultrasound of fetus is recommended to prevent hydrops fetalis.

Diseases in Immunocompromised Hosts

In immunocompromised individuals, clearance of virus gets delayed due to poor immune status of the host, which leads to persistence of the virus and chronic bone marrow suppression and manifested by chronic anemia. Symptoms of rash and arthralgia which occur mostly due to immune complex formation and are less noticed because of immune suppression. This condition is commonly observed in patients with hematological malignancies with chemotherapy and multiple blood transfusions.

Other diseases: B19 has also been associated with several other disease conditions like myocarditis, hepatitis, aplastic anemia, hemophagocytic syndrome, autoimmune disorders, glove and sock syndrome in both immunocompetent and immunocompromised individuals.

Diagnosis

Serology and detection of viral DNA are the common methods used for diagnosis of parvovirus infections.

Serology: Infection with HPVB19 induces development of IgM antibody after 10–12 days of infection, which then persists for a few months. IgG antibody starts appearing just after IgM and persists for life.

Parvovirus specific IgM: Detection of IgM antibody against recombinant VP2 antigen by ELISA is commonly used for diagnosis of acute infection. Majority of the immunocompetent individuals with erythema infectiosum or arthropathy shows IgM antibody positivity. The test is, however, not reliable in immunocompromised individuals.

Parvovirus specific IgG: IgG antibody to HPVB19 is detected by ELISA using VP1 and VP2 viral antigens. Presence of IgG in absence of IgM antibody indicates past infection and immunity to infection. As the IgG antibody persists for life, it is used for determination of seroprevalence. Seroconversion indicates acute infection.

Parvovirus DNA detection: Molecular tests like hybridization and polymerase chain reaction are used for virus DNA detection from blood or tissue. Detection of HPVB19 DNA in affected tissue generally indicates the association of virus with specific organ involvement. This strategy has been employed to show the association of HPVB19 in fetal tissue, bone marrow, myocarditis, liver cells, placental trophoblast, etc.

Interpretation of Parvovirus DNA Detection

- HPVB19 DNA can be detected in blood, bone marrow, and tissue samples. Detection of virus DNA in these samples can be either due to acute infection or due to persistence of DNA. Therefore, the qualitative detection of HPVB19 DNA alone cannot act as an indicator of acute infection in adults. Whereas, qualitative detection of HPVB19 DNA in infants or in placenta helps in diagnosing the intrauterine parvovirus infection. It also helps in detecting infection in immunocompromised individuals in whom IgM may be negative.
- Viral load estimation is helpful to differentiate between acute infection and persistence of DNA. Significant viral load during acute PVB19 infection can go up to 10^{12} **geq/mL**

blood. Though there is no consensus regarding the significant viral load, a **cut-off of 10⁴ vgc/mL blood** has been suggested as a diagnostic criterion by Maple et al.

Antigen detection assay: Parvovirus B19 antigen can be detected in the tissue specimen by immunohistochemistry using monoclonal antibody.

Enzyme immune assay for the detection of parvovirus B19 antigen in serum sample is available but because of its poor sensitivity this is not widely used.

Cytopathology: Bone marrow may show the presence of giant pronormoblast. Its presence is suggestive of PVB19 infection.

Treatment

There is no specific antiviral available against HPVB19. Most of the parvovirus diseases are self-limiting in nature and does not require any specific treatment. However, in severe cases, patients may be treated as mentioned below.

- Diseases in immunocompetent individuals are normally treated with anti-inflammatory drugs.
- Patients with aplastic crisis with increased erythrocyte turnover are usually managed with blood transfusions as required according to severity of anemia.
- Immunocompromised patients with chronic symptoms are treated with intravenous immunoglobulin **(IVIG)**. This reduces the anemia associated symptoms.

Bibliography

1. Broliden K, Tolfvenstam T, Norbeck O. Clinical aspects of parvovirus B19 infection. J Intern Med 2006;260(4):285–304.

2. Erik D. Heegaard and Kevin E. Brown. Clinical Microbiology Reviews 2002;15:485–505.

3. Maple PA, Hedman L, Dhanilall P, Kantola K, Nurmi V, Söderlund-Venermo M, Brown KE, Hedman K. Identification of past and recent parvovirus B19 infection in immunocompetent individuals by quantitative PCR and enzyme immunoassays: a dual-laboratory study. J Clin Microbiol 2014; 52:947–56.

4. Matthews PC, Malik A, Simmons R, Sharp C, Simmonds P, Klenerman P. PARV4: an emerging tetraparvovirus. PLoS Pathog 2014;10(5): e1004036.

5. Qiu J, Söderlund-Venermo M, Young NS. Human Parvoviruses. Clin Microbiol Rev 2017;30(1):43–113.

6. Servey JT, Reamy BV, Hodge J. Clinical presentations of parvovirus B19 infection. Am Fam Physician 2007;75(3):373–76.

7. Virus Taxonomy: 2015 Release. EC 47, London, UK, July 2015; Email ratification 2016 (MSL #30) accessed from http://www.ictvonline.org/virustaxonomy.asp.

8. Waldman M, Kopp JB. Parvovirus B19 and the kidney. Clin J Am Soc Nephrol 2007 Jul;2 Suppl 1:S47–56.

Adenovirus

Human adenovirus was discovered during early 1950s from adenoid tissue of patients with atypical pneumonia and has derived its name from its source of origin (adenoid). Thereafter the virus has been associated with several clinical diseases like acute respiratory disease, keratoconjunctivitis, pharyngitis, gastroenteritis, hemorrhagic cystitis, hepatitis, meningitis, myocarditis and disseminated disease. The ability to infect various types of cells, various organs and evidence of oncogenic potential in animals poses a challenge.

VIRUS

Human adenoviruses (HAdV) are nonenveloped, viruses with icosahedral symmetry and 70–100 nm diameter. The genome is linear double-stranded DNA of 34–37 kb. At the 3' and 5' end, genome has inverted terminal repeats. The unique feature of adenovirus genome is the presence of 1–2 non-coding virus associated RNA genes. The icosahedral capsid composed of 252 capsomers which includes 240 hexons and 12 pentons. Each penton is located at the 12 vertices of the icosahedrons. A spike-like projection is present at each vertex called fiber or **penton fiber**. Each of this fiber has three parts—base, shaft and terminal knob. The base or tail of the fiber binds to the pentons. The length of the shaft is different for different AdV types and the terminal knob acts as the ligand for attachment with the host cell receptor. Hexon contains the antigenic site that is common to all AdV types (Fig. 7.1).

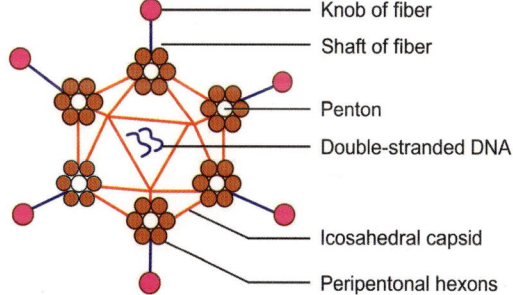

Fig. 7.1: Schematic diagram of adenovirus

Human adenoviruses (HAdV) belong to the family Adenoviridae and genus Mastadenovirus and are divided into seven species: A to G. Species are designated on the basis of phylogenetic distance of >5–15%, genome organization, length of the fiber, guanosine and cytosine percentage in the genome, host range, oncogenicity in rodents and hemagglutination pattern. Till date 67 HAdV types have been identified. HAdV 1–51 have been identified on the basis of traditional serum neutralization assay, whereas the remaining types 52–67 have been distinguished by using nucleotide analysis and bioinformatics analysis. Table 7.1 shows the classification of HAdV.

EPIDEMIOLOGY

Adenoviruses are ubiquitous. Most of the infections occur during childhood and by 10 years of age majority of the population is seropositive for the virus. This occurs due to several infections by different serotypes during childhood. In immunocompetent

Table 7.1: Classification of human adenovirus

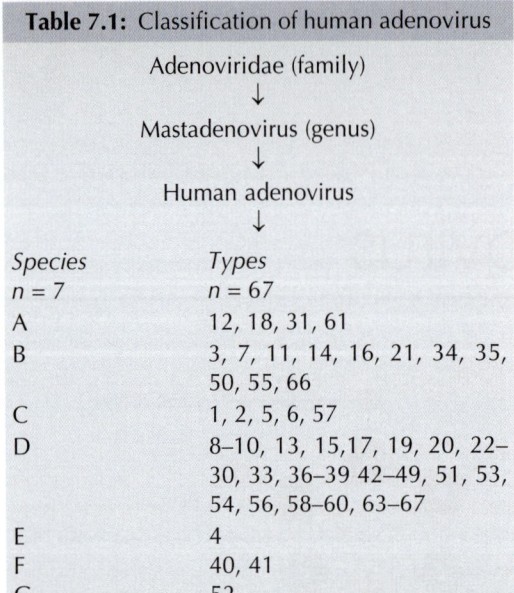

Adenoviridae (family)
↓
Mastadenovirus (genus)
↓
Human adenovirus
↓

Species	Types
n = 7	n = 67
A	12, 18, 31, 61
B	3, 7, 11, 14, 16, 21, 34, 35, 50, 55, 66
C	1, 2, 5, 6, 57
D	8–10, 13, 15,17, 19, 20, 22–30, 33, 36–39 42–49, 51, 53, 54, 56, 58–60, 63–67
E	4
F	40, 41
G	52

individuals, most of the infections are mild and self-limiting in nature. Thus infections in children go unnoticed. Infection with adenovirus in military recruits and in immunocompromised individuals can lead to severe infection. Adenovirus serotype 14 is one of the rare serotypes that has been reported to cause severe pneumonia in immunocompromised individuals particularly transplant recipients and patients with underlying chronic lung or cardiac disease.

HAdV can cause sporadic infection as well as outbreaks. Several outbreaks of acute respiratory tract infections have been reported in closed circuits like day care center and military recruits. Infection is mostly seasonal in immunocompetent hosts, whereas it occurs throughout the year in immunocompromised individuals.

Adenoviruses C, B, A and D are the common species associated with human infections. Table 7.2 shows the common adenovirus types associated with various infections.

The virus is transmitted commonly through aerosols, infected droplets and fomites, feco-oral route from individuals with acute infection. Infection through ocular instrument or swimming pool is a common mode of transmission during epidemic keratoconjunctivitis.

PATHOGENESIS

The incubation period of the disease is 2 days to 2 weeks and depends on the route of entry and AdV type. HAdV can infect various types of cells. The tissue tropism and the site of infection largely depend on the type of AdV and the route of infection (Table 7.2).

The **coxsackievirus and adenovirus receptor (CAR),** a transmembrane protein of immunoglobulin superfamily, is the primary receptor for HAdV. CAR is expressed in several organs, such as heart, pancreas, prostate, lungs, liver, central and peripheral nervous system. Several other proteins like CD46, CD80 and CD86 are also acts as receptor for certain adenovirus types.

The replication of adenovirus inside the host cell nucleus leads to accumulation of viral proteins which give rise to the formation of intranuclear inclusion bodies. **Penton can act like a toxin** and has been shown to cause

Table 7.2: Common adenovirus species and types associated with different sites of infection

Site of infection	Common AdV species	AdV types
Respiratory tract infection	C, B, A, E	1–7, 14, 21
Pharyngoconjunctival fever	C, B, A, E	1–7, 11–17, 19–21, 29
Epidemic keratoconjunctivitis	D	8, 19, 37
Gastrointestinal	F, G	40–41, 52
Genitourinary	B	11
Myocarditis	B	7, 21

detachment of the mononuclear cells in the cell culture. The virus also has the potential to shutdown the synthesis of mRNA synthesis in the host cell and produce its own proteins.

Adenovirus has been shown to remain in latent form after the primary infection. Tonsillar lymphocytes, central nervous system and gastrointestinal tract are the major sites of viral latency. Viral protein E3 plays the major role in maintaining latency by inhibiting the class I major histocompatibility complex (MHC) and affecting the antigen presentation. Reactivation of the endogenous latent virus is the main cause of infection in immunocompromised individuals.

CLINICAL SYNDROMES

Acute respiratory tract infection: Respiratory infection is more common in children. Upper respiratory tract infection is more common and usually mild in nature. In symptomatic cases, it manifests with pharyngitis or tracheitis accompanied with constitutional symptoms, such as fever, myalgia, headache, malaise. In infants, it may be associated with otitis media. Exudative tonsillitis caused by adenovirus is clinically indistinguishable from group A streptococcal infection. HAdV has also been associated with pertussis-like syndrome. Lower respiratory tract illness is characterized by fever, cough, and shortness of breath with bilateral ground glass patchy lung opacity.

The upper respiratory tract illnesses are commonly caused by AdV serotypes 1, 2, 5, and 6 of species C, whereas serotypes 3, 7, 14 and 21 of species B are mostly responsible for lower respiratory tract illness including severe complicated pneumonia.

Ocular infection: Eye infections of AdV can occur in the form of pharyngoconjunctival fever (PCF) or epidemic keratoconjunctivitis (EKC).

Pharyngoconjunctival fever (PCF) is a syndrome that is characterized by pharyngitis, follicular conjunctivitis, fever and cervical lymphadenopathy. Infection can occur as sporadic infection in children or as community outbreaks mostly due to improper chlorination in swimming pools. Disease is usually self-limiting in nature with rare corneal complications. HAdV types 3 and 7 are the two most common serotypes associated with PCF.

Epidemic keratoconjunctivitis (EKC) is characterized by severe unilateral or bilateral follicular conjunctivitis, edema of eyelids and subepithelial corneal infiltrates. Corneal involvement is usually painful and leads to blurring of vision. Incubation period is around 4–7 days. Symptoms may subside in 2 weeks to 1 month but corneal opacity may persist for a long time; months to years with reduced vision, photophobia and foreign body sensation (Fig. 7.2).

EKC commonly occurs in outbreaks in schools or hospital settings. The infection is highly contagious and commonly transmitted through ophthalmic instruments and lotions or directly through eye secretion. HAdV serotypes 8, 19 and 37 have been commonly associated with EKC.

Gastrointestinal disease: Acute gastrointestinal infection is caused by HAdV serotypes 40 and 41. The disease is common in infants and attribute to around 5% of acute diarrheal illness in children. It is manifested as acute diarrhea which is accompanied by fever, vomiting and abdominal pain. Symptoms usually last for about 8–10 days.

Hemorrhagic cystitis (HC): Acute hemorrhagic cystitis occurs in children and immunocompromised individuals. The disease is characterized by sudden onset of hematuria with dysuria. Hematuria lasts for around

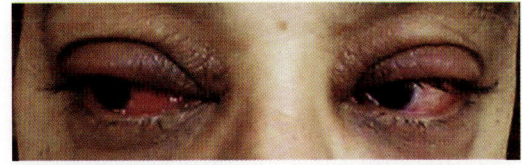

Fig. 7.2: Adenovirus keratoconjunctivitis

3 days, whereas dysuria and microscopic hematuria may persist for a few more days. Adenoviruses 11 and 21 have been commonly associated with HC.

Myocarditis: Adenovirus is considered to be one of the major causes of acute viral myocarditis, particularly in children. The disease process is believed to be immune mediated following the acute respiratory tract infection. The association of adenovirus has been shown by detection of viral DNA in myocardial tissue. Outbreaks of adenovirus-associated acute myocarditis have been reported.

Infections in immunocompromised patients: Amongst the immunocompromised patients, symptomatic adenoviral infection is more common in transplant recipients. Most of the infections occur within 3 months of transplantation. In solid organ transplant recipients (SOT), the transplanted organ is the primary site of infection. Hence, the manifestation is mostly related to the organ transplant such as hepatitis in liver transplant, hemorrhagic cystitis in renal transplant and pneumonia in lung transplant cases. In hematologic stem cell transplant (HSCT) recipients, symptoms can range from pneumonia, hepatitis, nephritis, enteritis, meningoencephalitis or disseminated disease.

The severity of infection is much more in pediatric patients as compared to the adults and more in patients receiving antilymphocyte antibodies. Presence of viral DNA in the blood has been associated with severity of disease and high rate of mortality.

DIAGNOSIS

Diagnosis of adenoviral infection is mainly relies on the direct methods like isolation of virus or its components such as viral antigen or viral genome. Indirect method like antibody detection is of less importance and not reliable due to its less sensitivity.

As the direct method of diagnosis is the mainstay of diagnosis, the clinical samples required for diagnosis are collected from the site of virus replication. Therefore, the type of sample depends on the symptoms of the patient or organs affected. For example, throat swab/nasopharyngeal aspirate or broncho-alveolar lavage in pneumonia; conjunctival swab in eye infection, urine in cystitis, cerebrospinal fluid in meningitis and blood in disseminated or severe disease and biopsy specimens from the affected organs.

Isolation of virus: Human adenoviruses can be isolated in various cell lines. Continuous cell lines of epithelial origin such as HeLA, Hep2, A549 and human embryonic kidney (HEK) are susceptible for adenovirus. Enteric adenoviruses; HAdV 40 and 41 are fastidious and grow in Graham 293 or adenovirus 5 transformed HEK cell lines. Cytopathic effect in epithelial cell line is produced mostly within seven days and is characterized by rounding of cells with clustering and intranuclear inclusion bodies.

Isolation of virus by cell culture is considered as the gold standard, but time consuming, and suffers from the disadvantages of poor sensitivity and liable for bacterial and fungal contamination.

Antigen detection: Indirect immunofluorescence (IIF) test using monoclonal antibody is used for samples like swabs and aspirates from different sites. Infected cells show apple green **intranuclear fluorescence** (Fig. 7.3). IIF is a rapid test and can be reported on the same day but poorly sensitive and needs fluorescence microscope.

Immunohistochemistry technique is used for demonstration of adenovirus antigen in the tissue.

Enzyme immunoassay (EIA) is used for antigen detection for AdV 40, 41 in fecal samples as the antigen is not intracellular in this sample, whereas EIA is not applicable for other samples where intracellular antigen has to be detected. Genus and type-specific EIA are available commercially with sensitivity and specificity is >90% and 95%, respectively.

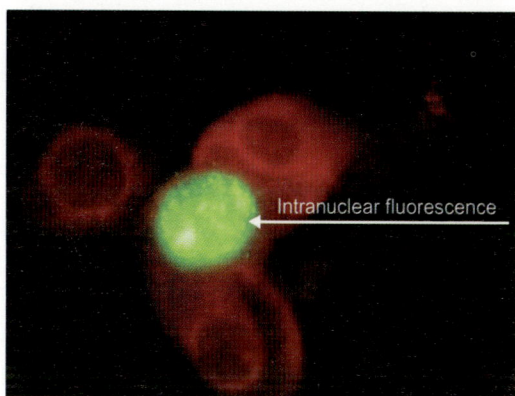

Fig. 7.3: Adenovirus antigen showing intranuclear fluorescence of infected conjunctival epithelial cell by indirect immunofluorescence

Histopathology is particularly helpful in case of pneumonia wherever biopsy is feasible. Mononuclear cell infiltration, formation of hyaline membrane and intranuclear inclusion bodies masking the nuclear membrane giving the appearance of **Smudge cell** are the characteristic findings.

Detection of viral genome: This is done most commonly by polymerase chain reaction (PCR). Hexon gene or fiber gene is commonly used as the target for amplification. The test can be done in all types of clinical samples with high sensitivity and specificity.

Commercially available real-time PCR assays are presently available for detection of panels of respiratory pathogens including adenovirus from respiratory samples.

Typing of adenovirus is traditionally done by serological assays like hemagglutination inhibition or serum neutralization assay. These tests have largely been replaced by molecular methods such as type specific PCR-RFLP or PCR followed by sequencing.

Presently, adenovirus DNA detection by PCR or RT-PCR is the commonly employed tests for diagnosis of adenoviral infection. Detection of AdV DNA in blood is considered as a marker of disease severity and is indicated in HSCT recipient children.

TREATMENT

There is no approved drug available for adenovirus therapy. Most of the immunocompromised patients recover or improve from the infection due to improvement in the T cell count.

Cidofovir has shown good *in vitro* antiviral effect and has been accepted for pre-emptive therapy. The drug is a cytosine analog with DNA polymerase activity. It is given pre-emptively as indicated in pediatric HSCT recipients during the post-transplant period. The dose is 5 mg/kg body weight, given weekly for 2 weeks and thereafter at 2 weeks interval. The duration of treatment is based on the clinical response or reduction in viral load. However, there is no consensus regarding the response of pre-emptive therapy. The drug needs to be monitored for its nephrotoxicity.

VACCINE

Adenovirus vaccine prepared from AdV type 4 and 7 was available for military recruits in USA. The vaccine was available in enteric capsule form for oral intake. The manufacture of the vaccine was stopped by its sole manufacturer during late 1990s and it was not available for few years. However, due to reemergence of severe form of adenovirus disease, the preparation of adenovirus vaccine has been restarted.

Bibliography

1. Adenovirus vaccine. http://www.cdc.gov/vaccines/hcp/vis/vis-statements/adenovirus.pdf.
2. Adhikary AK, Banik U. Human adenovirus type 8: the major agent of epidemic keratoconjunctivitis (EKC). J Clin Virol. 2014;61(4):477–86.
3. Lewis PF, Schmidt MA, Lu X, et al. A community-based outbreak of severe respiratory illness caused by human adenovirus serotype 14. J Infect Dis. 2009;199(10):1427–34.

4. Lion T. Adenovirus infections in immunocompetent and immunocompromised patients. Clin Microbiol Rev. 2014;27(3):441–62.

5. Louie JK, Kajon AE, Holodniy M, et al. Severe pneumonia due to adenovirus serotype 14: A new respiratory threat? Clin Infect Dis 2008; 46(3):421–25.

6. Pavia AT. Viral infections of the lower respiratory tract: Old viruses, new viruses, and the role of diagnosis. Clin Infect Dis 2011;52 Suppl 4:S284–89.

7. Ronchi A, Doern C, Brock E, Pugni L, Sánchez PJ. Neonatal adenoviral infection: A seventeen year experience and review of the literature. J Pediatr 2014;164(3):529–35.e1–4.

Poxviruses

The members of Poxviridae are large complex enveloped double-stranded viruses that infect both vertebrates and invertebrates causing diseases of medical and veterinary importance. Poxviridae is known for its most important member smallpox virus which was once one of the highly lethal and dreadful viruses of mankind. Smallpox virus was eradicated in 1977, but the other members of the family bearing the potential of causing similar manifestations in human and possible threat of using smallpox virus or similar engineered virus as bioweapons have regenerated interest in this virus family.

CLASSIFICATION

Poxviridae consists of two subfamilies—Chordopoxvirinae and Entomopoxvirinae.

Members of Chordopoxvirinae infect vertebrates and members of Entomopoxvirinae infect insects. According to the recent ICTV classification, Chordopoxvirinae has 11 genera of which members of 4 genera—Orthopoxvirus, Molluscipoxvirus, Parapoxvirus, and Yatapoxvirus infect humans. Amongst the members of these 4 genera, only variola virus (causative agent of smallpox) of genus Orthopoxvirus and molluscum contagiosum of genus Molluscipoxvirus are the sole human pathogens (Fig. 8.1).

Genus Orthopoxvirus has 10 species with the widest host range affecting several vertebrates. Of which variola virus is an obligate human pathogen, whereas cowpox, monkeypox and vaccinia viruses can also infect humans as zoonosis. The remaining viruses have restricted host range (Table 8.1).

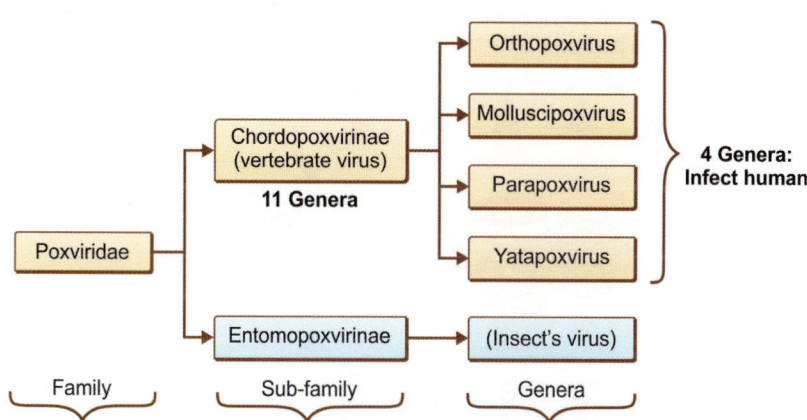

Fig. 8.1: Poxivrus classification showing genera that infects humans

Virus	Primary host	Other hosts
Table 8.1: Poxviruses and their host range		
Genus: Orthopoxvirus		
Camelpox virus	Camel	Nil
Cowpox virus	Bank vole rodent	Cattle, cat and human
Ectromelia virus	Rodents	Nil
Monkeypox virus	**Rodents**	Monkeys, prairie dogs and human
Raccoonpox virus	Raccoons	Nil
Skunkpox virus	Skunks	Nil
Taterapox virus	Gerbils	Nil
Vaccinia virus	**Human, cow**	Nil
Variola virus	**Human**	Nil
(smallpox virus)		
Volepox virus		
Genus: Parapoxvirus		
Milkers nodule	Cattle	Human
(Pseudo-cowpox virus)		
Orf virus	Sheep, Goat	Human
Genus: Molluscipoxvirus		
Molluscum contagiosum	Human	Nil
Genus: Yatapoxvirus		
Yabapox virus	Monkey	Human
Tanapox virus	Monkey	Human

VIRUS

Morphology: Poxviruses are amongst the largest of animal viruses. The shape of virion is brick or barrel shaped with an average dimension of 350 × 250 nm. The structure of virion consists of outer layer or envelope that encloses the internal structures (Fig. 8.2).

The outer layer or envelope consists of lipoproteins and has tubular structures giving a corrugated appearance. The outer membrane covers a dumbbell- or biconcave-shaped inner core and two lateral bodies of unknown function. The inner core contains the viral DNA and viral proteins. The genome is linear double-stranded DNA, the size of which in different poxviruses varies from 130–375 kbp. The ends of each DNA strand have inverted tandem repeats. The two DNA

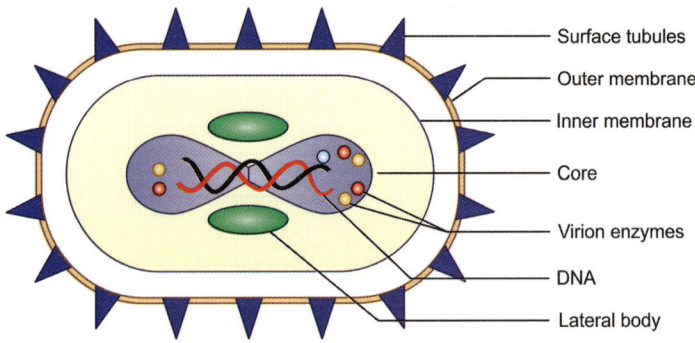

Surface tubules
Outer membrane
Inner membrane
Core
Virion enzymes
DNA
Lateral body

Fig. 8.2: Structure of poxvirus

strands are joined at both ends by terminal hairpin loops.

The nucleocapsid of poxvirus does not confirm either icosahedral or helical pattern and is termed as **"Complex"** symmetry.

The genome of poxvirus has the capacity to produce large number of proteins, up to 200 proteins. The proteins required for viral replication are coded by the central region of the genome and proteins that are responsible for host range, immunomodulation and virulence are coded by the terminal regions.

Viral Replication

Poxviruses are large viruses and are considered as self-sufficient regarding its replication. These viruses code the enzymes required for their replication. So, unlike other DNA viruses, poxviruses replicate inside the host cell cytoplasm instead of nucleus.

The transcription process occurs in a cascade manner. As first step, transcription begins by viral RNA polymerase that transcribes the viral genome into early mRNA and viral proteins. This process occurs inside the viral core. These proteins in turn complete the process of uncoating of viral core and transcribes around 100 early proteins. These are basically the enzymes required for replication of viral genome, viz. DNA polymerase, thymidine kinase. Early gene codes for the transcription factors required for intermediate genes and intermediate genes codes for the transcription factor required for the late genes.

The replication of viral genome occurs inside the cytoplasm by virally coded enzymes. The process starts after the second step of uncoating. Genome replication occurs in discrete sites of cytoplasm, called **"viroplasm"**, **"factories"** or **"inclusion bodies"**. Assembly of the progeny virus begins in these factories.

The morphogenesis and assembly of poxviruses is a complex process. Most of the **mature virions (MV)** remain in the cytoplasm and are released by cytolysis of the host cell.

Some of these viruses migrate to Golgi and get another coating of cell derived enveloped. These virions are called **"enveloped virions"** (EV) and are released from the host cell by exocytosis. Both the morphological forms of virus are infectious, however, EVs are known more for their efficient cell-to-cell spread of infection.

HUMAN POXVIRUS INFECTIONS

Smallpox

Smallpox is caused by variola virus a member of Orthopoxvirus genus. It is the first disease to be eradicated from the history of mankind. In 1977, the last case of endemic smallpox was reported from Somalia and in 1978 one laboratory acquired case was reported. **In 1980, WHO declared the world as free from smallpox.** However, the virus is now important for its potential threat as a biological weapon and in such scenario if the virus is released it will cause massive outbreak with expected high mortality as the major population of the world is susceptible. It is thus important to know the virus and be familiar with its clinical features.

Pathogenesis: The virus enters through respiratory tract from where it goes to the lymph nodes. Replication at these sites leads to primary viremia. During this period, virus is seeded in reticuloendothelial system and replicates at these sites leading to secondary viremia. This marks the onset of symptoms with prodromal symptoms. During this period, two different events occur.

1. Virus infects the mucosa of mouth and pharynx causing oropharyngeal lesions which appear before the skin rash. High concentration of virus is released in respiratory shedding. The infectivity is maximum during this period. Virus is also found in urine and several visceral organs such as spleen, kidney, liver, and bone marrow.

2. During the secondary viremia, the infected macrophage migrates to and infects epidermis, and then the virus invades the

dermal layer of the skin, which leads to the formation of vesicular lesion. Virus is also found in the lesions.

The histopathological feature of infected cells of all orthopoxviruses including smallpox shows typical cytoplasmic inclusions. Type A inclusions which are intracytoplasmic inclusion bodies containing the clusters of virions. The second one is type B inclusions, also called **"Guarneri bodies"** which are perinuclear intracytoplasmic inclusion bodies containing viroplasm and maturing virions. Table 8.2 describes the unique properties of poxviruses.

Clinical features: Incubation period ranges from 7–17 days. Symptoms start with the prodromal symptoms followed by lesions on oral and pharyngeal mucosa (enanthem) and on skin (exanthem).

Prodromal symptoms start with abrupt onset of high fever and severe constitutional symptoms. Most of the patients are severely ill during this period. It lasts for 2–3 days. This follows the appearance of lesions on oral mucosa and one day later skin lesions. The lesions on the skin are macules to start with and progresses to papules, vesicles and pustules by 7–10 days and formation of scab occurs by 2–3 weeks. The size of lesion also increases with the lesion progression from 2–3 mm to 7–10 mm in diameter. Lesions start

---

Table 8.2: Unique properties of poxviruses

Morphology:
- Largest brick-shaped virus
- Dumbbell shape core and lateral bodies
- Complex symmetry not fitting to either icosahedral or helical
- Two DNA strands are attached
- Virions produced are two types: Enveloped and non-enveloped

Replication: DNA virus that replicates inside cytoplasm

Transmission: Mostly by direct contact or respiratory tract

Clinical feature: All are associated with skin lesion

---

from face and are more condensed on face and limbs than trunk (centrifugal distribution) in contrast to chickenpox where the lesions are more on trunk with centripetal distribution. Also, in contrast to chickenpox, all lesions are in same stage. Hemorrhagic lesions are rare and associated with higher fatality.

WHO has classified the clinical features of smallpox into five types:
a. *Ordinary smallpox:* Most common (90%) with 30% case fatality rate.
b. *Flat type smallpox:* Seen in around 7% cases with high mortality of 97%.
c. *Hemorrhagic smallpox:* Seen in only 3% but mortality almost 100%.
d. *Modified smallpox:* Seen in only 2%, much less severe and rarely fatal.
e. *Variola sine eruption:* Asymptomatic and do not transmit the infection.

Differential diagnosis: Diagnosis of smallpox is based on its classical clinical presentation. However, it may be required to be differentiated from other diseases causing macular rash or papular or vesicular lesions. Table 8.3 gives the list of common causes of differential diagnosis of smallpox.

Vaccination

Variolation: The first preventive measure for smallpox was attempted by "variolation". This is a process in which material from smallpox lesion was directly introduced to healthy host with an aim to produce immunity against the virus by producing milder form of disease.

Two methods were practiced:
a. Inhalation of dried scab material.
b. Introduction of pus from active lesion by causing scratch.

In 1796, **Sir Edward Jenner** showed that intradermal introduction of human cowpox material to a healthy person provided protection against smallpox. This concept of Jenner created world's first vaccine.

Smallpox Vaccine used during Eradication

First generation smallpox vaccines were used during eradication. These were multiple

Table 8.3: Differential diagnosis of smallpox

Diseases with maculopapular rash	Diseases with papulovesicular lesions
Viral causes • Dengue • Chikungunya • Parvovirus B19 • Measles • Rubella • Enteroviruses	**Viral causes** *Other poxviruses:* • Monkeypox • Molluscum contagiosum *Viral causes other than poxviruses:* • Chickenpox (varicella) • Zoster • Hand-foot-mouth disease (HFMD)
Bacterial causes • Scarlet fever • Scalded skin syndrome • Rocky mounted spotted fever	**Bacterial causes** • Rickettsia pox • Yaws

strains of live vaccinia virus that were propagated on the skin of live animals. These vaccines had the chance of contamination with microorganisms, animal proteins and adventitious tissue materials. Lister-Elstree and New York City Health Board strains of vaccinia virus were the recommended strains by WHO during the eradication program.

Vaccine was given using a bifurcated needle with multiple punctures to introduce through abrasion as vaccinia virus does not grow on intact skin. The efficacy of first generation vaccine was 75%. After correct vaccination, a papule develops at the local site around 4th day and progresses to vesicle and pustule by 7th day.

Second generation smallpox vaccines were prepared from single clone of vaccinia virus grown in tissue culture. This was made to reduce the reactogenicity keeping the immunogenicity intact.

Third generation smallpox vaccines are based on replication competent attenuated vaccinia strains. Attenuation is based either on serial passage in non-human tissues or geneticmodification. LC 16m8 and Modified Vaccinia Ankara (MVA) are among the preferred vaccine strains. These vaccines are safer than previous generation vaccine, however, immunogenicity is probably less.

Eradication of Smallpox

The intensified smallpox eradication program began in 1967. Smallpox was declared to be eradicated from the globe in 1980.

The last case of naturally acquired smallpox by variola major was reported in 1975 in a 3 years old girl named Rahima Banu from Bangladesh. The last case of naturally acquired smallpox due to variola minor was reported in 1977 in Somalia. In 1978, the last case of smallpox was reported which was a lab-acquired case in England.

The factors which played a role in smallpox eradication are:
• Presence of effective vaccine and intensive vaccination program
• No animal reservoir
• Easily identifiable characteristic clinical features
• No subclinical infection.

Smallpox as Bioweapon

Currently, the stocks of variola virus is officially present only in two labs under direct supervision of WHO, Center for Disease Control and Prevention (CDC) Atlanta, and State Research Center for Virology and Biotechnology (VECTOR), Koltsovo, Russia. However, it is apprehended that the virus may be present unofficially in many more places and many groups are working on it to use it as a bioterrorism agent.

Smallpox is a suitable bioterrorism agent due to its following characteristics:

- Virus is easy to grow.
- Can be lyophilized.
- Survive in the aerosolized form and can be used for mass infection.
- Majority of the current population are susceptible.
- Disease has a long incubation period hence, may go unnoticed before it spreads.
- Clinicians of current generation are not familiar with the clinical presentation which may delay the diagnosis.
- Genetically modified virus can lead to atypical presentation leading to misdiagnosis and may evade vaccine-induced immunity.
- Limited number of stockpiles of vaccines.

Human Monkeypox

Monkeypox is caused by the monkeypox virus (MPXV), a member of the genus Orthopoxvirus. The name monkeypox was given as the disease was first recognized in an outbreak of vesicular lesion amongst captive monkeys in 1958. Human disease was first observed in early 1970s in West and Central Africa particularly Congo Basin countries after the eradication of smallpox from these regions. Monkeypox is presently the most important Orthopoxvirus for humans after the eradication of smallpox virus and cessation of smallpox vaccine. The disease in human resembles clinically with smallpox but in a milder form and its cross-protection by smallpox vaccine makes the virus more relevant clinically.

Epidemiology: The primary animal reservoir host for MPXV remains unknown, however, evidence of infection implicates squirrels and rodents. Infection in monkeys and humans is incidental.

The disease in humans is rare and occurs mostly as zoonosis. The **primary zoonotic infection** occurs by direct or indirect contact with the body fluids of the infected animals through handling, bites or scratches. The **secondary human to human** transmission occurs through infected respiratory droplets or direct or indirect contact with infected lesion material, body fluids or contaminated materials or surfaces. The virus enters through respiratory tract or through broken skin or mucosa (Fig. 8.3).

Monkeypox usually considered as a geographically restricted disease being confined to West and Central Africa particularly in the tropical rain forest areas. Two different

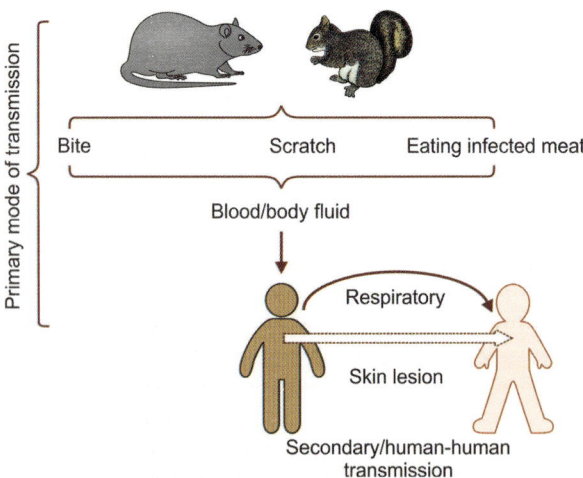

Fig. 8.3: Monkeypox virus: Modes of transmission

monkeypox virus clades exist: (1) Congo Basin clade which is prevalent in Central Africa, and (2) West Africa clade which is prevalent in West Africa. Both clades are found in Cameroon. Congo Basin clade has been found to be more virulent and transmissible.

The first human case was reported in the year 1970 in a 9-year-old boy in the Democratic Republic of Congo. Since 1970, several sporadic and outbreaks have been reported in central and West African regions. However, the number of human cases is rising in recent years with possible human-to-human transmission. Amongst the large outbreaks of human poxvirus infection, outbreaks in Democratic Republic of Congo (previously known as Zaire) in 1996–97 affecting >500 people and in 2017–18 in Nigeria affecting >100 people are of concern.

Imported human monkeypox outbreaks: The first report of MPXV infection in western hemisphere was reported in USA in 2003 where >80 people from different states were affected. The infection was traced to contact with prairie dogs that were kept along with the exotic rodents imported from Africa.

Imported cases were reported from Israel in 2018, from United Kingdom in 2018 and 2019 and from Singapore in 2019 in individuals who had travelled to Nigeria.

Clinical features in MPXV infection is similar to smallpox. The incubation period is 7–17 days. The symptoms start with prodrome of fever, headache, myalgia, followed by development of lesion on skin and mucosa within 1–3 days. Lesions are maculopapular to start with and become fluid filled vesicles and pustules and finally become crusted by 10 days. Crust disappears within 3 weeks.

Pronounced lymphadenopathy is common and is the clinical distinguishing feature than smallpox.

Smallpox and Monkeypox Vaccines

Currently two vaccines have been licensed in USA for prevention of smallpox, ACAM2000

and JYNNEOS™ and the later one JANNEOS™ is also licensed for prevention of monkeypox. Both are live virus preparation.

ACAM2000 is a replicating virus, given by skin prick. Successful inoculation leads to development of lesion at the site. As the virus propagates at the site, it can spread to other parts of the body and also to other persons. Individuals who receive this vaccine are advised to take adequate precaution to prevent spread of vaccine virus. The present recommendation by CDC and Advisory Committee for Immunization Practice (ACIP) is for individuals who are at high-risk for smallpox like laboratory personnel working on orthopoxviruses and military personnel.

JYNNEOS™ (also named Imvamune or Imvanex) is also a live vaccinia virus containing Modified Vaccinia Ankara (MVA). Unlike ACAM2000, it is non-replicating, given in two doses, at 4 weeks apart by subcutaneous route. There is no risk of spread of vaccine virus to other parts of body or to other people.

Efficacy of these vaccines is around 85% in preventing against monkeypox when given before exposure. In case of postexposure, vaccine is recommended within 4 days of exposure to prevent the disease. When given between 4 and 14 days of exposure, it may reduce the disease severity. Table 8.4 summarizes the important facts on human monkeypox virus infection.

Human Cowpox

Cowpox virus is a member of Orthopoxvirus. The virus has been found in Europe and Western parts of Russia. Rodents are the natural reservoir. The virus can infect various animals and humans who are the accidental host. Infection to animals is usually transmitted from the rodents whereas humans can acquire the infection either from infected animals or rodents. Cows and domestic cats are the common source of infection to man. This probably explains why the disease is

Table 8.4: Important facts of human monkeypox virus infection	
Causative agent	• Monkeypox virus • Genus: Orthopoxvirus • Family Poxviridae
Transmission	• Viral zoonotic disease • Primary mode to human is from wild animals such as rodents and primates • Secondary mode: Human-to-human
Case fatality rate	Up to 10%
Clinical presentation	Resembles that of smallpox fever, rash and enlarged lymph nodes
Vaccine	• JANNEOS™: Live, non-replicating • Two doses, 4 weeks apart • Route: Subcutaneous

common among milkers who get the infection by milking the cows suffering from lesions on the teats and udders and amongst children because of exposure to domestic cats. Human cowpox due to pet rats are also increasing in Europe. In India, a few case reports of human cowpox have been reported. However, the reports are based on clinical and histopathological diagnosis without virological confirmation.

Clinical manifestation in humans often presents with single maculopapular lesion with intense inflammatory reaction often on hand and face along with systemic signs of fever and lymphadenopathy.

Cowpox virus is immunologically similar to vaccinia virus and variola virus. It has been observed in the past that the person who had cowpox were immune to smallpox.

Human Buffalopox

The causative agent of buffalopox is considered as a **modified vaccinia virus**. The disease is found in domestic buffaloes in Indian subcontinent, Indonesia and Egypt. Clinical manifestation of buffalopox in buffaloes is similar to features of cowpox in cows where the buffaloes develop skin lesions. Man acquires the infection through contact with the infected buffaloes by manual milking the animals with lesions on teats and udders. Lesions are maculopapular to start with which progresses from macule to papule, vesicles and pustules.

Several virologically confirmed outbreaks of human buffalopox have been reported in India amongst the dairy workers. Laboratory acquired human buffalopox virus has also been reported.

Human Parapoxvirus Infections

Among the parapoxviruses, **milkers' nodule virus** (also called **pseudocowpox virus**) and **orf virus** can infect humans. Humans acquire the infection from infected cattle and sheep/goats, respectively. Disease mostly occurs in farm workers who work with livestock or their products. Like other zoonotic poxvirus infections, lesions occur on skin, but begin as papules which become granulomatous.

Human Yabapox and Tanapox

Yabapox virus and tanapox virus are members of genus Yatapoxvirus.

Yabapox virus infection occurs in Asian and African monkeys. It causes benign histiocytoma on hairless areas of face, palms and interdigital areas and mucosa of nose and mouth. Men acquire the infection through contact with infected monkeys and suffer from similar illness.

Human **tanapox** virus infection is common in Eastern Africa and Democratic Republic of Congo (DRC). Monkeys are believed to be the natural reservoir. The exact mode of transmission is not known. It is believed that the disease is mechanically transmitted by insect bite from infected monkeys. Skin lesion like other poxviruses is the common manifestation.

Molluscum Contagiosum

The disease molluscum contagiosum is caused by molluscum contagiosum virus, a member of genus Molluscipoxvirus. Virus is specific for humans and not seen in animals. The infection manifest as multiple pearly papular nodules, 2–5 mm diameter, dome shaped, with central umbilication containing yellowish cheesy material. Nodules are painless. Number of lesions may vary from 1–20 and disease lasts for 6–9 months. Lesion is confined to epidermis. The infected cells in the nodule are hypertrophied and contains acidophilic inclusion bodies called **Molluscum bodies** which consist of spongy matrix of multiple cavities containing clusters of viral particles.

Virus is transmitted through direct contact, abraded skin or sexual contact. In children, lesions are seen mostly on trunk and proximal extremities whereas in adults on trunk or pubic area. Molluscum contagiosum virus is

Table 8.5: Summary of human poxvirus infections			
Genus/virus	*Disease*	*Mode of transmission*	*Common clinical features*
Genus: Orthopoxvirus			
• Variola virus	Smallpox	Respiratory route: Human–human	Generalized pustular lesion Case fatality rate: 30%
• Vaccinia virus	1. As complication of smallpox vaccination 2. Human buffalopox	Direct contact with infected buffaloes	Localized/generalized pustular lesion Pustular lesion
• Monkeypox virus	Human monkeypox	Animal (rodents, squirrels, other animals)—human: Scratch/bite/direct contact with lesion Human–human: Respiratory/direct contact	Generalized pustular lesion Case fatality rate: up to 15%
• Cowpox virus	Human cowpox	Direct contact with infected cows/domestic cats	Localized maculopapular lesion
Genus: Parapoxvirus			
• Orf virus	Human parapox	Direct contact with infected cattle, sheep, goat	Papular lesion
• Milkers' nodule virus			
Genus: Molluscipoxvirus			
• Molluscum contagiosum virus	Molluscum contagiosum	Direct contact from infected human	Multiple pustular lesions
Genus: Yatapoxvirus			
• Yabapox virus	Human yabapox	Direct contact with infected monkeys	Histiocytoma on face
• Tanapox virus	Human tanapox	Insect bite from infected monkeys	Pustular lesion

worldwide, however, more cases are reported from African countries.

Diagnosis is often made by characteristic clinical features, presence of brick-shaped virus in the lesion material by transmission electron microscopy, characteristic histopathology showing molluscum bodies and PCR. Virus is not cultivable in tissue culture. Table 8.5 summarizes the different types of human poxviruses.

Laboratory diagnosis: Lab diagnosis can be made from material collected from lesions by transmission electron microscopy and PCR using specific primers. The typical brick-shaped morphology of the Orthopoxvirus can be observed by electron microscopy. PCR is usually done for further sequencing to track the source of infection. Isolation of virus can be done by inoculating the samples onto Vero cell line. Cytopathic effect can be observed by 24 hours. Presence of virus in the infected cells is confirmed by immunofluorescence using orthopoxvirus specific antibodies.

Bibliography

1. Breman JG, Henderson DA. Diagnosis and management of smallpox. N Engl J Med 2002 Apr 25;346(17):1300–08.

2. Damle AS, Gaikwad AA, Patwardhan NS, Duthade MM, Sheikh NS, Deshmukh DG. Outbreak of human buffalopox infection. J Glob Infect Dis 2011 Apr;3(2):187–88.

3. Doshi RH, Guagliardo SAJ, Doty JB, Babeaux AD, Matheny A, Burgado J, Townsend MB, Morgan CN, Satheshkumar PS, Ndakala N, Kanjingankolo T, Kitembo L, Malekani J, Kalemba L, Pukuta E, N'kaya T, Kangoula F, Moses C, McCollum AM, Reynolds MG, Mombouli JV, Nakazawa Y, Petersen BW. Epidemiologic and ecologic investigations of monkeypox, Likouala Department, Republic of the Congo, 2017. Emerg Infect Dis 2019 Feb; 25(2):281–89.

4. Erez N, Achdout H, Milrot E, Schwartz Y, Wiener-Well Y, Paran N, Politi B, Tamir H, Israely T, Weiss S, Beth-Din A, Shifman O, Israeli O, Yitzhaki S, Shapira SC, Melamed S, Schwartz E. Diagnosis of Imported Monkeypox, Israel, 2018. Emerg Infect Dis 2019 May;25(5): 980–83.

5. https://www.cdc.gov/poxvirus/monkeypox/clinicians/smallpox-vaccine.html

6. https://www.cdc.gov/smallpox/history/history.html

7. Kennedy RB, Ovsyannikova I, Poland GA. Smallpox vaccines for biodefense. Vaccine 2009 Nov 5;27 Suppl 4:D73-9. doi: 10.1016/j.vaccine.2009.07.103.

8. Moore ZS, Seward JF, Lane JM. Smallpox. Lancet 2006 Feb 4;367(9508):425–35.

9. Ogoina D, Izibewule JH, Ogunleye A, Ederiane E, Anebonam U, Neni A, Oyeyemi A, Etebu EN, Ihekweazu C. The 2017 human monkeypox outbreak in Nigeria-Report of outbreak experience and response in the Niger Delta University Teaching Hospital, Bayelsa State, Nigeria. PLoS One. 2019 Apr 17;14(4): e0214229.

10. Petersen BW, Kabamba J, McCollum AM, Lushima RS, Wemakoy EO, Muyembe Tamfum JJ, Nguete B, Hughes CM, Monroe BP, Reynolds MG. Vaccinating against monkeypox in the Democratic Republic of the Congo. Antiviral Res 2019 Feb;162:171–7.

11. Petersen E, Abubakar I, Ihekweazu C, Heymann D, Ntoumi F, Blumberg L, Asogun D, Mukonka V, Lule SA, Bates M, Honeyborne I, Mfinanga S, Mwaba P, Dar O, Vairo F, Mukhtar M, Kock R, McHugh TD, Ippolito G, Zumla A. Monkeypox–Enhancing public health preparedness for an emerging lethal human zoonotic epidemic threat in the wake of the smallpox post-eradication era. Int J Infect Dis 2019 Jan;78:78–84.

12. Petersen E, Kantele A, Koopmans M, Asogun D, Yinka-Ogunleye A, Ihekweazu C, Zumla A. Human Monkeypox: Epidemiologic and Clinical Characteristics, Diagnosis, and Prevention. Infect Dis Clin North Am 2019 Apr 10. pii: S0891-5520(19)30017-0.

13. Reynolds MG, Wauquier N, Li Y, Satheshkumar PS, Kanneh LD, Monroe B, Maikere J, Saffa G, Gonzalez JP, Fair J, Carroll DS, Jambai A, Dafae F, Khan SH, Moses LM. Human Monkeypox in Sierra Leone after 44-Year Absence of Reported Cases. Emerg Infect Dis 2019 May; 25(5):1023–25.

Chapter

9

Hepatitis Viruses

Hepatitis B is a global public health problem. It is caused by hepatitis B virus (HBV) which is one of the leading causes of chronic hepatitis, cirrhosis and hepatocellular carcinoma. One-third of the world's population have been infected with hepatitis B virus. More than 6,00,000 people die every year due to HBV-related diseases.

HISTORY

Epidemiologically two types of agents were suspected as the cause of hepatitis: Feco-oral transmission and serum transmitted. Hepatitis A virus (HAV) was known to be the major cause of the first one. Several outbreaks of hepatitis were also linked to vaccines prepared from convalescent human serum, use of contaminated syringe and sharing of needle. These outbreaks of hepatitis were considered as another form of hepatitis which is different from HAV and was termed "serum hepatitis".

The first link of association of HBV came from the experiment of **Blumberg**, who discovered a human antigen in the serum of Aborigine of Australia while studying for the genetic marker. Subsequently this **Australia antigen** was associated with hepatitis in several groups of patients including patients with multiple transfusions and was found to be a major cause of serum hepatitis including post-transfusion hepatitis. Further study revealed the antigen as a component of a DNA virus, which was named hepatitis B virus (HBV). Blumberg received Nobel Prize in 1976 for his discovery.

CLASSIFICATION

HBV belongs to the family Hepadnaviridae (hepar, liver and DNA is according to the type of virus genome or hepatitis DNA virus or hepatotropic DNA virus).

The family Hepadnaviridae consists of hepatitis B like viruses. It has two genera:

1. *Orthohepadnavirus:* Consists of viruses that infect mammals. HBV belongs to the genus Orthohepadnavirus.
2. *Avihepadnavirus:* Consists of viruses that infect birds.

The members of Hepadnaviridae share the unique properties of: (i) Having small size of viral genome, (ii) unique arrangement of ORF, (iii) unique replication strategy using reverse transcription for replication of viral DNA, and (iv) hepatotropism.

Members within each genera share strong genome similarities but the same is not observed among the groups.

VIRUS

Three forms of virus particles are present in the serum of the infected patient.

Two are smaller incomplete form of virus particles (subvirus particles), present in spherical form with 20 nm diameter and filamentous form with variable length with a width of 22 nm. Both these forms contain hepatitis B surface antigen (HBsAg) and host derived lipid but do not contain viral nucleic acid, hence are non-infectious particles. These are highly immunogenic, produce neutralizing antibodies, and in purified form can be used for vaccine purpose (Fig. 9.1a and b).

The infectious virion is the complete virus particle which is also known as **"Dane particle"** named after the pathologist Dane. HBV is a small spherical virus having two shells—outer and inner. Outer one is the lipid envelop of 42 nm in diameter, consisting of the HBV surface antigen (HBsAg) together with the host derived lipids. The inner shell is the nucleocapsid of 22 nm in diameter. It consists of the hepatitis B core protein (HBcAg) complexed with the viral genome and virally encoded polymerase (Fig. 9.1c).

The genome of HBV is a small, circular, partially double-stranded DNA of 3–3.2 kilo-base size (3200 kb), consisting of one complete and one incomplete strand. The complete strand is of negative polarity, so known as "minus strand". The incomplete strand is called "plus strand" because of its positive polarity. It contains viral enzymes; reverse transcriptase (RT) and a cellular protein serine kinase.

Genome organization: Genome contains four partially overlapping open reading frames (ORFs)—S, C, P and X (Fig. 9.2).

S ORF encodes for the surface antigen (HBsAg). Structurally and functionally it can be divided into pre-S1, pre-S2 and S that code for the three components of hepatitis surface antigen L, M and S, respectively.

- *S (small or major) protein:* Small or short surface antigen. This is the major component of HBsAg. Present in abundant both in virion and subviral particles. It contains the major antigenic determinant called Australia antigen.
- *M (middle) protein:* It represents around 10–15% of the envelop protein.
- *L (large) protein:* In virion, L protein represents around 17% of envelop protein and acts as the main ligand for the viral receptor.

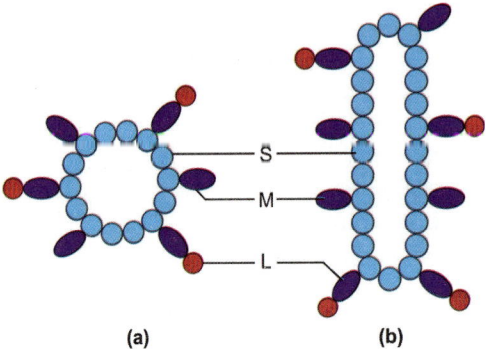

(a)   **(b)**

Fig. 9.1a and b: Non-infectious secreted spherical and filamentous particle

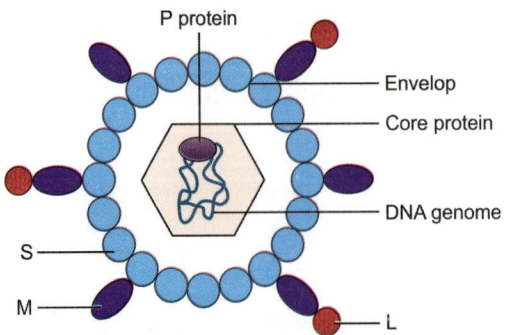

Fig. 9.1c: Infectious Dane particle

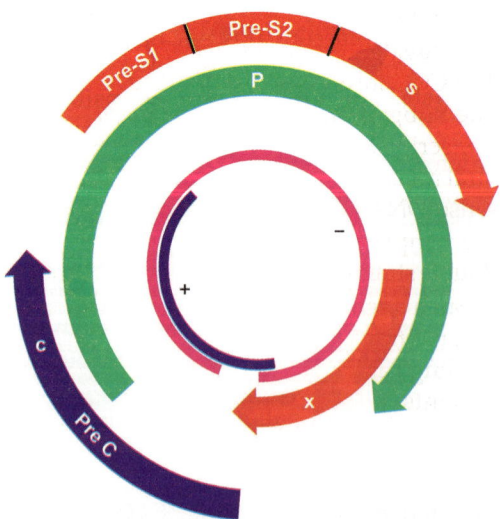

Fig. 9.2: Schematic representation of HBV genome

C ORF codes for nucleocapsid or core gene (HBcAg) and precore region (pre C) encodes for a secretory core protein called hepatitis E antigen (HBeAg). The formation of HBcAg or HBe antigen depends on the site of translation initiation from core or precore region.

P ORF codes for the polymerase gene. It is functionally divided into three parts: (i) Terminal domain is responsible for encapsidation and formation of minus strand, (ii) reverse transcriptase domain is responsible for genome synthesis, and (iii) ribonuclease H domain is responsible for degradation of pregenomic RNA and facilitates replication.

X ORF codes for HBx gene, which is essential for productive HBV infection and has the oncogenic potential.

Replication

The replication of HBV is unique from other DNA viruses. In spite of being a DNA virus, it utilizes an RNA intermediate and enzyme reverse transcriptase.

HBV enters the susceptible hepatocyte through receptor. The Pre-S1 component of L1 surface antigen of the virus plays an important role in entry of virus into the host cell. Inside the cytoplasm, capsid is removed; viral genome is released and transported to the nucleus. Inside the nucleus, viral genome gets converted into **covalently closed circular (CCC)** form. The CCC form of viral DNA is transcribed to one subgenomic and one full length RNA transcripts. The latter act as the template for viral reverse transcriptase from which a minus sense DNA is produced. This then serves as the template for synthesis of positive sense double strand DNA by viral DNA polymerase.

Such a replication system is error prone as reverse transcriptase does not have a proofreading capacity. This leads to generation of mutants.

MODES OF TRANSMISSION

Infected blood and body fluids including saliva, semen, cervical secretions, are the source of infection. HBV can survive for a long period in the environmental surface. Thus exposure to the contaminated surfaces can transmit the virus. The transmission of virus can occur through following modes:

1. *Perinatal from infected mother to fetus:* Predominant mode in high prevalence locality.
2. *Horizontal transmission during childhood:* Occurs through close contact between family members. Predominant in intermediate prevalence area.
3. *Percutaneous route:* Transmission through sharing of contaminated needle, tattoo, razor, acupuncture, etc. predominant in low prevalence area.
4. *Sexual transmission:* Infection is mostly acquired in adolescent and adults in low endemic areas. More common in homosexuals and individuals with multiple partners.
5. *Blood transfusion:* The risk through this route has substantially decreased worldwide after implementation of HBV screening of donors.
6. *Hemodialysis:* Patients undergoing hemodialysis are high-risk group for acquiring HBV infection.
7. *Health care setting:* Exposure to needlestick injury or contaminated blood sample possesses risk in a health care setting.

EPIDEMIOLOGY

Hepatitis B is a public health problem across the globe. According to the World Health Organization's (WHO) estimation, more than 2 billion people (one-third of world's population) are infected with HBV. Approximately 240 million people suffer from chronic infection. Amongst these carriers, nearly 25% develop serious form of liver diseases like chronic hepatitis, cirrhosis and hepatocellular carcinoma leading to one million deaths every year.

The prevalence of HBV infection is based on the HBV surface antigen (HBsAg) positivity in the population which varies from 0.1 to 20% in different geographic regions. Three types of endemic regions have been classified.

1. *High endemic:* ≥8% HBsAg positivity in population.
2. *Intermediate endemic:* >2 to <8% HBsAg positivity in population.
3. *Low endemic:* ≤2% HBsAg positivity in population.

The endemicity of different regions primarily depends on the age of acquisition of infection.

Table 9.1 shows the age of HBV acquisition and mode of transmission in different grades of endemic areas.

Burden in India: Most of the studies have reported the prevalence of 2–8% HBsAg positivity. The average HBsAg carrier rate being 5%, India falls in the category of intermediate endemic region. This contributes to an estimated 40 million HBsAg carrier which is around 15% of the world's HBV carrier pool. The point prevalence of HBV among non-tribal and tribal population has been calculated using population-weights as 3.07% and 11.85% respectively. However, some of the tribal islands of Andaman and Nicobar have been found to be hyperendemic with the prevalence of 23 to >65%.

Genotypes: Genotypes of HBV are divided on the basis of ≥8% divergence in the complete genome sequence or >4% in the S gene sequence. Presently HBV has been divided into 10 genotypes: A–J. Initially four genotypes A–D were identified. Genotypes E and F were then added. Genotype G was identified in the year 2000 from French and American patients who were co-infected with genotype A. Genotype H which was identified in central America and Mexico and is closely related to genotype F. Genotype I was first identified in Vietnam and is based on 7% divergence in the complete genome sequence. Genotype J is the latest genotype found in Japan and is close to the strains found in primates.

Subgenotypes: These classifications are based on the divergence of >4% of the nucleotide sequence in the complete genome sequence. Currently up to 40 subgenotypes have been described amongst the genotypes A–D and F.

Clades: Within the subgenotypes, divergence of <4% are described as clades.

Genotypes and subgenotypes vary in their geographic location, clinical severity and drug response.

Geographic distribution of genotypes is given in Table 9.2.

Genotypes in India: Genotype D is the predominant genotype reported from all parts of India. In North India, genotype D is the most common followed by genotype A. Recently, genotype F also has been reported from Aligarh, Uttar Pradesh, and North India.

In southern India, genotype D is the most prevalent one with >90% prevalence, whereas genotype A and C are occasional.

Table 9.1: Relation between HBV endemicity with age and mode of infection

Endemicity	Area affected	Age of infection acquisition	Common mode of transmission
High	Southeast Asia, China, Pacific islands, Middle East	Preschool age	Vertical/horizontal through close contact
Intermediate	Africa, South Asia, Russia, Eastern and South Europe, South and Central America		Horizontal through close contact
Low	USA, West Europe, Australia	Adolescent/Adult	Parenteral/sexual

Continent	Countries	Predominant genotype
Table 9.2: Continent-wise distribution of HBV genotypes		
Africa	East and South Africa	A
	North Africa	D
	West Africa	E
Asia	Western, South, Central Asia and Middle East	D
	Southeast Asian countries	B and C
	China and Japan	C
Europe	Mediterranean countries (Spain, Greece, Italy)	D
	Eastern Europe	D
	Russian countries	D
	Other parts	A and D
America	Canada, Greenland	B
	South and Central America	F
	Central America	H
Australia		A, B, C

In Eastern India, genotype D was found in around 50% cases followed by genotypes C and A in near 30% and 26%, respectively.

In north east part of the country, both genotypes D and C were found to be common with only around 5% prevalence of genotype A.

Figure 9.3 depicts the schematic representation of HBV genotypes distribution in India.

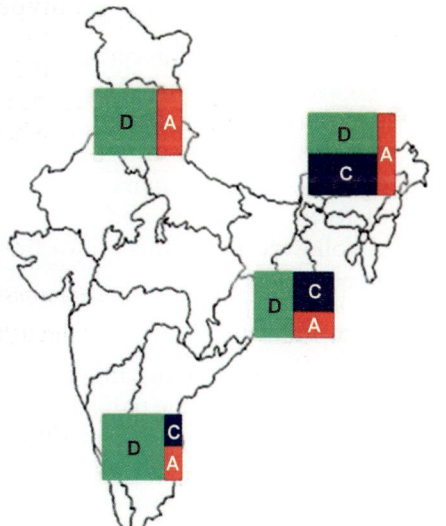

Fig. 9.3: HBV genotype distribution in India (schematic)

Clinical relevance of genotypes: Genotypes and subgenotypes of HBV have been associated with the clinical outcome of the disease by influencing the disease chronicity, severity, rate of HBe antigen clearance, viral load, development of chronic liver disease and hepatocellular carcinoma and response to antiviral treatment. It has been observed that different genotypes have been associated with different modes of transmission.

Studies on association of various genotypes with clinical outcome have reported the comparative result between two prevalent genotypes in a specified region.

Risk of progression to chronicity: **HBV genotype C (C2)** has been independently associated with progression to chronicity. As compared to genotype B, genotype A has a higher risk of progression to chronicity following acute infection in adults.

Rate of HBe antigen clearance: Failure or delayed HBe antigen clearance has been associated with genotype C to the tune of decades in comparison to another genotypes. Rate of spontaneous HBe antigen clearance is higher in genotype B as compared to genotype C.

Role in viral load: Genotypes C and D have shown high viral load as compared to genotypes

B and A, respectively. In a comparison between genotypes E and D in Africa, genotype E was associated with more HBe positive disease and high viral load as compared to D. High viral load has been linked to more chance of vertical transmission.

Development of chronic liver disease: Genotypes C and D have been associated with cirrhosis and development of fibrosis as compared to genotypes B and A respectively. Whereas genotype H acquired during adulthood, develop low viral load with less risk for development of chronic liver disease or HCC.

Association with HCC: Several HBV genotypes have been associated with heptocellular carcinoma (HCC) more commonly with genotypes C, B, A and F. Genotype C has been associated with higher risk of developing HCC as well as with tumor recurrence. Whereas, genotypes B, A and F have been associated with HCC developed at a young age in their respective places.

Treatment response: HBV genotypes differ in their response to PegIFN therapy. Response to therapy after 52 weeks in HBeAg positive cases has been shown to be maximum with genotype A followed by B, C and D (47%, 44%, 28% and 25%, respectively). Poor response also has been reported in genotypes E and G.

Genotypes and modes of transmission: The distribution of different HBV genotypes is geographically and ethnically restricted. It has also been noted that the predominant mode of transmission varies in different population and locality.

In hyperendemic eastern Asian countries where genotypes B and C are predominant, vertical transmission is quite common. Whereas, in Africa, Europe and India where genotypes D and A are prevalent, horizontal transmission is common. Genotype G has been reported mostly in MSM (men having sex with men) from America and Europe.

CLINICAL FEATURES

The infection can remain asymptomatic. Clinical manifestation may vary from anicteric hepatitis to icteric hepatitis to fulminant hepatitis. Chronic stage of the disease can be asymptomatic carrier to cirrhosis or hepatocellular carcinoma.

Acute Hepatitis

Incubation period for development of acute hepatitis varies from 6 weeks to 6 months. Acute HBV infection is mostly asymptomatic in nature and clinical manifestation occurs only in one-third of infected adults.

The manifestation can range from mild constitutional symptoms, jaundice to fulminant liver failure. It has three phases: **Pre-icteric, icteric and convalescent**. Preicteric phase lasts from days to a week. Icteric phase lasts from 1 to 2 weeks and convalescent phase lasts from weeks to months.

Preicteric phase is manifested by generalized constitutional symptoms, such as fever, malaise, loss of appetite, nausea, vomiting which is followed by the icteric phase with signs of jaundice and liver tenderness. Jaundice is more seen in adults as compared to children. Severity is more in case of co-infection with hepatitis D virus infection and in alcoholics. The symptoms subside within 1–3 months. Majority (95%) of immunocompetent adults recover from the disease and the rate of going towards chronicity decreases with increase in age group. Majority of the infected children develop chronicity.

Liver failure occurs in around 1% of acute HBV hepatitis cases. It is marked by sudden onset of fever, vomiting, abdominal pain, jaundice followed by confusion, disorientation, and coma.

Chronic Hepatitis

Persistence of HBsAg for more than six months in the blood of the patient is defined as chronic hepatitis. Patient usually suffers from non-specific symptoms and diagnosed incidentally for elevated aminotransferase unless develops cirrhosis or HCC.

The course of chronic hepatitis has three basic phases of disease process which occurs

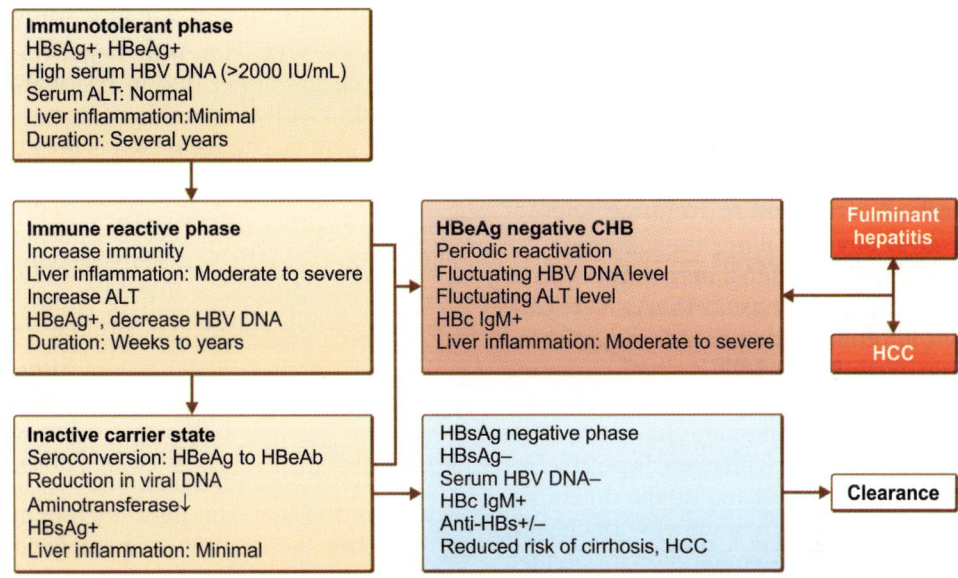

Fig. 9.4: Natural course of chronic HBV infection

due to interaction between virus and host. The progress course of chronic HBV infection is depicted in Fig. 9.4.

Immune tolerant or replicative phase: This is the initial phase when the infection is transmitted during perinatal period. This is so named because the immune system of the host tolerates the high level of virus replication. This phase is reflected by presence of virus in terms of HBsAg, high levels of HBV DNA in serum **(often >10⁷ IU/mL)**, HBe

antigen positivity, but minimal liver inflammation manifested by mild disease, normal serum ALT and minimum change in liver biopsy. This phase **may last for 10–30 years** and there is a low rate of spontaneous HBeAg clearance during this phase.

Immune active/immune clearance phase: The immune tolerant phase gradually switches over to immune active phase where, the tolerance to HBV is lost and host immune system mounts antibody against the virus and

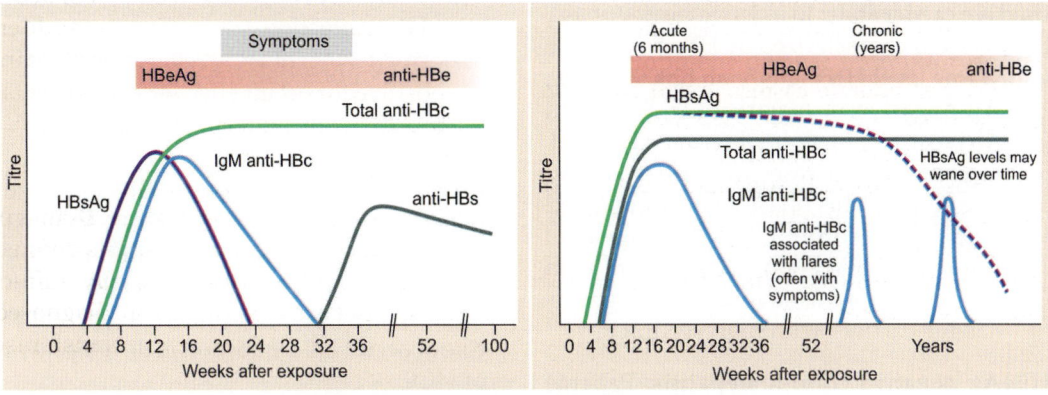

Source: Guidelines on Hepatitis B and C testing, WHO, 2017

Fig. 9.5a and b: (a) Acute HBV infection; (b) Chronic HBV infection damages the hepatocyte. Therefore, immune active phase is also called *immune clearance phase*. This is the **most aggressive phase** of chronic HBV infection. During this phase, HBeAg is usually positive indicating active viral replication. HBV DNA level in this phase is lower as compared to immune tolerant phase.

The chance of **HBeAg seroconversion** is high during this phase. HBeAg seroconversion is often accompanied with sudden increase in serum ALT level due to immune mediated lysis of hepatocytes. Usually this exacerbation is asymptomatic, however, some patients may present with acute hepatitis-like symptom. This may misguide the diagnosis, if the patient is not a known case of chronic hepatitis B. Misdiagnosis may get compounded by positive anti-HBc IgM which might flare up during exacerbation.

In some cases, exacerbation does not lead to HBe seroconversion and HBV DNA clearance. This is called **abortive immune clearance**. Repeated episodes of exacerbation may lead to intermittent HBeAg disappearance, HBV DNA clearance and fluctuating ALT level. Such repeated episodes increase the risk of fulminant hepatitis or HCC.

In majority of the chronic HBV patients, seroconversion of HBeAg occurs to HBeAb within 10 years of diagnosis. Therefore, HBeAg negative indicates late phase of chronic HB.

Inactive carrier state: In inactive carrier phase, the disease is characterized by HBsAg positive with a low level **HBV DNA of <2000 IU/mL, HBeAg negative** in majority (75–80%) and **HBeAb positive** with **normal serum ALT** level. Small percentage of cases may clear HBsAg over the course. This phase may be confused with HBeAg negative CHB. HBsAg level <1000 IU/mL, HBV DNA <2000 IU/ml with HBeAb positive status favours the diagnosis of "inactive carrier" with high accuracy.

HBeAg negative chronic hepatitis: Patients of HbeAg negative CHB category have moderate level of HBV DNA replication with increased ALT level and chronic liver inflammation with HBaAg is negative. This probably occurs either due to residual wild virus or mutant virus with precore or core promoter genetic change which is not able to produce HBe antigen.

Phase of resolution of CHB: Some CHB patients becomes HBsAg negative and usually carries a good prognosis unless there is some liver injury persists. However, the risk of HCC remains.

Occult Hepatitis B

Occult hepatitis B is defined as presence of HBV DNA in liver and/or serum in an HBsAg negative individual. This may be presented as two varieties: Seropositive and seronegative.

- **Seropositive occult HBV infection**: Positive for anti-HBs, anti-HBc.
- **Seronegative occult HBV infection**: Negative for anti-HBs, anti-HBc.

Mutation in the 'a' variant of HBsAg or mutation in pre-S gene leading to sequence variation in HBV genome has been implicated for occult HBV infection. The prevalence of occult hepatitis B has been reported in up to 30% cases.

Clinical Relevance

- Occult HBV in presence of chronic HCV infection has been associated more commonly with cirrhosis in comparison to only HCV infection.
- Majority (> 70%) of HCV related HCC cases are positive for occult HBV which indicates the possible contributory role of occult HBV in development of HCC.
- Core promoter mutation in HBV genome is frequently detected in HBV associated HCC. Presence of same mutation in HBV genome in occult HBV associated with HCV related HCC supports the possible role of occult HBV infection in development of HCC.
- However, clinical relevance of lone occult hepatitis B infection is still not clear.

- *False occult B*: Failure to detect HBsAg by available ELISA system may be falsely labeled as occult HBV infection. This is associated with the mutation in the "a" determinant of the surface antigen.

Hepatocellular Carcinoma (HCC)

HCC is one of the important causes of cancer-related deaths. High incidence of HCC is seen in areas with high HBV endemicity. Chronic HBV infection plays a major role in development of HCC. Various host and viral factors have been associated with higher risk of development of HCC in chronic HBV infection (Table 9.3).

Mechanisms of Carcinogenesis in HBV Related HCC

The process of carcinogenesis in HCC is primarily driven by three mechanisms. The first, being the chronic inflammatory process in the liver leading to genetic damage. Second is the integration of HBV DNA into the host chromosome inducing chromosomal instability and third is the modulation of cellular proliferation by HBx protein.

Role of chronic inflammation: Chronic active hepatitis leads to repeated cycle of inflammation induced apoptosis and liver cell regeneration. This in turn leads to accumulation of genetic damage which facilitates oncogenesis.

Table 9.3: Risk factors in chronic HBV infection for development of HCC

Host factors:	Male sex
	Alcoholics
	Smoking
	Diabetes
Viral factors:	High HBV DNA load (>2000 IU/mL)
	HBeAg positive
	Genotype C
	Core promoter mutation
	Prolong duration for seroconversion from HBeAg to HBe antibody
	Co-infection with HCV/HIV/HDV

Role of HBV DNA integration: Integration of viral DNA into the host cell chromosome occurs during the acute phase of disease. The number of integrations are higher in chronic hepatitis and integration is seen in 80 to 90% of HBV-related HCC.

Most of the DNA integration occurs at such sites that are prone to genetic instability and involved in cell proliferation. The integration of DNA into host chromosome leads to disruption of cell control machinery that are involved in cell survival, proliferation and immortalization.

Role of HBx protein in HCC carcinogenesis: The primary mechanism of HBx protein is to act through various cell signaling pathways that are involved in cellular proliferation, oncogenesis and apoptosis.

The trans-activation of HBx protein activates certain genes that are associated with cell proliferation such as tumor necrosis factor α (TNF-α), interleukin 8 (IL-8), transforming growth factor (TGF-β1), epidermal growth factor receptor (EGFR) and transcription factor.

It activates several oncogenes like c-myc, c-jun, c-fos. It inactivates the p53 protein by binding to it and thereby inhibits the p53 mediated apoptosis pathway.

Diagnosis of HBV Infection

Laboratory Diagnosis

A panel of HBV markers are present in the blood during various phases of acute and chronic HBV infection which can be tested to detect the infection, differentiate acute and chronic HBV infection, to ascertain the stage of disease, monitor the disease progression and also to monitor the treatment response.

Amongst them, HBV DNA is the first to appear, followed by HBV surface antigen (HBsAg) which appears 1–10 weeks after exposure. HBe antigen is the second viral antigen to appear after HBs antigen. Acute infection seroconversion of HBe antigen occurs early to HBe antibody and both the HBs and HBe antigen disappears and the infection resolve by itself. The core antigen of HBV does not appear in the blood as this is not secreted but can be detected in liver cells. Serum enzymes; ALT and AST appear after a few weeks of viral markers. Jaundice may appear during this time.

Persistence of HBs antigen leads to chronicity of infection. In chronic HBV infection, HBe antigen persists for a long period, seroconversion may get delayed for years.

Antibody to core antigen HBc is the first amongst the antibodies to appear. It appears after 1–2 weeks of appearance of HBs antigen and may be the only viral marker during the window period (after disappearance of HBsAg and before appearance of anti HBs antibody). During the initial period of infection, HBc antibody is IgM in nature which disappears within 6 months which is followed by the appearance of IgG to HBc. Figure 9.5a and b shows the schematic representation of acute and chronic HBV infection.

Marker of recovery: In majority of the cases, HBe Ag disappears early, whereas HBV DNA and HBs Ag persist throughout the symptomatic phase.

Disappearance of HBsAg, HBV DNA, HBeAg, anti HBc IgM along with presence of antibody to HBsAg and HBeAg are indicator recovery. Antibody to HBeAg appears after the disappearance of HBeAg, thus acts as a favourable indicator. Anti-HBs antibody appears late and acts as a marker of immunity.

Kinetics of chronic hepatitis B: The initial phase of chronic hepatitis B is positive for HBV DNA, HBsAg, HBeAg and anti-HBc antibody. As the viral replication persists, HBV DNA, HBsAg and HBeAg continue to remain positive in majority of the cases. The further course of chronic hepatitis however differs in different hosts depending on the pattern of chronicity (Fig. 9.4).

Role of HBV infection markers in diagnosis and management of disease is given in Table 9.4.

Types of assays used in HBV diagnosis: Serological tests are the main stay of diagnosis of HBV infection. Detection of various serological markers are often used to know the stage and phase of disease and its progression. Molecular test is done to detect the HBV DNA. It can be performed to detect the infection particularly in absence of HBsAg in a suspected case, but more important is the quantitation of HBV DNA to determine the viral load which is an important marker to assess the need of antiviral treatment and also to monitor the disease progression.

Serological assays: Serological tests for HBV markers are available in various formats such as, rapid format, enzyme immunoassay, chemiluminescence assay, etc.

Rapid diagnostic tests (RDT) are single use, disposable format. Available for detection of antigen and antibody. It involves simple procedure and the result is read visually without need of any other equipment and available in less than 30 minutes. RDTs are based on the principle of immunochromatographic assay (lateral flow), immunoconcentration, particle agglutination, etc. The test can be performed in serum, plasma, whole blood or finger prick capillary blood samples.

The pooled sensitivity and specific of RDT as compared to EIA is reported as 90% and 99.5% respectively. The pooled sensitivity of EIA is 88.9% and specificity 98.4%. Whereas the overall analytical sensitivity of EIA is 50–100 times better than RDT. However, high analytical sensitivity does not affect the diagnostic performance of RDT much because the HBsAg concentration in most of the chronic HBV patients is well above 10 IU/ml.

Enzyme immunoassay (EIA), chemiluminescence assay (CLIA) are the other common serological test methods. In most of the medium and high throughput labs, enzyme immunoassay is preferred as it can test the samples in batches. CLIA is a platform where large number of samples can be tested in a short time with higher sensitivity and specificity than

Table 9.4: Clinical implications of HBV markers

Viral marker	Clinical implication
HBV Surface antigen (HBsAg)	
Positive: Persistence for >6 months: Loss of HBsAg: Undetectable	• HBV infection and person is infectious • Chronic HBV infection • Elimination of HBV • Occult HBV infection (HBsAg negative, HBV DNA positive)
HBcAg: Capsid protein	• Expressed in the liver tissue, not secreted in serum
HBeAg	• Marker of active viral replication • Associated with high level of viremia, hence "Indicator of severity" • Prolong positivity associated with progressive liver disease
Anti-HBc IgM Ab	• Primarily used when there is a suspicion that the patient is in early convalescence during "window period" (disappearance of HBsAg and before appearance of anti-HBsAb). • Mostly present from 3 to 12 months • Positivity indicates acute infection in preceeding 4–6 months • Used to differentiate between acute and chronic HBV infection, but its reappearance during "flares" in chronic HBV infection make it an unreliable indicator of recent primary HBV infection
Anti-HBc Total Ab	• Develops around 3 months after infection • Most constant marker of infection • Together with anti-HBs, indicates resolved infection • Anti-HBc, with or without anti-HBs, also indicates individuals who may reactivate in the context of immunosuppression • Absent in individuals immune due to vaccination
Anti-HBeAb	• Appears due to seroconversion from HBeAg to anti-HBeAb • Indicator of end of active viral replication • Indicator of clinical resolution • Present in the immune control and immune escape phases • May co-exist with HBeAg towards the end of immune tolerant phase
Hepatitis B surface antibody (anti-HBsAb)	• Marker of protection to HBV (10 IU/mL is considered as protective) • Determination of Ab level is required to find the need of booster dose of vaccine or HBIG in case of exposure • Develops after natural infection or vaccination • May coexist with HBsAg, so presence cannot be used to exclude current infection
HBV DNA	• Detection of HBV DNA can act as a marker of infection. May be detectable in early infection before HBsAg, and therefore useful in early diagnosis of at-risk individuals before HBsAg appears • Also present at low levels in the absence of HBsAg in the context of occult infection • Used to differentiate active from inactive HBeAg negative CHB infection. • Quantitation of HBV DNA is used to determine need for antiviral therapy in conjunction with ALT levels and degree of liver fibrosis

(Contd.)

Table 9.4: Clinical implications of HBV markers (contd.)

Viral marker	Clinical implication
	• To monitor treatment response: Increase in HBV DNA load may indicate inadequate adherence or the emergence of resistant variants
	• Quantitation of HBV DNA is used as a more direct and accurate measure of active viral replication, which correlates with disease progression

Table 9.5: Interpretation of serological and viral markers in various HBV infections/immune status

Clinical status	HBsAg	Anti-HBs	HBeAg	Anti-HBe	Anti-HBc	IgM anti-HBc	HBV DNA	ALT
Susceptible	–	–	–	–	–	–	–	N
Acute HBV hepatitis	+	–	+	–	+	+	+	I
Chronic HBV hepatitis	+	–	+/–	+/–	+	–	+	N/I
Immune due to natural infection	–	+	–	+/–	+	–	–	N
Immune due to HBV vaccination	–	+	–	–	–	–	–	N
Inactive carrier	+	–	–	+	+	–	–/N	N
Inconclusive (resolved/resolving/low level chronic)	–	–	–	–	+	–	–	N

N: Normal; I: Increased

ELISA. Hence, should be used in a high throughput lab where large number of samples are tested every day. RDT is preferred in labs where few samples are tested and when rapid result is required such as in emergence labs.

Testing for HBV Infection is Required in Several Scenario

• **Seroprevalence:** This is estimated by estimating the rate of HBV infection in a community by HBs antigen detection.
• **Screening:** In blood/organ donors to prevent the transmission and before surgical procedure.
• **Antenatal screening for HBsAg** is done to detect infection in mother so that transmission to baby can be prevented.
• **Diagnosis:** In suspected patients with acute or chronic hepatitis.

Molecular test: Detection of HBV DNA is important both for diagnosis as well as for assessment and monitoring of antiviral treatment. Several molecular techniques including point of care tests are commercially available.

Qualitative HBV DNA detection is important in following clinical scenario:
• To distinguish between window period from chronic HBV infection when only anti-HBc is positive: HBV DNA is positive in the first scenario and usually negative in the second situation.
• Patients with features of fulminant hepatitis who are negative for HBsAg on presentation.
• Serum HBV DNA negative in a previously positive patient indicates virus clearance.

Quantitative HBV DNA (viral load) estimation is indicated in the following clinical conditions:

- To assess the candidacy for antiviral treatment
- To assess the therapeutic response
- Planning of antiviral management: Infection with high HBV DNA usually refractory to interferon therapy.

Staging of chronic HBV infection: Staging of CHB infection is done by clinical assessment along with various laboratory parameters:

- Virological:
  - HBsAg positive >6 months
  - HBeAg and HBeAb
  - HBV DNA load estimation
- Biochemical: ALT level
- Pathological: Degree of fibrosis

As per the WHO and Indian National Guideline Antiviral Therapy is recommended in:

- Patient of any age group with CHB, with evidence of cirrhosis should receive antiviral therapy irrespective of HBeAg positivity, HBV DNA load and ALT level.
- In non-cirrhotic CHB patients, those aged >30 years who have persistently abnormal ALT and HBV DNA levels >20,000 IU/mL, irrespective of HBeAg status.

Monitoring of disease progression: Annual monitoring is recommended for HBsAg, HBeAg, HBV DNA load, ALT level. Monitoring at a frequent interval may be required if levels of HBV DNA or ALT are fluctuating or in patients who are on treatment.

Treatment is not recommended in patients with minimum liver disease, persistently normal ALT level and HBV DNA <2000 IU/mL.

TREATMENT

The objective of antiviral treatment in chronic HBV infection is to reduce the morbidity and mortality. True "cure" or eradication of virus is not attainable because of the persistence of **covalently closed circular DNA (CCC DNA)** in the host hepatocytes which poses the possibility of reactivation even after serological clearance.

Therefore, the **goal of treatment** is to achieve:

- HbsAg clearance
- Sustained suppression of HBV DNA
- Normal ALT level
- Loss of HBeAg
- Improvement in liver pathology.

According to American Association of Study of Liver Disease (AASLD), six antiviral agents have been approved for treatment of chronic HBV infection in adults and five for children.

These drugs belong to two groups: 1. Interferon (IFN) and pegylated interferon (Peg IFN), and 2. Nucleosides/tides analogs (NUC).

Interferon: Interferon-α is the recombinant form of natural interferon protein. It has antiviral, immunomodulatory and antifibrotic effects. The major effect through which it acts against HBV is possibly immunological mediated. In both HbeAg positive and negative groups of patients, it has shown to decrease HBV DNA load, loss of HBeAg and seroconversion from HBeAg to anti-HBe (in HBeAg positive patients), normalization of ALT and loss of HBsAg (in HBeAg negative patients), the risk of progression to cirrhosis and HCC becomes less. Currently Pegylated form of IFN is preferred.

Two forms of Peg IFN are available:

- PEG IFN 2α (adult): For adults, recommended for 48 weeks, 180 µg/week
- PEG IFN-α 2b: For children × >1 year.

Nucleoside/tide analogs (NUC): Several drugs belong to this group, of which lamivudine, telbivudine, antecavir, adefovir and tenofovir are used against HBV. These are polymerase inhibitors and inhibit viral synthesis. These are the main drugs used for chronic HBV infection and are preferred over IFN therapy because of better tolerance. The rate of virus clearance is, however, slow for which NUC therapy is given for a prolonged period. This also leads to the chance of drug resistance.

AASLD recommends antiviral treatment for the following groups of chronic HBV patients:

i. Immunoactive CHB in adults with HBeAg +/−
ii. HBeAg positive children with increased ALT and HBV DNA.

VACCINE

Two types of HBV vaccines are currently available:

1. Plasma derived
2. Recombinant HBV vaccine

Plasma derived: This is prepared by purification of HBs antigen from plasma of HBV carriers and contains S protein. They are safe, effective and cheaper in comparison to the recombinant vaccine. Concern related to plasma derived preparation has led to the use of recombinant vaccine.

Recombinant vaccine: These vaccines contain immunodominant "a" determinant of HBsAg in S protein. Vaccines using two types of expression system are available in the market: *Saccharomyces cervisiae* yeast and Chinese hamster ovary cell line. The former one is used to express only S protein from the recombinant DNA, whereas the later is used to express both S and M surface proteins.

Dose and schedule of vaccine: Vaccines are available in 0.5 mL (standard pediatric dose) or in multiple dose vials.

Vaccine is given either in three doses or four doses schedule. In both the cases, first dose is given at birth (within 24 hours of birth). The subsequent doses can be either monovalent or multivalent {along with Diphtheria, Pertussis and Tetanus (DPT)}.

In three-dose schedule, the second and third doses of HBV vaccine are given at the time of 1st and 3rd doses of DPT (i.e. 6 weeks and 14 weeks after birth). In four-dose schedule, after the first dose at birth, the remaining three doses are given with the three doses of DPT (i.e. 6 weeks, 10 weeks and 14 weeks after birth).

The recommended site of vaccine in infants is the anterolateral aspect of thigh and in older children and adults it is given in deltoid region.

Majority of the vaccines (90–95%) develop **protective titer (anti-HBs titer of 10 IU/mL).** 5–10% may be non-responders. Alcoholics, smoking, old age, immunosuppression are the major factors for being non-responders.

The protective level of antibody remains for 20 years or more. Hence, booster dose is not recommended by the World Health Organization for individuals who have completed the primary three-/four-dose vaccine schedule.

Vaccine Indication

Routine immunization: WHO recommends inclusion of HBV vaccine in national immunisation program for every infant in every country. This is based on the observation that early acquisition of HBV during infancy usually leads to chronicity because of poor viral clearance capability in infants which in turn poses high-risk of development of cirrhosis and HCC.

Prevention of perinatal HBV transmission: This is particularly recommended in countries with high seroprevalence like Southeast Asia in order to prevent HBV transmission from mothers to baby.

Catch-up vaccination: This is applicable in low endemic countries where infection may be acquired at a later age. Vaccination is recommended for older children (11–12 years), adolescents and adults.

Other conditions: Individuals who are at high-risk of acquiring HBV infection like frequent blood transfusion recipients, dialysis patients, healthcare personnel who are frequently exposed to HBV patients or samples are also recommended for HBV vaccination.

Vaccine escape mutants: Infection with HBV mutants has been observed in infants who have received HBIG and HBV vaccine containing S protein. The most common type of mutation is glycine to arginine substitution at 145 positions **(G145R)** of "a" determinant of HBsAg. The mutant-HBsAg does not bind to the Hbs antibody and leads to infection even in the presence of anti-HBs. Infection due to mutants is able to establish itself when the anti-HBs response in the vaccine is solely against the region between 139 and 147, as the anti-HBs is not able to neutralize the mutants which lies within this region. But when the anti-HBs response is outside the 139–147 region then mutants get neutralized by anti-HBs, so no infection occurs.

The surface antigen mutants also pose problems in diagnosis, if the recombinant antibody used in the detection system fails to bind it. This leads to false negative test.

Bibliography

1. Alexopoulou A, Karayiannis P. HBeAg negative variants and their role in the natural history of chronic hepatitis B virus infection. World J Gastroenterol 2014;20(24):7644–52.

2. Ali A, Abdel-Hafiz H, Suhail M, Al-Mars A, Zakaria MK, Fatima K, Ahmad S, Azhar E, Chaudhary A, Qadri I. Hepatitis B virus, HBx mutants and their role in hepatocellular carcinoma. World J Gastroenterol 2014;14; 20(30):10238–48.

3. Besharat S, Katoonizadeh A, Moradi A. Potential mutations associated with occult hepatitis B virus status. Hepat Mon 2014;14(5):e15275.

4. Croagh CM, Desmond PV, Bell SJ. Genotypes and viral variants in chronic hepatitis B: A review of epidemiology and clinical relevance. World J Hepatol 2015;7(3):289–303.

5. Francis J. Mahoney Update on Diagnosis, Management, and Prevention of Hepatitis B Virus Infection CMR 1999; 12: 351–366.

6. Gao S, Duan ZP, Coffin CS. Clinical relevance of hepatitis B virus variants. World J Hepatol 2015;18;7(8):1086–96.

7. Gomes MA, Priolli DG, Tralhão JG, Botelho MF. Hepatocellular carcinoma: Epidemiology, biology, diagnosis, and therapies. Rev Assoc Med Bras (1992). 2013;59(5):514–24.

8. Hai H, Tamori A, Kawada N. Role of hepatitis B virus DNA integration in human hepatocarcinogenesis. World J Gastroenterol 2014;28;20(20): 6236–43.

9. https://www.cdc.gov/vaccines/pubs/pinkbook/downloads/hepb.pdf

10. http://www.who.int/mediacentre/factsheets/fs204/en/

11. Knipe DM, Howley PM. Fields virology. 6th edition. Philadelphia, USA. Wolter Kluwer. Lippincott Williams & Wilkins. 2013.

12. Morales-Romero J, Vargas G, García-Román R. Occult HBV infection: a faceless enemy in liver cancer development. Viruses 2014;8;6(4): 1590–611.

13. Paul F. Coleman. Detecting Hepatitis B Surface Antigen Mutants. Emerging Infectious Diseases 2006;12:198–203.

14. Puri P. Tackling the Hepatitis B Disease Burden in India. J Clin Exp Hepatol 2014;4(4):312–19.

15. Samal J, Kandpal M, Vivekanandan P. Molecular Mechanisms Underlying Occult Hepatitis B Virus Infection. CMR 2012;25:142–63.

16. Shuaibu A, Hudu SA, Malik YA, Niazlin MT, Harmal NS, Sekawi Z. An Overview of Hepatitis B Virus Surface Antigen Mutant in the Asia Pacific. Curr Issues Mol Biol 2014;16: 69–78.

17. Tarocchi M, Polvani S, Marroncini G, Galli A. Molecular mechanism of hepatitis virus-induced hepatocarcinogenesis. World J Gastroenterol 2014;20(33):11630–40.

18. Terrault NA, Bzowej NH, Chang K-M, et al. AASLD Guidelines for Treatment of Chronic Hepatitis B. Hepatology 2016; 63:261–83.

B. HEPATITIS C VIRUS

Hepatitis C virus is one of the important agents of chronic hepatitis, cirrhosis and hepatocellular carcinoma worldwide. More than 185 million people worldwide have been affected by it leading to around 35,000 deaths every year. The seropositivity rate across the globe has increased over the years from 2.3 to 2.8%.

HISTORY

After the development of serological tests for hepatitis A and hepatitis B virus, it was realized that majority of the post-transfusion hepatitis are caused by agent other than hepatitis A or B virus. This led to coining of the term **"non-A non-B hepatitis"**, which emphasized the existence of the entity hepatitis that was neither due to A or B hepatitis virus. Further, studies in chimpanzee with blood of non-A non-B hepatitis patients led to the discovery of a small enveloped virus as the causative agent of non-A non-B hepatitis, and named hepatitis C virus (HCV).

VIRUS

Hepatitis C virus (HCV, recently renamed Hepacivirus C) is a small, spherical virus of 55 nm diameter with a 9.6 kb, single stranded positive sense RNA genome. The genome composed of a long open reading frame flanked by two untranslated region at both ends. Due to similarities of its structure, genomic organization and replicative cycle, HCV is a member of Flaviviridae family. But it is distinct enough to be in a separate genus that is Hepacivirus.

The genome of HCV encodes ten proteins—three are structural, six non-structural and one membrane associated ion channel.

- The structural proteins are core or capsid protein (c) and two envelop proteins E1 and E2.
- The non-structural proteins are NS2, NS3, NS4A, NS4B, NS5A and NS5B.

Genotypes: Presently there are seven phylogenetically distinct genotypes or clades. The seventh one has been added recently.

The **nucleotide sequence similarity is <50% between the genotypes**. With this much differences in identity, other RNA viruses usually show different serotypes. However, in HCV, this is not noticed so far. Different HCV genotypes show similarity in transmissibility, replication and rate of disease progression, but they differ in their response to interferon therapy.

Subtypes or subgenotypes: Within each genotype, strains that share nucleotide sequence

identity within core-E1 and NS5B region of **75 to 85% are termed subtype** and **91 to 99% are termed Quasi species**. The generation of quasi species occurs due to the lack of proofreading capacity of the viral polymerase and high rate of replication.

In contrast to HBV and HIV where recombination between subtypes is one of the major modes of evolution, recombination is a relatively rare event in HCV.

Geographic distribution of genotypes:
- Genotype 1a, 1b: North America, Europe
- Genotype 2: West Africa
- Genotype 3: South East Asia and other geographic area with IV drug abuse
- Genotype 4: Egypt, Africa, Middle East
- Genotype 5 and 6: South Africa and South East Asia

HCV cell culture: In 2005, first time HCV was grown on the cell culture system. The HCV **strain JFH1of genotype 2a** that was derived from a patient with fulminant hepatitis was able to infect **hepatoma cell line Huh7**. The virus strain was found to replicate efficiently with production of infectious virus particles. The successful *in vitro* culture of HCV has facilitated the studies on virus replication cycle, structural analysis of mature virion and has helped in the development of directly acting antivirals (DAA) against HCV.

Burden of HCV infection: The estimated global prevalence of HCV infection is 2.2% amounting to 185 million infected population worldwide. The rate of prevalence varies countrywise. Based on the seroprevalence of anti-HCV antibody, different geographic areas have been categorized into:
  i. High prevalence: >3.5%
 ii. Intermediate prevalence: 1.5 to 3.5%
iii. Low prevalence: <1.5% prevalence rate (Table 9.6).

Modes of transmission: HCV is mainly transmitted through infected blood. The risk of transmission depends on the size of inoculums and type of exposure.

Table 9.6: HCV seroprevalence	
Rate of prevalence	*Countries*
High: >3.5%	Central and East Asia
	North Africa
	Middle East
Intermediate: 1.5–3.5%	Eastern and Western Europe
Low <1.5%	India

- Percutaneous transmission
  - Blood transfusion
  - Injection drug use
  - Contaminate needle use
  - Organ transplantation from seropositive donor
  - Others: Needles-stick injury, tattooing, etc.
- Sexual transmission
- Mother to fetus (vertical)

CLINICAL FEATURES

Acute hepatitis: Majority of the HCV infections occur silently and manifest only when become chronic. A few percentage of cases may manifest during the acute phase of infection with malaise, nausea, anorexia, fatigue, dark urine and sometimes with pale stool and jaundice. Incubation period (exposure to development of symptoms or increase liver enzymes) is 1–7 weeks. HCV RNA becomes positive in blood within a few days and may reach up to 10^5 to 10^7 IU/mL. Serum level of liver enzymes (alanine aminotransferase and aspartate aminotransferase) becomes high. High initial viral load is usually associated with higher rate of spontaneous clearance.

Fulminant hepatitis: HCV related fulminant hepatitis is a rare entity, but has been reported in around 50% of NANB fulminant hepatitis in Japan.

Chronic hepatitis: Majority of HCV infected patients (near 85%) develop chronic infection due to persistence of virus (Fig. 9.6). Chronic HCV is indicated by high or fluctuating level of hepatic transaminases (ALT, AST), high serum HCV RNA and presence of liver inflammation. HCV RNA level usually remains constant around 10^6 IU/mL (Table 9.7). The degree of these parameters is variable in different hosts and also does not correlate with the disease progression.

Table 9.7: Laboratory parameters of chronic HCV infection

HCV RNA load: ≈10^6 IU/mL
Hepatic transaminase: High and fluctuating
Liver pathology: Evidence of inflammation (infiltration of lymphocytes near portal triad, then forms bridge to central vein and other triads and ultimately damages the liver architecture). Histopathology of liver is the best marker of stage of disease.

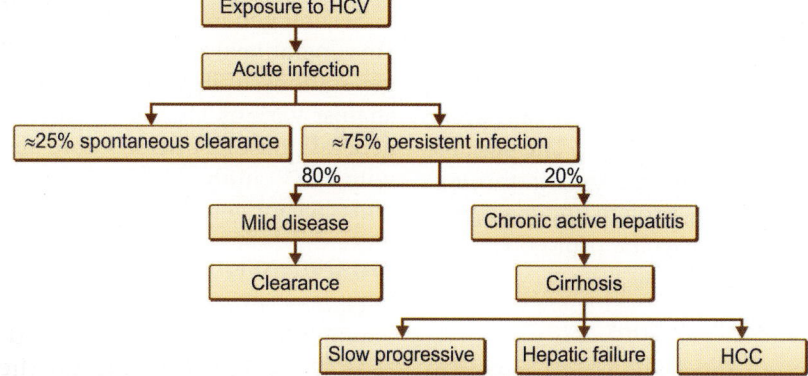

Fig. 9.6: Disease course of HCV infection

Some patients develop cirrhosis during the course of disease process. Once cirrhosis develops, decompensation occurs within years, which is manifested by coagulopathy, esophageal varices, ascites, encephalopathy and hepatocellular carcinoma. High HCV viral RNA does not show strong correlation with the rate of disease progression or extent of fibrosis. Some patients with persistent high level of HCV RNA also do not develop fibrosis.

Chronic hepatitis is also associated with several metabolic dysfunctions like insulin resistance, type II diabetes mellitus and steatosis. Insulin resistance develops because of downregulation of hepatic insulin receptor. Steatosis is more commonly associated with HCV genotype 3.

Hepatocellular carcinoma: HCC typically occurs during the last stage of chronic HCV infection and more particularly after development of cirrhosis. HCC is more commonly seen in Egypt and Japan where outbreak of HCV has occurred 10–20 years back.

Clinical manifestation occurs with sudden worsening of symptoms and signs of cirrhosis that are ascites, jaundice and fatigue with pain in right upper quadrant. No effective screening test is available. High level serum **alpha-fetoprotein (AFP) biomarker (>500 ng/mL) is indicative of HCC** irrespective of etiology. However, the AFP levels do not correlate with the severity of HCC. Imaging technique helps in detecting the intrahepatic mass, but confirmation is made with liver biopsy.

Unlike the direct oncogenic role of HBV in development of HCC, the direct oncogenic role of HCV is less, and not clear. The possible mechanism of HCV oncogenesis is through overexpression of core protein under a strong promoter which induces the proto-oncogene and suppresses apoptosis.

Extrahepatic Manifestations

- Extrahepatic manifestations are common in chronic HCV infection.
- Weakness, purpuric rash and joint pain are the common manifestations. Vasculitis may develop in some patients.
- Essential mixed cryoglobulinemia, membranous glomerulonephritis and porphyria cutanea tarda are strongly associated with chronic HCV infection.
- HCV infection leads to chronic stimulation of B cell and, therefore, associated with non-Hodgkin's lymphoma.
- Thyroiditis in women.
- Evidences of HCV infection: Anti-HCV antibodies and HCV RNA positive, response to antiviral therapy.

DIAGNOSIS

Laboratory diagnosis of HCV is based on the detection of anti-HCV antibody by enzyme immune assay or chemiluminescence assay. Detection of HCV RNA is measured to establish ongoing infection.

Detection of Anti-HCV Antibody

HCV antibody can be detected by enzyme immune assay (EIA) or recombinant immune blot assay (RIBA).

In enzyme immune assay (EIA) method, antibodies to HCV is tested against various recombinant HCV antigens where antigens are used together. In recombinant immune blot assay (RIBA), antibodies are detected separately against each antigen. So, specificity of antibody detected by EIA is confirmed by RIBA.

Several generations of EIA and RIBA (1st–4th generations) are available. 1st, 2nd and 3rd generations ELISA detect antibody to HCV against various HCV proteins (Fig. 9.7), 4th generation ELISA detects core antigen along with HCV antibody against core, NS3, NS4 and NS5 antigens.

First Generation EIA

- Detects the antibody to c100-3 antigen which is the larger clone of the original HCV specific clone c100, and derived from the chimpanzee plasma and expressed in

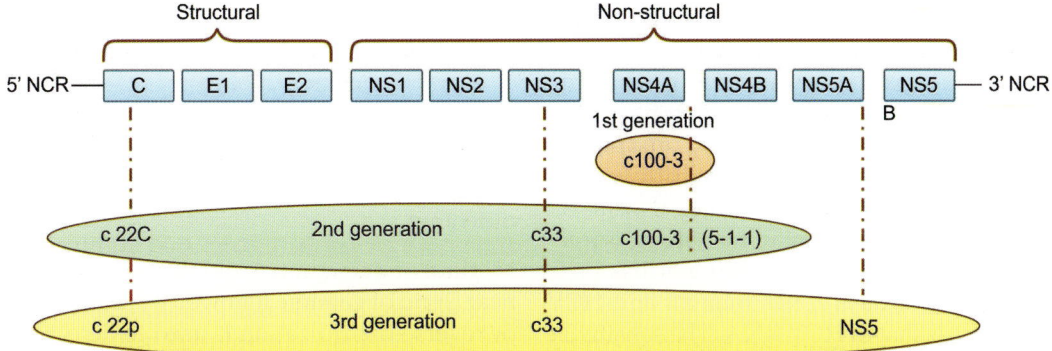

Fig. 9.7: Genome organization of HCV showing proteins associated with different generations of serological test

yeast. The derived protein is encoded by NS4 region of the genome.

- Using the 1st generation EIA, the causative role of HCV in post-transfusion non-A, non- B hepatitis was discovered.
- Average window period with 1st generation EIA: 16 weeks.

Second Generation EIA

- Detects antibodies against the antigen from core and NS3, NS4 non-structural proteins.
- Antigens used are: Core: c22c; NS3: c33; NS4: c100-3/5-1-1.
- In RIBA, antibody to two of these antigens (two bands) is confirmatory and antibody to one antigen (one band) is indeterminate.
- Sensitivity and specificity is higher than first generation counterpart.
- Average window period: 10 weeks.
- Use of second generation tests for screening of blood donors practically eliminates the possibility of HCV transmission through blood transfusion.

Third Generation EIA

- Detects the antibodies against the antigen from core: c22p (reconfigured core antigen); NS3:c33; and NS5.
- Average window period: 8 weeks.

Fourth Generation EIA

Detects HCV core antigen and antibodies to core, NS3, NS4 and NS5 region.

Comparative evaluation of different generation of ELISA is shown in Table 9.8.

Current guideline for serology: Independent RIBA test is not recommended for confirmation.

Previously, positive EIA was recommended for confirmation by RIBA whenever required. RIBA is no more in use. Presently confirmation is done by molecular test for HCV RNA.

Signal-to-cutoff ratio: Titer of antibody in EIA by measuring signal-to-cutoff ratio (S/CO) is also now recommended for confirmation. S/CO of 3.8 and 8 in ELISA and chemiluminescence assay (CLIA) respectively has been shown to predict true HCV viremia in 95–98% cases.

Table 9.8: Comparison of different generation of serological tests

Status	1st generation	2nd generation	3rd generation	4th generation
Detects	HCV antibody	HCV antibody	HCV antibody	HCV antibody and antigen
Antigen used	NS4	Core, NS3, NS4	Core, NS3, NS5	Core, NS3, NS4 and NS5
Window period	16 weeks	10 weeks	8 weeks	2 weeks
False positive	More	Less than 1st generation	Rare	Practically nil

Detection of HCV core antigen: HCV protein in blood is produced during viral replication and thus a reliable marker of ongoing infection. ELISA and CLIA formats have been developed for detection of HCV core antigen. HCV core antigen presence follows the dynamics of HCV RNA and thus used as an alternative to molecular test in resource-limited setting. HCV core antigen has been detected 40–50 days earlier than the anti-HCV antibody as that of HCV RNA. The sensitivity of the test has been reported to be lower than nucleic acid amplification technique (NAT). New chemiluminescence quantitative assay for HCV core antigen has reported almost equal sensitivity as that of PCR.

Detection of HCV RNA: This can be detected by various molecular techniques (nucleic acid amplification technique or NAT) like reverse transcriptase polymerase chain reaction (RT-PCR), real-time PCR, transcription mediated amplification (TMA) and branched DNA assay (bDNA), etc.

The presence of HCV RNA in plasma or serum indicates viremia and thus ongoing infection. The NAT is, therefore, considered as the gold standard test for active HCV replication.

Indications of Qualitative NAT

- Confirmation of HCV infection that is positive for anti-HCV antibody.
- Detection of acute infection after around 1 week of post needle-stick injury.
- Screening of blood for transfusion
- Confirmation of infection in babies born to HCV positive mother.

Indications of Quantitative NAT

Real-time PCR (qRTPCR) or branched DNA (bDNA) assay is commonly used for this purpose.

- To diagnose chronic HCV
- To measure the baseline viral load prior to starting of treatment
- Monitoring of treatment response (same assay should be performed both pre- and post-treatment)

- Viral load estimation for treatment monitoring is generally recommended after every 12 weeks.

HCV genotype determination: HCV genotype determination is important clinically, as this is the best possible predictor of treatment outcome.

Method of detection: Amplification of region near 5'end of genome followed by sequencing or reverse hybridization is used to detect the genotype.

Figure 9.8 depicts the diagnostic algorithm of suspected acute and chronic HCV infection.

TREATMENT

The goal of treating HCV is to prevent the complications associated with chronic HCV, reduce spread of infection and also to eradicate the infection.

Goals of Treatment

- Prevent cirrhosis and decompensation due to cirrhosis
- Prevent development of HCC
- Reduce the need of liver transplantation
- Increase the quality of life.
- Reduce transmission

Treatment recommendations: The current recommendations of WHO for treatment of HCV infection are:

- All HCV infected individuals of age 18 years and above should receive pangenotypic direct acting antivirals (DAA) treatment regimen. Pangenotypic DAA regimen is defined as those which achieve >85% SVR rate across all the major six genotypes.
- All HCV infected individuals of age 12 years and above should receive treatment irrespective of their stage of disease.

Treatment Response

Various types of treatment response in HCV infection is described in Table 9.9.

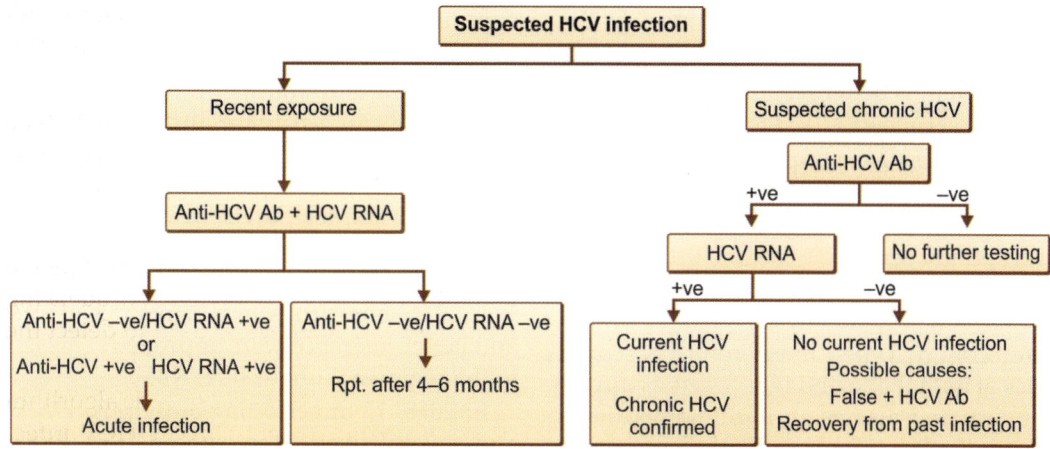

Fig. 9.8: Diagnostic algorithm of HCV infection

Table 9.9: Treatment response with their clinical implications		
Outcome	Criteria	Implications
Rapid virologic response (RVR)	HCV RNA undetectable (50 IU/mL) after 4 weeks of starting of treatment	RVR is a strong predictor of treatment success
Early virologic response (EVR)	HCV RNA level decreases at least by 2 log or undetectable after 12 weeks of starting of treatment	Non-achievement of EVR is a strong predictor of treatment failure. Chances of achieving SVR in these cases are almost nil
End of treatment response (ETR)	Undetectable HCV RNA level at the end of treatment regimen.	
Sustained virologic response (SVR 12)	Undetectable HCV RNA 12 weeks after treatment completion	SVR acts as the marker of cure and associated with improvement of liver pathology and clinical response. With interferon therapy, almost all patients who achieved SVR remain negative for HCV RNA for 4 years
Null non-response	Consistent HCV RNA level with <2 log 10 reduction by 24 weeks	
Partial response	Consistent HCV RNA level with >2 log10 reduction by 24 weeks	
Relapse	Detectable HCV RNA after achieving ETR	
Breakthrough	Suppression of HCV RNA to undetectable level followed by detectable level during therapy which is not due to new HCV infection	

Predictors of Treatment Outcome

- *Viral load:* Low viral load at the starting of treatment is better responsive.
- *Co-infection with HIV:* Less responsive.
- *Liver fibrosis:* Better response with less fibrosis.
- *Host factors:* Patients with young age (<40 yrs), female sex and low body weight responds better.

Antiviral Agents

Interferon α: Type I Interferon (IFN) constitutes of both IFN-α and IFN-β, acts as directly antiviral and also as an immunomodulator. Currently Pegylated (polyethylene glycol) form of IFN (Peginterferon) is preferred which increases its efficacy and pharmacokinetics. Half-life of PegIFN is more. Recombinant form of PegIFN has been approved for use in HCV infection. Dose schedule of PegIFN is once weekly as compared to 3 times per week for standard IFN.

Ribavirin (RBV): Ribavirin is a guanosine analog. It is usually given in combination with pegylated IFN for HCV treatment. It has broad antiviral activity and also acts as an immunomodulator. It increases the rate of error of replication to a level that induces lethal mutagenesis. It increases the rate of SVR by reducing viremic relapse.

The common adverse effects of IFN and ribavirin are fatigue, headache and psychiatric disorders. There may be bone marrow suppression leading to anemia, neutropenia and thrombocytopenia.

Direct acting antiviral agents (DAA): In 2011, two agents of protease inhibitors were approved for use along with IFN and ribavirin for the treatment of genotype I. These are boceprevir and telaprevir. Addition of these drugs increases the rate of SVR, also the toxicity and cost of therapy.

In 2014, more drugs with higher rate of SVR and better tolerance were approved. This led to the regimen without IFN/RBV.

In 2018, 13 direct acting antivirals (DAA) and several fixed dose combination (FDC) drugs got the FDA approval for treatment of HCV infection. The FDA approved DAA according to their class (mechanism of action) is shown in Table 9.10.

The comprehensive algorithm of diagnosis and treatment of HCV infection as per WHO recommendation is given in Fig. 9.9.

VACCINE

Presently there is no licensed vaccine available for HCV. Two basic approaches have been tried for HCV vaccine:

- **Prophylactic vaccine**
- **Therapeutic vaccine**.

Effective **prophylactic vaccine** is expected to produce both humoral and cellular immune responses to prevent infection against all HCV genotypes. Recombinant vaccine using

Table 9.10: FDA approved direct acting antiviral agents for HCV therapy according to their class

NS3/4A protease inhibitor	NS5A inhibitor	NS5B polymerase inhibitors (nucleotide analogue)	NS5B polymerase inhibitors (non-nucleoside analogue)
Glecaprevir	Daclatasvir	Sofosbuvir	Dasabuvir
Voxilaprevir	Velpatasvir		
Grazoprevir	Prebitasvir		
Paritaprevir	Ledipasvir		
Simeprevir	Ombitasvir		
	Pibrentasvir		
	Elbasvir		

(Ref: WHO. Guidelines for the care and treatment of persons diagnosed with chronic hepatitis C virus infection, July 2018)

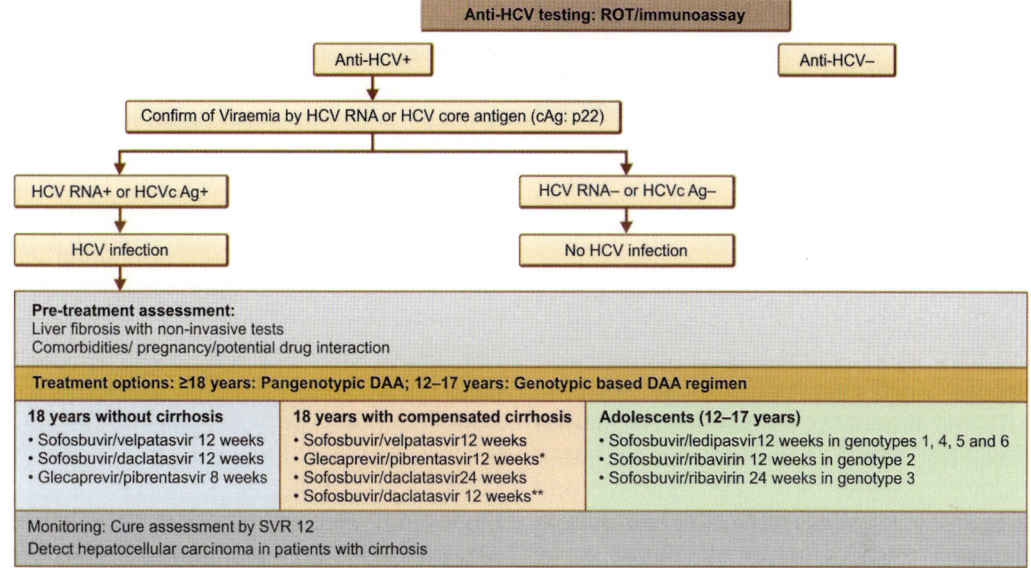

Fig. 9.9: Comprehensive algorithm of HCV testing and treatment

structural proteins E1 and E2 has been shown to elicit neutralizing antibody response and to reduce the chance of chronic HCV infection.

Therapeutic vaccine is meant for use during chronic infection along with the treatment in order to modulate the course of chronic infection and reduce the failure rate, duration and dose of antivirals.

Bibliography

1. AASLD, IDSA. Recommendations for Testing, Managing, and Treating Hepatitis C. Downloaded from http://www.hcvguidelines.org

2. Akihiro Tamori, Masaru Enomoto, Norifumi Kawada. Recent Advances in Antiviral Therapy for Chronic Hepatitis C. Mediators of inflammation. 2016. Volume 2016 (2016), Article ID 6841628. http://dx.doi.org/10.1155/2016/6841628.

3. Bashar M Attar, David H Van Thiel. Hepatitis C virus: A time for decisions. Who should be treated and when? World J Gastrointest Pharmacol Ther 2016; 7(1): 33–40.

4. Catanese MT, Dorner M. Advances in experimental systems to study hepatitis C virus *in vitro* and *in vivo*. Virology 2015;479–80: 221–33.

5. Cloherty G, Talal A, Coller K, et al. Role of Serologic and Molecular Diagnostic Assays in Identification and Management of Hepatitis C Virus Infection. J Clin Microbiol 2016;54(2): 265–73.

6. Cuthbert JA. Hepatitis C: Progress and problems. Clinical Microbiology Reviews 1994;7: 505–32.

7. Kouka Saadeldin Abdelwahab, Zeinab Nabil Ahmed Said. Status of hepatitis C virus vaccination: Recent update. World J Gastroenterol 2016; 22(2): 862–73.

8. Lee J, Conniff J, Kraus C, Schrager S. A Brief Clinical Update on Hepatitis C. The Essentials. WMJ. 2015;114(6):263–69.

9. Rocío González-Grande, Miguel Jiménez-Pérez, Carolina González Arjona, José Mostazo Torres. New approaches in the treatment of hepatitis C. World J Gastroenterol 2016;22(4): 1421–32.

10. Sung PS, Shin EC, Yoon SK. Interferon Response in Hepatitis C Virus (HCV) Infection: Lessons from Cell Culture Systems of HCV Infection. Int J Mol Sci 2015;16(10):23683–94.

11. Webster DP, Klenerman P, Dusheiko GM. Hepatitis C. Lancet 2015;385(9973):1124–35.

C. HEPATITIS A VIRUS

Hepatitis A virus (HAV, recently renamed Hepatovirus A) is one of the common agents of sporadic as well as epidemic jaundice causing acute self-limiting infection of the liver.

VIRUS

It belongs to Picornaviridae family which consists of 31 genera, of which *Enterovirus* and *Hepatovirus* are of importance to humans. HAV is the only member of the genus Hepatovirus.

It is a small round virus with 27–30 nm diameter, non-enveloped with icosahedral symmetry. The genome is single stranded positive sense RNA of 7.5 kb length. As compared to other members of the same family, HAV takes long time to adapt to cell culture, grows slowly without producing cytopathic effect.

HAV has **seven genotypes**, I to VII, of these four (genotypes I, II, III and VII) are of human origin and genotypes IV, V and VI are simian origin. **In humans, I and III genotypes** are more prevalent.

In general, HAV is considered to be more resistant to low pH, temperature, disinfectants and harsh environmental conditions in comparison to enteroviruses (Table 9.11).

These properties of HAV make it suitable for transmission through food, water, environment and person-to-person contact.

EPIDEMIOLOGY

Modes of Transmission

HAV is excreted in stool, bile and also present in blood. Transmission of virus occurs mainly through feco-oral route.

Person-to-person contact: Most of the transmission occurs through close household contact. Asymptomatic infection along with long fecal shedding in children makes them important source of infection. Outbreaks in day care centers are common due to this.

Food and water: Transmission through contaminated food and water usually leads to single source outbreak. Contamination of food can occur during cultivation, preparation, processing or handling.

- Food items that are served raw like salads can get contaminated during handling by infected person. Salad items like raspberry, strawberry, lettuce, green onion, sandwiches, fruits, pastries have been implicated as the source of HAV infection in various outbreaks.
- Improperly or inadequately cooked oysters or clam obtained from sewerage contaminated water acts as the source of HAV. These bivalves can filter large amount of water within short span of time and thus concentrates the virus up to 100 times within them, therefore, can act as a good source of infection when consumed with

Properties	HAV	Enterovirus
Stability at pH 1.0	Stable	Not stable
Temperature of inactivation	85°C × 1 minute	72°C × 15 seconds
Environmental stability	Stable at room temperature for 4 weeks Resistant to drying Survive in sea water, waste water, soil, contaminated fresh water	Not stable
Disinfection with sodium hypochlorite (household bleach)	1:100 × 1 min	1:125 × 30 seconds
Growth characteristics on cell culture	Slow growth No cytopathic effect	Fast growth Produces cytopathic effect

Table 9.11: Differences between HAV and enteroviruses

inadequate cooking preparation like steaming.

- Unlike HEV, waterborne outbreaks of HAV are less common as compared to that of foodborne outbreaks. Poor management of sewage is mostly responsible for water contamination of ground water, river or drinking water sources.

MSM and IVDU: High incidence of HAV infection has been reported amongst men who had sex with men (MSM) and IV drug users. This could be due to oral-anal contamination and poor hygiene.

Parenteral transmission: Parenteral transmission of HAV has been reported mostly through blood or clotting factor. Asymptomatic viremic individuals and inadequate virus inactivation due to solvent-detergent has been implicated as source of infection for blood and clotting factor, respectively.

Health care setting: As the virus spreads through contaminated feces, it can pose risk in various hospital situations.

Patterns of Endemicity

Patterns of endemicity of HAV infection vary in countries with different socioeconomic status and living conditions.

High endemicity: In developing and underdeveloped countries with crowding, poor sanitation and hygiene, infection is acquired during childhood and almost 100% children became immune by first decade of life. This reflects high endemicity.

Moderate or low endemicity: In developed countries and developing countries where sanitation condition is better, infection is not acquired during childhood. Thus majority of the population remain susceptible to infection and acquire the infection when get exposed from various sources. Single source outbreaks are reported more commonly in these settings.

HAV infection in children is mild and rarely apparent; whereas in adults or adolescent children, the infection usually manifest in overt symptomatic form. Therefore, HAV infection is more noticed in developed countries or low endemic settings as it is symptomatic. With the improved sanitary condition and living style in developing countries, the age of acquisition is getting delayed and more cases of HAV are reported in adults.

PATHOGENESIS

HAV primarily transmitted through feco-oral route. HAV being resistant to acid survives the acidic environment of stomach and finally reaches liver because of hepatotropic nature. Virus replicates in liver, excreted in bile also produces viremia. Through bile canaliculi virus passes into the intestine and excreted in the stool (Fig. 9.10).

- Fecal shedding of virus starts from 1 to 3 weeks before onset of symptoms. Maximum fecal concentration of virus up to 10^9/g stool occurs just before appearance of symptoms (jaundice and elevated alanine transferase). Fecal shedding continues for around a week after disappearance of symptoms (Fig. 9.11).
- Virus concentration in blood is approximately 2–3 log10 less than stool, persists for longer duration than stool. Concentration of HAV substantially decreases after appearance of symptoms. HAV RNA in blood can be detected for several weeks during the convalescent period (Fig. 9.11).

CLINICAL FEATURES

Incubation period can vary from 2 to 8 weeks with a mean of 4 weeks.

Infection in children is mostly asymptomatic and more overt symptoms occur with increasing age.

Symptoms in HAV infection is classically described in three stages—prodromal, icteric and convalescent.

Prodormal phase: Prodromal phase occurs before the jaundice starts. It lasts around one week. The beginning of this phase often occurs

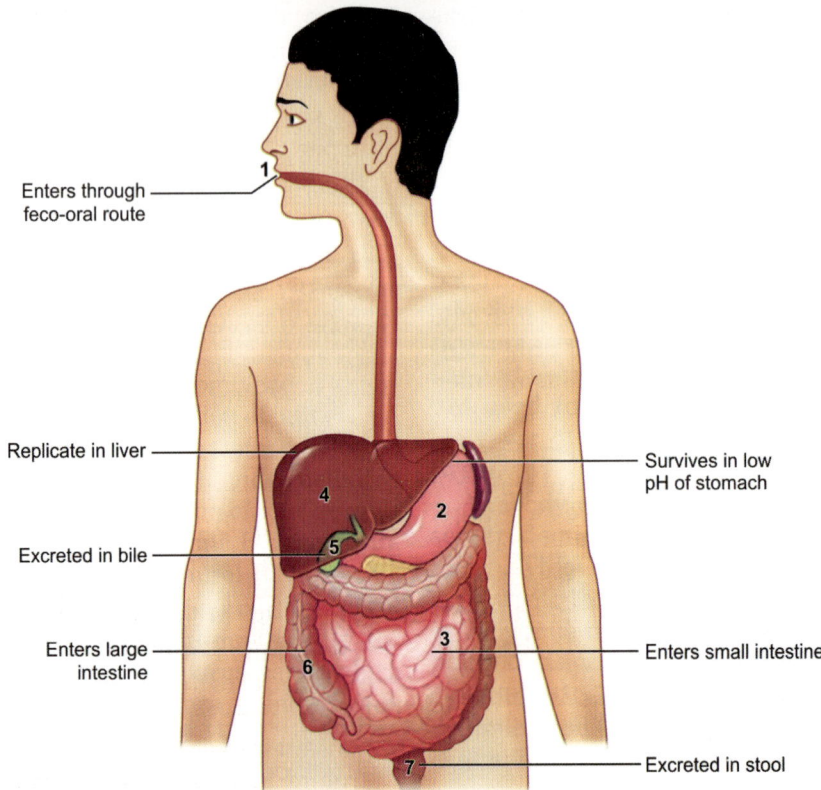

Enters through feco-oral route — 1

Replicate in liver — 4

Survives in low pH of stomach — 2

Excreted in bile — 5

Enters large intestine — 6

Enters small intestine — 3

Excreted in stool — 7

Fig. 9.10: Natural course of HAV infection

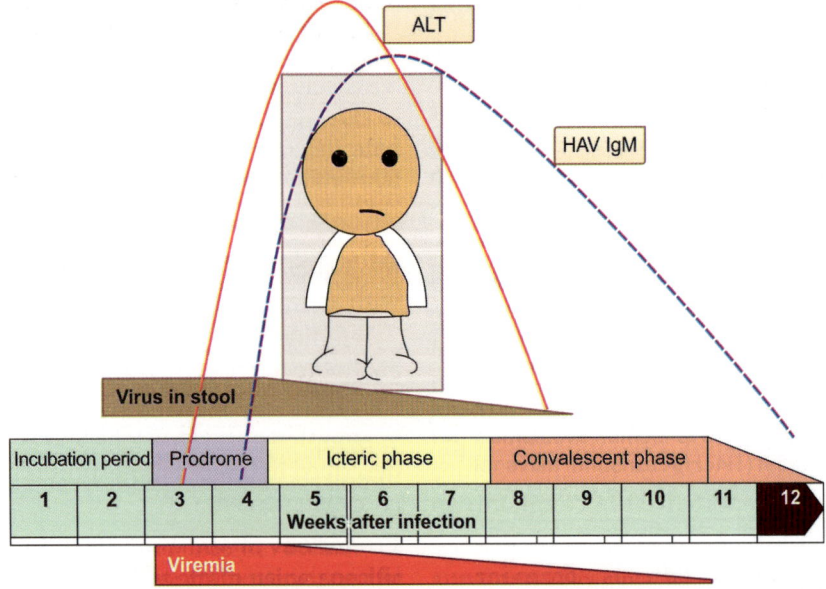

Fig. 9.11: Kinetics of different parameters of HAV infection (schematic)

abruptly with development of anorexia, fever, malaise, fatigue and disinterest for smoking and alcohol in chronic users. Nauseating feeling occurs on smell or sight of food, more so for fatty food.

Icteric phase: This phase begins with development of yellow coloration of skin, conjunctiva, sclera, etc. High bilirubin causes dark yellow to brown coloration of urine and pale stool. Itching may occur due to cholestasis in 50% cases.

Signs: Enlarged tender liver is a common finding. Splenomegaly may be therein 10–15% cases. Presence of spider nevi on trunk and palmar erythema. Symptoms and signs often last for two weeks.

Complications

- Cholestatic hepatitis with features of pruritus, fever and diarrhea. Once occurs, persists for a longer period. Recovery may take several months.
- Autoimmune hepatitis
- Relapsing hepatitis
- Acalculus hepatitis with jaundice
- Fulminant hepatitis: Severe form of acute hepatitis with hepatic encephalopathy. There is impairment of liver functions which are manifested as increased prothrombin time, bleeding diathesis, increased serum bilirubin and creatinine, and ascites. Symptoms of fulminant hepatitis usually develop after 1–4 weeks after jaundice. In around 0.5% of acute HAV infection, fulminant hepatitis occurs.
- Transverse myelitis, Guillain-Barré syndrome: Rarely reported.

LABORATORY DIAGNOSIS

As the clinical features of acute hepatitis are more or less similar, laboratory confirmation is essential in suspected cases. Detection of antibodies in serum is the mainstay of diagnosis.

HAV IgM: Detection of HAV IgM antibody in serum is the most common method for confirmation of diagnosis. Titer of HAV IgM rises in 4–6 weeks and disappears in 3–6 months after subsidence of symptoms. Patient is positive for IgM antibody during the symptomatic phase and a few weeks thereafter. Rarely IgM antibody may be negative during the early part of illness, then test should be repeated after 7–10 days for confirmation. Thus, anti-HAV IgM acts as the marker of current or recent infection.

HAV IgG: IgG antibody to HAV once appears persists for lifelong and protects from further infection. Presence of IgG is the marker of past infection or vaccination.

Total anti-HAV antibody: This detects both IgG and IgM to HAV together. So, in absence of IgM HAV, it indicates past infection or immunity. Along with IgM HAV, it indicates current or recent HAV, infection. Absence of total HAV antibody in presence of HAV IgM could indicate false positive IgM.

Interpretation of various HAV antibody tests is depicted in Table 9.12.

HAV RNA detection and sequencing: This test per se is not important for confirmation of diagnosis. Detection of HAV RNA is required for nucleotide sequencing which is required to find the relatedness of strains. This test is

Table 9.12: Interpretation of HAV antibodies in various phases of infection				
Infection	Symptoms	HAV IgM	Total HAV Ab	HAV IgG
Acute	+	+	+	+/–
Recent past	–	+	+	+
Past	–	–	+	+
Vaccination	–	–	+	+

usually employed to find out the source of infection or outbreaks.

HAV antigen detection: Detection of HAV antigen can be done in infected liver tissue by using immunohistochemistry or enzyme immunoassay method. It is employed to study the HAV infection and not required for patient diagnosis.

Detection of HAV from food and water: Detection of HAV from food and water is done by molecular method using polymerase chain reaction.

HAV from water can be detected after concentrating the virus from water samples. This is traditionally done by filtration or ultracentrifugation of water sample.

Food samples are usually not used for virus detection in HAV as they are either discarded or consumed because of long incubation period of HAV infection. HAV can be detected from food materials like muscles or oysters. Extraction of viral genome from food samples is done by using various precipitation and concentration techniques like polyethylene glycol, ultracentrifugation and magnetic beads.

VACCINE

Two types of vaccines are available for prevention of hepatitis A virus infection.
1. Inactivated vaccine
2. Live attenuated vaccine

Inactivated vaccine: Formaldehyde inactivated vaccine prepared from HAV grown on cell culture is widely used in many countries. Vaccine is available in **monovalent form** (Havrix, Vaqta, Avaxim) and **polyvalent form** with HBV vaccine (Twinrix) or typhoid vaccine (ViATIM, ViVAXIM), worldwide. Two doses of vaccine recommended to individuals above 1 year of age to travellers to endemic area or immunosuppressed individuals. In children of 1–15 years age, the dose is 0.5 mL each and 1 mL in adults, given intramuscularly with 6–12 months gap. Protective level

(20 IU/mL) persists for >20 years after completion of two doses.

Live attenuated vaccine: Live attenuated HAV vaccines have been prepared in China from **H2 and LA-1 strains** of virus isolated in cell culture from the fecal sample of HAV-infected patients.

H2 strain vaccine was prepared by attenuation of wild virus strain by serial passage in new monkey kidney cell line followed by human diploid fibroblast at lower temperature. LA-1 strain vaccine was prepared by serial passage in human diploid fibroblast. Protective immunity persists for 3 years. However, detail information related to safety, adverse effects and genetic stability are still required. Both the vaccine preparations have got the approval for use in China and India.

Recommendation: Vaccine is recommended to individuals above 1 year age, given as single dose intramuscularly. In China, HAV vaccine has been included in the National Immunisation Programme since 2008. In India, vaccine is not included in the National Immunisation Programme. Indian Academy of Pediatrics has recommended two doses of HAV vaccination (inactivated vaccine) at 12 months and 18 months of age and single dose live attenuated vaccine at 12 months.

Bibliography

1. Chonmaitree P, Methawasin K. Transverse myelitis in acute hepatitis A infection: The rare co-occurrence of hepatology and neurology. Case Rep Gastroenterol 2016;10(1):44–9.

2. Cuthbert JA. Hepatitis A: Old and New. Clinical Microbiology Reviews 2001;14:38–58.

3. Dahanayaka NJ, Kiyohara T, Agampodi SB, et al. Clinical features and transmission pattern of hepatitis A: An experience from a hepatitis A outbreak caused by two cocirculating genotypes in Sri Lanka. Am J Trop Med Hyg 2016; Jul 5. pii: 16-0221. [Epub ahead of print]

4. http://www.who.int/vaccine_safety/committee/topics/hepatitisa/Jun_2010/en/ downloaded on 10/10/2016.

5. http://www.who.int/ith/vaccines/hepatitisA/en/ downloaded on 10/10/2016.

6. Mathur P, Arora NK. Epidemiological transition of hepatitis A in India: Issues for vaccination in developing countries. Indian J Med Res 2008; 128: 699–704.

7. Nainan OV, Xia G, Vaughan G, Margolis HS. Diagnosis of hepatitis a virus infection: A molecular approach. Clin Microbiol Rev 2006; 19(1):63–79.

8. Xu ZY, Wang XY. Live attenuated hepatitis A vaccines developed in China. Hum Vaccine Immunother 2014;10(3):659–66.

9. Yılmaz-Çiftdoğan D, Köse E, Aslan S, Gayyurhan E. Atypical clinical manifestations of hepatitis A among children aged 1–16 years in South-Eastern Region of Turkey. Turk J Pediatr 2015; 57(4):339–44.

D. HEPATITIS E VIRUS

Hepatitis E virus is the causative agent of hepatitis E and is the major cause of enterically transmitted hepatitis worldwide. It is mainly associated with outbreak associated hepatitis in developing countries and sporadic cases in both developing as well as developed countries. It can lead to fulminant hepatitis with high mortality particularly in pregnant women.

DISCOVERY

The existence of a new pathogen in causing hepatitis was first suspected during 1978–79 when a waterborne outbreak of hepatitis in Kashmir valley affecting more than 16,000 population was found not to be due to hepatitis A virus or hepatitis B virus, the two hepatitis viruses known that time. The clinical features were also not similar to non-A and non-B transfusion transmitted virus which was later named hepatitis C virus. During the same period, sera collected and stored from a previous outbreak in north India during 1956 was also found negative for hepatitis A or B virus. Finally a new hepatitis virus was identified by immune electron microscopy in a stool sample of a volunteer who had ingested the pooled fecal suspension of patients of acute hepatitis in Afghanistan.

VIRUS

Hepatitis E virus (HEV) belongs to the family Hepeviridae and genus Orthohepevirus. HEV is a small spherical virus of 27–30 nm diameter, non-enveloped with icosahedral symmetry. It contains a single stranded positive sense RNA genome of 7.2 kb length. It contains three open reading frames (ORF)—ORF1 to ORF3 (Fig. 9.12).

- **ORF1** is the largest coding unit, located at the 5′ end, consists of 5000 nucleotides. It encodes a protein which contains several enzymes that are required for viral replication and protein processing.
- **ORF2** encodes for capsid proteins which help in viral assembly, attachment with host cell and immunogenicity.
- **ORF3** overlaps ORF2, encodes a small protein that is responsible for virus morphogenesis and release.

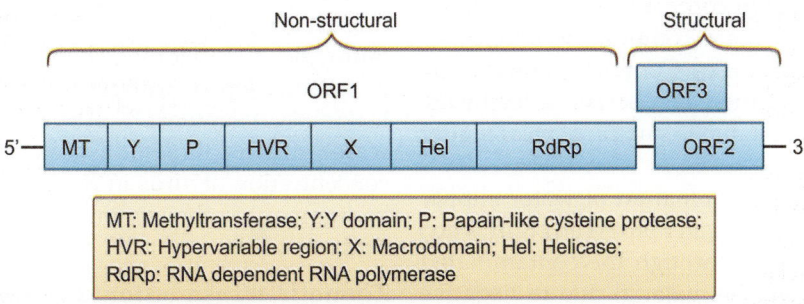

Fig. 9.12: Genomic organization of HEV

HEV can be cultured in cell lines derived from hepatocellular carcinoma (PLC/PRF/5) and human lung cancer (A549).

Classification: According to the current taxonomy of International Committee for Taxonomy for Viruses (ICTV), the family Hepeviridae comprises two genera: Ortho-hepevirus and Piscihepevirus. The genus Orthohepevirus contains four species: Ortho-hepevirus A–D.

Orthohepevirus A has seven genotypes: HEV-1 to 7, of which humans can be infected with HEV1–4. The distribution of four HEV genotypes in humans and animals is as follows:

- **HEV-1 and HEV-2:** Occur only in humans and are found mainly in developing countries.
- **HEV-3** has been isolated from humans and several animal species including pigs and is widely distributed worldwide.
- **HEV-4** has been isolated from humans and pigs and is found mainly in Asia.

EPIDEMIOLOGY

Epidemiology of hepatitis E virus is different in developing and developed countries. Developing countries are mostly endemic for HEV infection. **Feco-oral route** is the major mode of transmission which can be either waterborne or foodborne. Sporadic infections are reported throughout the year, whereas outbreaks are more commonly observed during summer and rainy seasons. Outbreaks due to sewerage contamination of drinking water supply are a common occurrence in developing countries. Contamination usually occurs during monsoon season after heavy rainfall or flood due to mixing of fecal matter with water.

Sewerage contamination of water supply due to old leaking water pipes can occur when the pipes passes through the contaminated soil and sucks the material due to negative pressure created during no flow.

In developing countries, bathing and fecal matter disposal in river is a common practice. During summer season, river water gets dried up and the concentration of HEV becomes high. Drinking of contaminated river water during this period poses a high-risk of HEV infection. Most of the HEV infections in **developing countries are due to genotype 1 or 2**, former being more common than the later.

The peak seroprevalence is seen in young adults of 15–30 years age group. The disease is more prevalent in South Asian countries like India, Nepal, Bangladesh and Africa.

Indigenous HEV infection in developed countries is mostly **zoonotic origin** and associated with eating undercooked or impro-perly cooked meat of wild boar, pork and game animals. Because of zoonotic source of trans-mission, genotypes 3 and 4 are more commonly seen in developed countries which are basically animal variants virus and have been found in domestic pigs, wild boar, deer and many other animals. Pigs are considered as the reservoir because of high seroprevalence and asympto-matic nature of infection. **Genotype 3 is seen worldwide**, whereas **genotype 4 is mostly reported from China, Japan and Europe**.

Amongst the developed countries, the disease is prevalent in North America, European countries, New Zealand and Japan. Seroprevalence in different Western countries varies from <1 to ≈15%.

TRANSMISSION

Waterborne transmission: Contaminated drinking water supply plays the most important mode of infection and cause of outbreaks in developing countries. Sewerage contamination of water supply due to old leaking water pipes, fecal contamination of water during monsoon season and high concentration of virus in river due to drying in summer are the common causes of water contamination.

Zoonotic transmission: Consumption of undercooked meat of infected wild boar, deer, pork, or rabbit has been associated with HEV transmission to human. Consumption of shell fish has also been associated with HEV infection. Virus can resist temperature up to 60°C and thus survives steaming process.

Vertical transmission: In endemic countries, HEV is the most common cause of hepatitis during pregnancy and has been associated with mother to fetus transmission. Transmission can be intrauterine or perinatal. HEV infection in fetus causes severe hepatitis and can lead to high mortality.

Person-to-person transmission: There is very little direct evidence in favor of person-to-person mode of transmission of HEV. However, in developing countries with poor hygiene practices, it is believed to play an important role in transmission.

Parenteral transmission: Due to high seroprevalence in some localities and asymptomatic nature of infection, it is thought that HEV transmission is possible through blood transfusion during silent viremic phase. Genotypes 1 and 3 have been reported with parenteral transmission.

CLINICAL FEATURES

In high endemic areas: In developing countries with high endemicity, HEV is the most common cause of acute viral hepatitis. In around 70–80% of cases, HEV causes asymptomatic infection. In the remaining symptomatic cases, the range of severity varies widely from self-limiting illness to fulminant hepatic failure. Symptomatic infection is commonly seen in young adults of 15–50 years age group.

Symptoms manifest as fever, anorexia, nausea, vomiting, and weakness. Jaundice is seen in around 40% cases with dark colorations of urine and pale stool. This phase is accompanied with elevated liver enzymes; aspartate transaminase (AST) and alanine transaminase (ALT). Symptoms last for a few weeks.

Severe infection is usually seen in individuals with underlying chronic liver disease and during pregnancy.

In low endemic areas: In developed countries with low endemicity, HEV causes similar manifestations as that of high endemic region. Most of the infections occur as sporadic case or in small clusters due to foodborne outbreak.

Jaundice is seen in majority of cases (75%) and with high level of liver enzymes. Patients with prior underlying chronic liver disease show poor prognosis. Infection with HEV4 has been reported to cause more severe infection with higher level of liver enzyme.

Extrahepatic Manifestations

HEV has been associated with several extrahepatic manifestations. They are acute pancreatitis, neurological disorders, renal and musculoskeletal injury, hematological disorders and several autoimmune disorders.

Neurological disorders: In general, neurologic manifestations are rare in HEV infection. It is more commonly seen in immunosuppressed individuals, though has also been reported in immunocompetent patients. Most of the infections are associated with genotypes 1 and 3 infections. Guillain-Barré syndromes (GBS), transverse myelitis, Bell's palsy, neurologic amyotrophy are the common manifestations.

HEV IgM and/or HEV RNA positivity along with elevated hepatic transaminases are keys to diagnose HEV associated neural manifestations.

Exact pathogenesis of neurological manifestation is not known. In some cases, HEV RNA has been detected in CSF samples has shown variant form of virus suggesting the presence of neurotropic variants. **MO3.13 oligodendrite cells** have been shown to be permissive and support HEV replication.

Acute pancreatitis: Usually associated with fulminant hepatitis E virus infection due to **HEV1.** So far it has been reported from Asian countries: India and Nepal.

Renal: Impaired renal function has been observed mostly during acute infection with HEV genotype 3 and also in chronic HEV infection. Two types of glomerular pathology; membranoproliferative glomerulonephritis and membranous glomerulonephritis are seen with decreased glomerular filtration rate. Renal complications are mostly seen with **HEV1 and 3 genotypes**. Patients with glomerulonephritis with elevated transaminase should be investigated for HEV. Treatment with Ribavirin can be started for rapid viral clearance.

HEV in Pregnancy

HEV infection in pregnant women of developing countries is usually severe in comparison to their non-pregnant counterparts. Severity is more in second and third trimesters. **Mortality up to 20–25%** is seen during third trimester. Common complications in pregnant women are eclampsia, hemorrhage and fulminant hepatic failure. Complications in fetus are prematurity and stillbirth.

Disease pathogenesis is mostly immune mediated as evident from increased level of cytokines. High viral load and HEV replication in placenta have also been attributed to the severity. Severe HEV cases during pregnancy have been observed in developing countries and associated with **genotypes 1 and 2**.

Chronic Hepatitis E Infection

In general, HEV does not lead to chronic hepatitis. Chronic hepatitis E infection is seen in **immunocompromised** patients such as solid organ transplant recipients, HIV and hematological stem cell transplant (HSCT) patients.

Chronic hepatitis E virus infection is defined as elevated liver aminotransferase level, persistence of HEV RNA and related liver pathology for **more than three months**. The disease has so far been reported from developed countries and is associated with **genotype 3**.

Chronic HEV infection (CHE) develops in near 60% of HEV infected immunocompromised patients. 10–15% of CHE cases develop extrahepatic manifestations.

Diagnosis of CHE is made by demonstration of persistence of HEV IgM or presence of HEV RNA in blood or stool for more than three months.

Evolution of chronic HEV cases is shown in Fig. 9.13.

PATHOGENESIS

HEV enters the body through feco-oral route. It replicates in the small intestine and reaches liver where it replicates in the cytoplasm of the hepatocytes. Release of virus then occurs in the blood and bile. Finally the virus is excreted in the stool. Incubation period is 2–6 weeks.

Exact pathogenesis of HEV associated hepatitis is not clear. HEV being non-cytopathic does not cause direct damage to liver. The damage to liver is mostly immune mediated by cytotoxic T cell and NK cells.

Pathogenesis of fulminant hepatitis: In general, the disease is mostly self-limiting

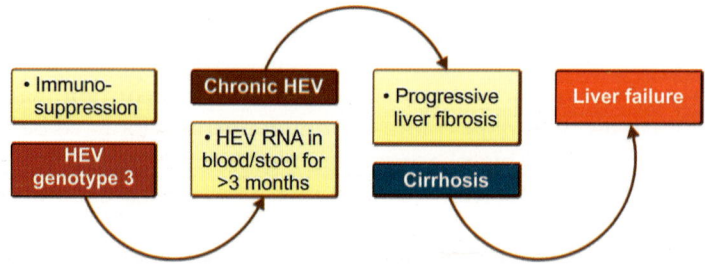

Fig. 9.13: Evolution of chronic HEV infection

in nature. Fulminant hepatitis may occur in around 4% of HEV cases. Several studies have been conducted to study the role of various viral and host factors in disease severity.

Viral factors such as genotypes and viral load have been associated with severe disease. HEV 3 and 4 genotypes are generally considered to be less pathogenic than HEV1 and 2. HEV 4 genotype has been shown to be associated with severe illness as compared to HEV genotype 3. In a recent study on re-examination of published evidences of various factors associated with severe HEV disease has concluded that it is the host factors than the virus factors that are associated with fulminant hepatitis.

In studies conducted to find out the role of cellular as well as humoral immune response in fulminant HEV hepatitis, heightened humoral response has been found to be associated with fulminant hepatitis more consistently than cellular immune response. Higher anti-HEV IgM and IgG titer have been found in fulminant HEV cases than those of uncomplicated cases.

Pathogenesis of HEV infection in pregnancy: Pregnant women with fulminant HEV have shown decreased expression of toll-like receptors that play key role in pattern associated recognition and inhibition of monocyte-macrophage function. These factors are thought to be the important contributing factors for pathogenesis by reducing the innate immunity.

Increased level of cytokines, such as TNF-α, IFN-β, interleukin-6 have been shown to be associated with severity. Increased level of pregnancy-related hormones (estrogen, progesterone) is thought to increase the viral replication.

LAB DIAGNOSIS

During the course of natural infection, HEV RNA can be detected in serum mostly during the prodrome period and disappears after appearance of symptoms. HEV RNA in stool of patient persists at a higher level than in serum and can be detected throughout the symptomatic period. Anti-HEV IgM is the first antibody to appear and usually present in symptomatic patient. This is followed by anti-HEV IgG which persists for many years.

HEV infection is diagnosed mainly by serologic assay by detection of IgM or IgG antibody, or molecular assay for detection of HEV RNA. Recently HEV antigen detection tests are also available as an additional diagnostic marker. The course of HEV markers is shown in Fig. 9.14.

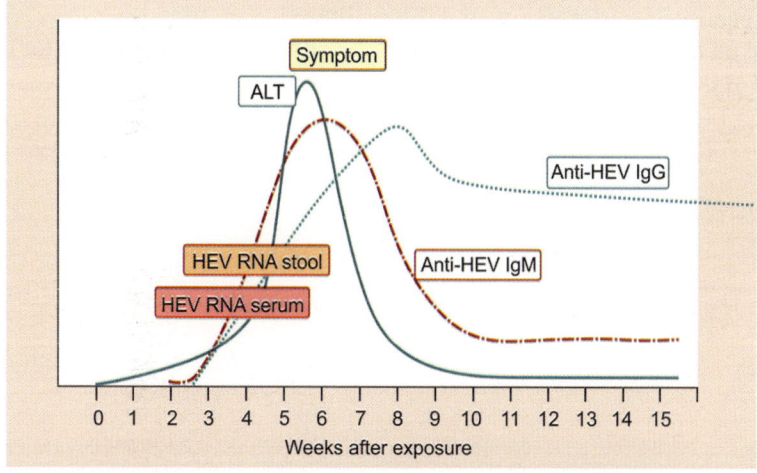

Fig. 9.14: Course of HEV diagnostic markers (schematic)

Anti-HEV antibody: HEV IgM or IgG detection can be done by ELISA or immunochromatographic tests (point of care test).

HEV IgM is the marker of acute or recent infection. It is usually present at the time when symptom appears or within a few days after jaundice ensues. IgM antibody persists for around 8 weeks after the onset of symptoms.

- The tests for detection of HEV IgM are common for all genotypes.
- Most of the assays include recombinant antigens from **ORF2 and ORF3**. Inclusion of antigens from ORF2 is important to increase the sensitivity as it contains the region shared amongst genotypes which includes minimum neutralizing domain.
- Sensitivity and specificity of various commercially available tests show 70–95% and 80–95%, respectively. Sensitivity in immunocompetent patients is higher in comparison to immunosuppressed individuals (97% *vs* 87%).

HEV IgG persists for a long time and detection of HEV IgG alone indicates past infection.

- IgG detection is used to study the seroprevalence of HEV.
- Determination of HEV IgG titer is done to check the protective titer after natural infection or for vaccine trial.

HEV antigen: HEV antigen detection test is developed recently for detection of HEV antigen in serum and stool sample by ELISA. HEV antigen appears earlier and persists for a shorter period as compared to HEV IgM. The kinetic of HEV antigen has been shown as similar as that of HEV RNA. Thus HEV antigen can act as an additional marker for diagnosis of acute HEV infection.

The test is helpful in following clinical conditions:

- In immunosuppressed patients where sensitivity of HEV IgM is less.
- Initial few days of jaundice when HEV IgM may be negative.
- To confirm acute HEV infection with HEV RNA in absence of HEV IgM.

HEV RNA: HEV RNA can be detected from serum and stool samples. Because of uncertainty of IgM antibody detection due to less sensitivity of certain commercial tests, RNA detection is recommended to confirm the infection. It is also considered as the gold standard for HEV diagnosis.

The test is recommended in the following situations:

- Immunosuppressed patients who are negative for HEV IgM.
- To confirm acute HEV infection in absence of HEV IgM.
- Nucleotide sequencing for genotype assay.

Table 9.13 mentions the association of HEV markers with different clinical stages of illness.

The diagnostic algorithm for HEV diagnosis is shown in Fig. 9.15.

Table 9.13: Diagnostic markers of HEV at different stages of illness

Stage of illness	ALT	HEV RNA in stool	HEV RNA in serum	Anti-HEV IgM	Anti-HEV IgG
Early acute	+	+	+	–	–
Acute	+	+	+	+	+/–
Convalescent	–	–	–	+	+
Past	–	–	–	–	+
Chronic	+ve for >3 months	+ve for >3 months	+ve for >3 months	+ve for >3 months	+ve for >3 months
Vaccination	–	–	–	–	+

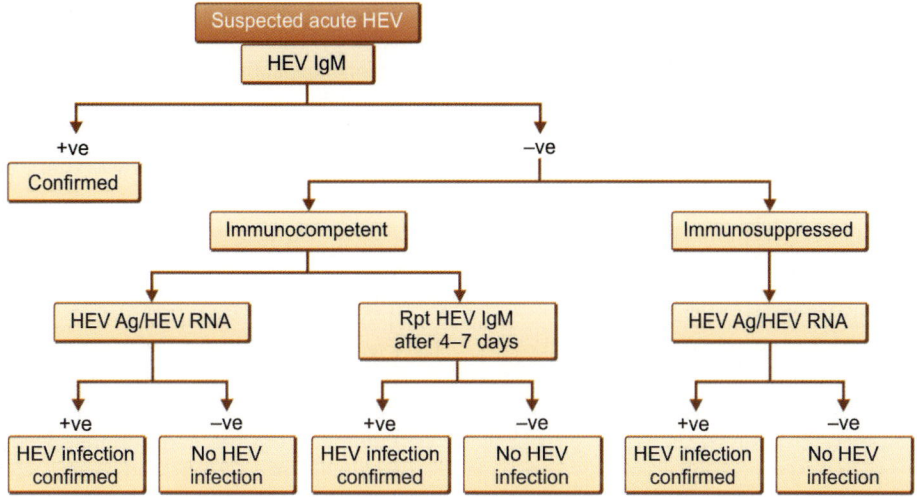

Fig. 9.15: Diagnostic algorithm for acute HEV infection

TREATMENT

Currently, antivirals are recommended in chronic HEV in immunosuppressed patients who do not respond to reduction of immuno-suppressive therapy or patients with extra-hepatic manifestations. Pegylated interferon or ribavirin has been tried in these patients.

VACCINE

Recently one vaccine, **HEV 239**, has been approved in China for HEV infection. The vaccine was expressed in *Escherichia coli* as a virus-like particle (VLP) which has been purified and adsorbed on aluminum hydroxide and suspended in buffered saline. It contains E2 domain of the HEV genotype 1 capsid protein. In a large community based phase III clinical trial in China, enrolling 1,13,000 individuals of 16–50 years age group, 100% efficacy was shown in the first year and 87% in the next 4.5 years. Vaccine has shown protection against HEV genotype 4 but protective status against genotype 3 is not known. Vaccine has been approved in China since 2012 and commercialized by Innovax (Xiaman Innovax Biotec) in the name of Hecolin. Three doses of vaccine are recommended at 0, 1, and 6 months.

Another HEV vaccine expressed in insect cell infected with a baculovirus expressing a truncated capsid protein of HEV 1. In a phase 2–3 clinical trials among Nepali soldiers, the vaccine was found to be safe, immunogenic and showed 95% efficacy after three doses at 0, 1, and 6 months. The vaccine is not yet commercialized.

Bibliography

1. Aggarwal A, Perumpail RB, Tummala S, Ahmed A. Hepatitis E virus infection in the liver transplant recipients: Clinical presentation and management. World J Hepatol 2016;8(2): 117–22.

2. Ahmed A, Ali IA, Ghazal H, Fazili J, Nusrat S. Mystery of hepatitis E virus: recent advances in its diagnosis and management. Int J Hepatol 2015;2015:872431.

3. Bazerbachi F, Haffar S, Garg SK, Lake JR. Extra-hepatic manifestations associated with hepatitis E virus infection: A comprehensive review of the literature. Gastroenterol Rep (Oxf) 2016;4(1):1–15.

4. Behrendt P, Steinmann E, Manns MP, Wedemeyer H. The impact of hepatitis E in the liver trans-plant setting. J Hepatol 2014; 61(6):1418–29.

5. Doyle JS, Thompson AJ. Local transmission of hepatitis E virus in Australia: Implications for

clinicians and public health. Med J Aust 2016; 204(7):254.

6. El Sayed Zaki M, El Razek MM, El Razek HM. Maternal-fetal hepatitis E transmission: Is it underestimated? J Clin Transl Hepatol 2014; 2(2):117–23.

7. Jeblaoui A, Haim-Boukobza S, Marchadier E, Mokhtari C, Roque-Afonso AM. Genotype 4 hepatitis E virus in France: An autochthonous infection with a more severe presentation. Clin Infect Dis 2013; 57,e122–6.

8. Kamar N, Dalton HR, Abravanel F, Izopet J. Hepatitis E virus infection. Clin Microbiol Rev 2014; 27(1):116–38.

9. Kim JH, Nelson KE, Panzner U, Kasture Y, Labrique AB, Wierzba TF. A systematic review of the epidemiology of hepatitis E virus in Africa. BMC Infect Dis 2014; 14:308.

10. Krain LJ, Nelson KE, Labrique AB. Host immune status and response to hepatitis E virus infection. Clin Microbiol Rev 2014; 27(1):139–65.

11. Lapa D, Capobianchi MR, Garbuglia AR. Epidemiology of Hepatitis E Virus in European Countries. Int J Mol Sci 2015; 16(10):25711–43.

12. Lee GY, Poovorawan K, Intharasongkroh D, Sa-Nguanmoo P, Vongpunsawad S, Chirathaworn C, Poovorawan Y. Hepatitis E virus infection: Epidemiology and treatment implications. World J Virol 2015; 4(4):343–55.

13. Lhomme S, Marion O, Abravanel F, Chapuy-Regaud S, Kamar N, Izopet J. Hepatitis E pathogenesis. Viruses 2016; Aug 5;8(8).

14. Li SW, Zhao Q, Wu T, Chen S, Zhang J, Xia NS. The development of a recombinant hepatitis E vaccine HEV 239. Hum Vaccin Immunother 2015;11(4):908–14.

15. Mirazo S, Ramos N, Mainardi V, Gerona S, Arbiza J. Transmission, diagnosis, and management of hepatitis E: an update. Hepat Med 2014; 6:45–59.

16. Nan Y, Zhang YJ. Molecular biology and infection of hepatitis E virus. Front Microbiol 2016; 7:1419.

17. Nelson KE, Heaney CD, Labrique AB, Kmush BL, Krain LJ. Hepatitis E: Prevention and treatment. Curr Opin Infect Dis 2016 Oct;29(5):478–85.

18. Okello AL, Burniston S, Conlan JV, et al. Prevalence of Endemic Pig-Associated Zoonoses in Southeast Asia: A Review of Findings from the Lao People's Democratic Republic. Am J Trop Med Hyg 2015; 92(5):1059–66.

19. Park WJ, Park BJ, Ahn HS, Lee JB, Park SY, Song CS, Lee SW, Yoo HS, Choi IS. Hepatitis E virus as an emerging zoonotic pathogen. J Vet Sci 2016; 17(1):1–11.

20. Pérez-Gracia MT, García M, Suay B, Mateos-Lindemann ML. Current Knowledge on Hepatitis E. J Clin Transl Hepatol. 2015; 3(2):117–26.

21. Smith DB, Simmonds P. Hepatitis E virus and fulminant hepatitis A virus or host-specific pathology? Liver Int 2015; 35: 1334–40.

22. Song YJ, Park WJ, Park BJ, Lee JB, Park SY, Song CS, Lee NH, Seo KH, Kang YS, Choi IS. Hepatitis E virus infections in humans and animals. Clin Exp Vaccine Res 2014; 3(1):29–36.

23. Sridhar S, Lau SK, Woo PC. Hepatitis E: A disease of reemerging importance. J Formos Med Assoc 2015; 114(8):681–90.

24. Verghese VP, Robinson JL. A systematic review of hepatitis E virus infection in children. Clin Infect Dis 2014; 59(5):689–97.

Arboviruses

A. DENGUE VIRUS

INTRODUCTION

Presently dengue is the most widespread vector-borne viral disease worldwide. Dengue virus is transmitted by *Aedes* mosquito. Nearly half of the world's population reside in areas which are at risk of dengue infection. According to World Health Organization (WHO), 50–100 million people are infected with dengue virus every year with 50 lakhs cases are of dengue hemorrhagic fever.

VIRUS

Dengue viruses belong to the family *Flaviviridae* and genus *Flavivirus*. There are four dengue serotypes: DEN-1, DEN-2, DEN-3 and DEN-4 based on neutralization assay. These are single-stranded RNA viruses of positive polarity, 40–50 µm in size and each has a lipid envelop surrounding it. The viral genome is approximately 11,000 base pairs long containing a single open reading frame (ORF) that encodes three structural [capsid (C), premembrane (prM) and envelop (E)] and seven non-structural proteins (NS1, NS2A, NS2B, NS3, NS4A,

NS4B, NS5) flanked by 5' and 3' non-translated regions (Fig. 10.1).

Dengue serotypes: Dengue virus has four serotypes, known as dengue 1, 2, 3 and 4. These serotypes have been classified on the basis of neutralization assay. The genetic similarity between the serotypes is 65%. Each serotype can mount distinct immune response in the host.

Recently another serotype, dengue-5, has been found in 2013 in a sample of a patient collected during the outbreak in Sarawak state of Malaysia in 2007. The full genetic analysis of DEN-5 showed it to be different from the existing four dengue serotypes. Infection of monkeys with the new DEN-5 serotype also developed antibody that were different from the previous type of antibody confirming the isolate as a distinct serotype. It was hypo-thesized that DEN-5 has been circulating amongst the monkeys for centuries in the South East Asian forests which has now crossed the species and has infected the human host. So far DEN-5 has been detected in only a single sample from a patient suffering from mild dengue fever and a single outbreak since 2007. So far there is no evidence to suggest that it

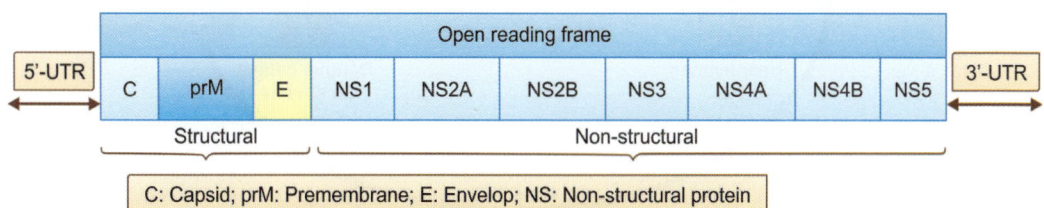

Fig. 10.1: Dengue virus genome (schematic)

can cause severe infection. However, if this serotype would establish itself in the human host, the formulation of dengue vaccine may require modification.

TRANSMISSION CYCLE

Dengue virus is maintained in two types of transmission cycles.

Endemic/epidemic/urban cycle: The viruses are maintained mainly between *Aedes aegypti* mosquitoes and humans. Human beings act as both reservoir and amplifying host. *Aedes aegypti* acts as the principal vector which grows in the water storage containers available in and around the human dwellings. *Aedes albopictus* and other *Aedes* species also act as secondary vectors. This is the most important form of transmission cycle from public health point of view (Fig. 10.2).

Man gets the infection through the bite of infected mosquitoes. The virus undergoes an incubation period of 4–7 days after which the viremia phase begins. This phase lasts for about 2–5 days during which the virus circulates in the blood. *Aedes* mosquitoes get the infection from infected human during its blood meal in viremia phase and subsequently transmit the virus after an extrinsic incubation period of 8–12 days. The virus is also maintained in mosquitoes through trans-ovarial and transstadial transmission.

Sylvatic/forest cycle: This occurs between the non-human primates and canopy dwelling *Aedes* mosquitoes, mostly in the rain forests of Asia and Africa. *Aedes niveus*, the canopy dwelling *Aedes*, is the main vector in Asia. *Aedes africanus*, *Aedes leuteocephalus* and *Aedes stegomyia* are the species responsible in Africa.

EPIDEMIOLOGY

Dengue virus is currently endemic in more than 100 countries worldwide. Almost two-thirds of dengue global burden are borne by the Southeast Asian countries. During the last three decades, the virus has spread all over the world and presently developed countries are also at risk.

History of Dengue Emergence

The emergence of dengue as a global problem started during the World War II and continued to spread in the post-war period.

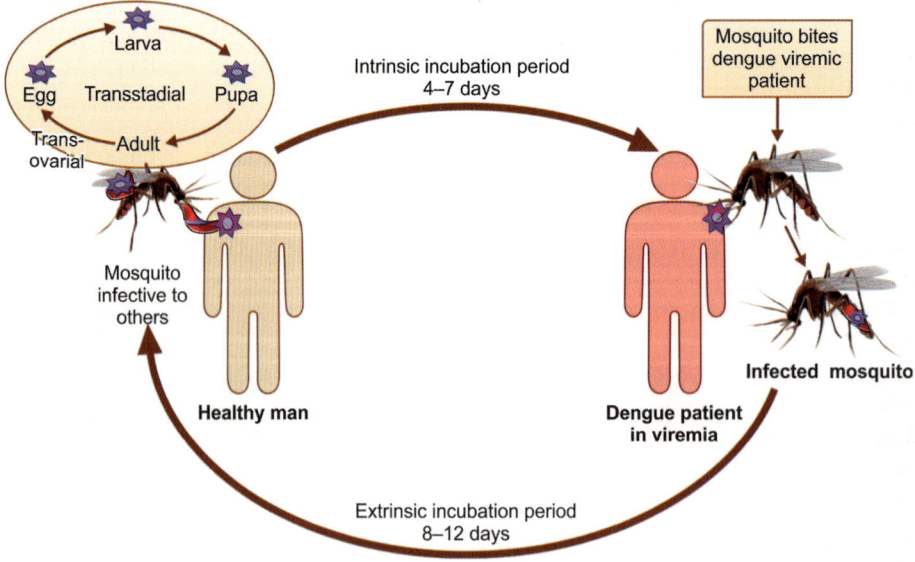

Fig. 10.2: Dengue virus transmission cycle

Before World War II, the virus was mostly endemic in nature and maintained in the sylvatic cycle. The ecological disturbance in the Southeast Asian region due to World War II led to the disruption of dengue cycle and increased transmission of the virus was observed. More number of epidemic cycle (man–mosquito–man) and circulation of multiple dengue serotypes led to the occurrence of dengue hemorrhagic fever (DHF), the serious form of the disease. First epidemic of DHF occurred in Philippines during 1953–54 and thereafter the virus spread throughout the Southeast Asian countries and also to west side affecting other Asian countries including India.

Global Spread

The endemicity of dengue virus has expanded to the entire Southeast Asian countries and as well as to several other Asian countries. The number of cases as well as the number of epidemics has increased many folds in Africa. The disease epidemics are increasingly being reported from countries with temperate climate like northern Australia, southern Europe and USA since last decade.

Factors responsible for global spread: Multiple factors are responsible for global spread of dengue virus which is transmitted by *Aedes* mosquito.

Increased urbanization is an important factor particularly applicable for southeast Asian countries. Unplanned or poorly planned urbanization process involving water laying provides suitable breeding place for the vector. *Aedes aegypti* which breeds in small pool of clean water finds the urban area suitable for its breeding and facilitates the spread of dengue virus. Increased number of construction work all over also acts as excellent breeding place for *Aedes aegypti*.

Increase in trade and transport leading to extensive air travel facilitates the spread of dengue virus from one place to another which includes city-to-city and from one continent to the other.

Global warming and climate change are also thought to facilitate the vector survival and breeding.

Absence of effective antiviral and vaccine against dengue virus helps the virus to survive in the host.

Spread of dengue virus can occur also from asymptomatic and mild dengue fever cases during their viremic phase.

Persistence of dengue virus in mosquito helps its survival in nature. Mosquito vector once infected with dengue virus remains infective throughout its life. Dengue virus gets transmitted to the next generation through transovarial transmission and persists in different developmental stages through transmission (transstadial).

Genetic diversity of dengue virus along with continuous genetic evolution helps the virus to find a naïve population.

Trends of Dengue Epidemics

Seasonality: Dengue being a mosquito-borne disease usually occurs as a seasonal disease in endemic areas. The cases of dengue start occurring during rainy season, i.e. July–August, and take its peak in post-monsoon period, i.e. October–November.

Cyclical epidemics: Epidemics of dengue usually occur in cyclical manner with epidemics in every 2–3 years. The exact reason of the cyclical trend is not yet known. However, it is thought to be due to intricate relationship among host, viral and environmental factors.

Development of host immune response against the circulating dengue type and genotype along with emergence of new dengue virus genotype in a particular location and replacement of prevalent serotype with another are the main reasons for cyclical trend.

Global warming and El Nino–Southern Oscillation (ENSO) have also been suggested as the factors responsible for cyclical epidemics.

Age: Dengue fever predominantly affects young adults and adolescents, whereas dengue hemorrhagic fever predominantly affects children below 15 years of age.

Gender: Male predominance is a more frequent observation. The difference in gender distribution has a doubtful clinical significance. Some studies also have reported equal gender distribution.

Dengue Outbreaks in India

The first documented dengue cases in India dates back to 1946. Thereafter, there were no reports of dengue virus activity for near two decades. In 1963–64, dengue epidemics were reported from eastern coastal states of the country. Subsequently several dengue outbreaks have been reported from various northern and southern states of the country.

Most of the epidemics have occurred due to DEN-2 and DEN-3, followed by DEN-1 in all over the country.

During 1950 and 1960, DEN-2 isolates were of genotype V. This was replaced gradually by DEN-2 genotype IV, which was probably silently circulating in northern states. In 1996, first outbreak of DHF occurred in Delhi and northern part of the country.

DEN-4 has been reported mostly from southern, western and central India.

It has been observed that one serotype predominates during a season. This is followed by entry of another serotype in the subsequent year and both types circulate together for 1–3 years. The second serotype gradually replaces the first one and subsequently becoming the predominant one. So, the predominance of circulating serotype keeps rotating in a particular geographic location. This reflects the role of herd immunity acting as the selection pressure for emergence of new serotype and disappearance of previously circulating serotype.

Circulations of all four serotypes and concurrent infection with two serotypes have been reported from various parts of the country.

PATHOGENESIS

Several hypotheses have been explained as pathogenetic mechanism for dengue hemorrhagic fever (DHF) or dengue shock syndrome (DSS).

Antibody dependent enhancement (ADE): The phenomenon of ADE was observed with dengue virus when it was found that dengue virus grows better in culture of peripheral blood mononuclear cells (PBMC) from human or monkeys who were immune to dengue, as compared to that from non-immune hosts.

The role of ADE in disease pathogenesis is evidenced by increased risk of developing DHF/DSS after a secondary dengue viral infection with a heterologous dengue virus serotype.

Mechanism of ADE: During the primary dengue infection, both homotypic neutralizing antibodies as well as cross-protective heterotypic neutralizing antibodies are developed. Homotypic antibodies persist for a long time but the level of heterotypic antibodies decreases to sub-neutralizing level after a few months to years. Therefore, secondary infection with a different serotype when occurs within a short span of time significant cross-protection occurs due to the presence of heterotypic antibodies developed during the previous infection.

Thus, during the initial period, heterotypic antibodies provide cross-protection against the infection and for a further more period against the severe disease. However, the level of cross-protection decreases with increase in gap between the primary and secondary infections.

When sequential infection occurs after a few years with a different serotype, the pre-existing heterotypic sub-neutralizing antibodies bind to the virus but not able to neutralize it. This antibody–virus (antigen) complex attaches to the Fcγ receptor present on the circulating monocytes, facilitating the entry of virus to its target cells. This in turn leads to infection of more number of target cells, high viral replication leading to high viral load and thereby increase in NS1 antigen level, release of cytokines and contributing to pathogenesis of severity (Table 10.1).

Table 10.1: Mechanism of antibody dependent enhancement

During primary dengue infection in the host, dengue specific IgG antibodies are developed
↓
(Development of homotypic neutralizing antibody and heterotypic neutralizing antibody)
↓
Homotypic neutralizing antibodies persist for a long time
Concentration of heterotypic antibodies lowers down to
sub-neutralizing level after months to years
↓
After a few years
Infection with different dengue virus serotype
↓
Pre-existing dengue heterotypic antibody binds to the virus
but unable to neutralize the virus
This antibody–virus (Ab–Ag) complex binds to the Fc receptor present on
the surface of the monocyte/macrophage, and enters inside the cells
↓
This leads to increase of viral load, increase in NS1 antigen level, and
release of cytokines, consumption of complement.
↓
Severe disease (DHF/DSS)

In addition to the ADE mechanism, it is also thought that entry of virus through Fcγ receptor inhibits the innate antiviral response.

Cross-reactive memory T cell response: Cross-reactive memory CD8 cells are present in previously infected individual. These memory cells recognize the viral antigen and get activated but not capable of clearing the virus. CD8 memory T cells may have high avidity or low avidity to the heterologous antigen.

CD8 T cells having low avidity for the heterologous dengue virus epitopes, produce high level of proinflammatory cytokines such as TNF-α, IL-6 and other soluble factors. These cytokines cause endothelial damage and lead to increase of vascular permeability.

Complement mediated damage: During secondary infection with heterologous dengue virus, reaction between viral antigen and pre-existing antibody activates the complement through C3 activator and initiating C1, C2 and C4 cascade. This leads to increase in the level of C3a and C5a anaphylotoxins which mediates the endothelial damage and leads to increase in vascular permeability.

Heterophile immune response: Antibody produced against the viral protein NS1 cross-reacts with several host cell proteins such as endothelial cell, liver cell and blood clotting proteins. The level of this antibody reaches pathologic level during secondary infection and is believed to cause damage to the cross-reactive host cells.

Soluble factors: Several studies have shown association of severity with increased level of various proinflammatory cytokines. However, the role of TNF-α, IL-10 and IL-6 has been extensively elucidated. These cytokines lead to increase in endothelial cell permeability, thrombocytopenia, derangement of coagulation and fibrinolysis.

Viral factors: No specific dengue virus serotype or genotype has been exclusively associated with severe disease. A recent meta-analysis study has shown secondary infection with DEN-2, DEN-3 and DEN-4 and primary infection with DEN-3 serotypes are associated with more severe disease in South East Asian (SEA) countries and secondary infection with

DEN-2 and DEN-3 are associated more with severe diseases in non-SEA countries.

Heterotypic infection with Asian genotype of DEN2 has been observed to cause more severe disease of DHF and DSS in Southeast Asia as compared to the American genotype of DEN-2 which causes mild disease.

The sequential infection with different dengue virus serotypes is responsible for development of severe dengue viral disease. Primary infection with DENV-1 followed by secondary infection with DEN-2 or DEN-3 has been associated with DHF epidemics. The evolution of dengue virus strain during a particular epidemic (intraepidemic) has also been associated with severe dengue disease.

Role of dengue virus NS1 antigen: High level of dengue virus NS1 antigenemia has been correlated with disease severity. NS1 antigen along with virus envelop (E) protein binds to the heparin sulfate which is the major glycosaminoglycan of the endothelial cell layer. This adherence alters the endothelial cell permeability leading to vascular leakage.

Dengue NS1 antigen has also been shown to interact with toll-like receptor 4 (TLR4) present on the surface of monocyte, macrophage and endothelial cells. This induces the release of cytokines responsible for disease severity and endothelial cell damage.

CLINICAL FEATURES

Dengue virus infection can remain asymptomatic or may be symptomatic. The symptoms of dengue range from mild fever to severe illness like dengue hemorrhagic fever or dengue shock syndrome. Severe form of dengue infection (DHF/DSS) can lead to circulatory failure and death.

Classic dengue fever: This starts with sudden onset of fever with severe headache, malaise, retro-orbital pain, severe back ache, myalgia and there may be pain in the joints. Fever is of high grade in nature, lasts for about 2–7 days.

In near half of the patients, macular or maculopapular rash appears at the time of defervescence after 3–7 days of illness. Mild respiratory or gastrointestinal symptoms may be there. Anorexia, nausea and vomiting are commonly seen.

Hemorrhagic manifestations may appear as purpura, petechiae and positive tourniquet test (**≥20 petechiae in 1 inch^2** on forearm after deflation of blood pressure cuff). Bleeding from various sites in the form of bleeding gum, epistaxis, vaginal bleeding may occur at this stage.

Tables 10.2 and 10.3 describe the criteria of dengue hemorrhagic fever and dengue shock syndrome according to the WHO definition.

The previous WHO classification of symptomatic dengue into DF, DHF and DSS suffers from the problem of classifying DHF in clinical setting. This led to revised criteria by WHO in 2009 which classify the dengue cases into severe dengue and non-severe dengue. The latter group is again divided into patients with and without warning signs. However, patient without warning signs may also develop severe disease. The warning signs as described by WHO are given in Table 10.4.

Table 10.2: Dengue hemorrhagic fever (DHF) (according to WHO criteria)

- Fever for 2–7 days
- Features of hemorrhagic manifestation
- Thrombocytopenia (<1,00,000/cumm)
- Evidence of plasma leakage by hemoconcentration (≥20% above average for age), pleural effusion or ascites, hypoproteinemia.

Table 10.3: Dengue shock syndrome (DSS) (according to WHO criteria)

- All features of DHF and features of circulatory failure
- Rapid and weak pulse and narrow pulse pressure (<20 mm Hg)
- Severe hypotension (systolic pressure <90 mm Hg in <5 yaers old)
- Cold clammy skin and restlessness.

Table 10.4: Warning signs (according to WHO classification)

- Abdominal pain and tenderness
- Persistent vomiting
- Clinical fluid accumulation
- Lethargy and restlessness
- Enlargement of liver >2 cm
- Increase in hematocrit
- Decrease in platelet count

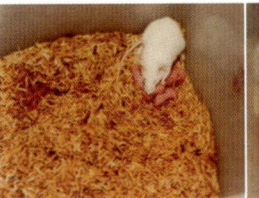

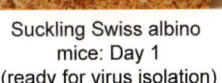

Suckling Swiss albino mice: Day 1 (ready for virus isolation)	Dengue infected suckling mice: Day 10 (inoculated mice about to die)

Fig. 10.4: Dengue virus isolation in mice

LAB DIAGNOSIS

Blood is the most important sample for dengue diagnosis. As dengue virus is sensitive to heat, pH, lipid solvents and proteolytic enzymes, sample for virus isolation and viral RNA detection should be transported in ice to the lab. The samples can be stored at 4°C for not more than 24 hours at peripheral hospital and should be transported in ice to the lab within that period. For prolonged storage, samples are stored at below –70°C.

Figure 10.3 depicts the timeline of dengue NS1 antigen, viremia, IgM and IgG antibody with symptoms.

Methods of Lab Diagnosis

Virus Isolation

Following three methods are used for dengue virus isolation.

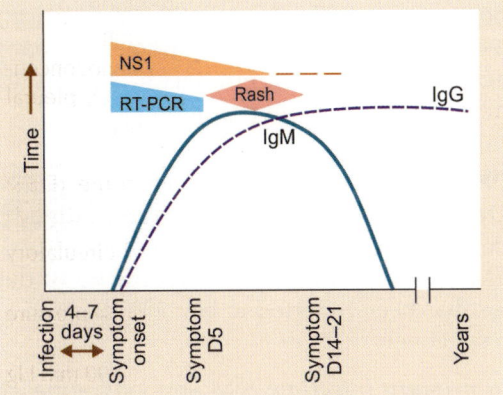

Fig. 10.3: Kinetics of dengue virus infection (schematic)

1. Mice inoculation: 1 to 3 days old newborn swiss albino mice is used for this purpose. Patient's serum is inoculated through intracerebral route. In positive cases, mice develops hunch back, less intake of food, fine tremor, and paralysis by 8–10 days post-inoculation (Fig. 10.4). Mice needs to be observed for development of symptoms for 2–3 weeks before declaring negative. The test is slow, time consuming, requires animal facility, and poorly sensitive. Other arboviruses that cause dengue-like illness can also be isolated in the mice host system. Because of restriction in use of lab animals and poor sensitivity, the technique is no more recommended.

Identification of virus can be done from the harvested mice brain.

- Dengue antigen can be detected in mice brain by direct or indirect fluorescent staining using specific polyclonal or monoclonal antibody.
- RT-PCR for dengue virus RNA detection.

2. Mosquito cell culture: This is presently the most common method and considered as the gold standard for dengue virus diagnosis. Various mosquito cell lines like C6/36 clone of *Aedes albopictus, AP-61 from Aedes pseudocutellaris* and *TRA-284 from Toxorhynchities amboinensis* are used for this. Rounding and clustering of infected cells may be observed as cytopathic effect. Cytopathic effects may be apparent in second or third passage.

Identification of virus:

- Infected cells are harvested for identification by direct or indirect fluorescent staining using specific polyclonal or monoclonal antibody.

- RT-PCR from cell culture supernatant.

3. Mosquito inoculation: Intrathoracic route of inoculation in mosquito is considered as the most sensitive method for dengue virus isolation. Four mosquito species have been used for this test; *Aedes aegypti, Aedes albopictus, Toxorhynchities amboinensis* and *Toxorhynchities splendens*. Virus grows rapidly to high titer in this method as compared to other methods of isolation.

Identification of virus: Antigen detection or RT-PCR from mosquito head-squash preparation.

Serological Tests

Hemagglutination inhibition (HI), complement fixation, neutralization, enzyme-linked immunosorbent assay (ELISA), and immunochromatographic tests are used for detection of dengue antibody.

Hemagglutination Inhibition (HI)

Principle: This is based on the principle of agglutination of dengue virus with chick red blood cells. The presence of dengue specific antibody in the patient's serum inhibits the agglutination. The antigen employed in HI test is prepared either from dengue virus infected suckling mouse brain with sucrose and acetone extraction or infected mosquito cell culture concentrate.

Primary infection: HI antibody begins to appear from five days of illness onwards and the titer goes up to 640 in convalescent phase serum and persists for 2–3 months.

Secondary infection: The titer rises rapidly and becomes 5120 or more.

Interpretation: HI titer more than 1280 in acute phase serum is often considered as the presumptive diagnosis of acute dengue infection.

Rise in fourfold titer between acute and convalescent sample is confirmative.

Use: As HI antibody persists for a long time, the test is often used for serosurveillance. The test is sensitive but lacks specificity.

Complement Fixation Tests (CFT)

The test is based on the principle of consumption of complement during antigen antibody reactions.

CF antibodies appear later than HI antibody and do not persist for long periods.

Therefore, positive CF antibody titer indicates acute or recent infection and not used for sero-surveillance.

The CF antibody is more specific and often monotypic as compared to broad specificity of HI antibody and secondary dengue infection.

However, CFT requires vigorous standardization and difficult to perform.

Neutralization

The principle is based on the production of cytopathic effect or plaques in susceptible cell lines. When serum containing dengue antibody is mixed with the virus, it causes inhibition in the number of plaques produced by the virus.

Neutralizing antibody appears later than HI antibody but earlier than CF antibody and persists for a long time. This test is **most sensitive and specific** for dengue virus.

These above mentioned serological tests are not used these dates for patients diagnosis because of difficulty in standardization, tedious to perform and requirement of both acute and convalescent serum sample to demonstrate fourfold rise of titer.

Newer Serological Tests

Detection of IgM antibody: IgM antibody detection by capture ELISA is known as MAC ELISA. The sensitivity and specificity of the test has been reported to be >90% and >95% respectively.

In primary infection: IgM antibody appears from 5 days of illness and persists for about 3 months.

In secondary infection, IgM appears slightly earlier as compared to primary infection.

Detection of specific IgM antibody in single sample indicates on-going or recent infection.

Seroconversion of IgM antibody in the convalescent sample depicts acute infection (negative in acute sample and positive in convalescent sample).

Because of requirement of single serum sample, ease of the procedure, wide availability, capability of testing a large number of samples and low cost, **MAC ELISA** is the recommended method for IgM antibody detection.

Detection of dengue IgG antibody: This can be done by various format of ELISA like IgG capture ELISA (GAC ELISA), indirect ELISA, etc.

Demonstration of seroconversion or four-fold rise in titer in acute and convalescent sample indicates acute infection.

Presence of IgG indicates exposure to the virus and thus used for serosurveillance.

IgG in acute sample indicates previous infection and helps to differentiate between primary and secondary infections.

Kinetics of dengue IgM and IgG antibody:
- Primary dengue infection is characterized by a low and delayed antibody response.
- IgM antibody is the first antibody to appear.
- Anti-dengue IgG appears after IgM at the end of first week of illness.
- During secondary dengue infection, both IgG and IgM antibodies are detected in the first week of illness in most of the cases.

Differentiation between primary and secondary infection: As per the WHO recommendation, the ratio between IgM and IgG (IgM/IgG) antibodies against dengue virus E/M protein is considered to differentiate between primary and secondary infections as below:
- *Primary infection:* **Ratio of IgM/IgG OD is >1.2** (serum dilution 1/100) or >1.4 (serum dilution 1/20).

- *Secondary infection:* **Ratio of IgM/IgG OD is <1.2** (serum dilution at 1/100) or <1.4 (serum dilution at 1/20).

Detection of dengue NS1 antigen: Dengue NS1 is a highly conserved glycoprotein. During infection, NS1 is associated with intracellular organelles or transported to the cell surface through secretory pathway. The antigen has been found to be circulated in patients' blood during acute phase of illness and presently considered as an **acute marker** of dengue virus infection.

The ELISA and rapid immunochromatographic tests are commercially available for detection of NS1 antigen in blood sample.

Dengue NS1 can be detected in the blood of the patient from day 1 to day 9 or more and used as early diagnostic marker.

Sensitivity: 75%; Specificity: 100%.

In secondary infection, it may disappear early due to early appearance of dengue IgM.

Rapid immunochromatographic test: These test formats are available for detection of dengue IgM, dengue IgG antibody as well as for dengue NS1 antigen. The combined antigen–antibody format is available by several manufacturers.

Rapidity and ease of use are its main advantages for which these tests are widely used.

Molecular Tests

Various formats of molecular tests are available for the detection of dengue virus RNA from clinical sample, like reverse transcriptase PCR (RT-PCR) (Fig. 10.5), real-time reverse transcriptase PCR, nested RT-PCR, single tube RT-PCR, and loop mediated isothermal amplification (LAMP). All these formats are used both for detection as well as for dengue virus type determination. Pre-membrane **(prM)** gene is most commonly targeted for detection of dengue virus RNA. Test is applicable only during the first 3–5 days of illness, due to short period of viremia. The sensitivity and specificity of molecular tests is near 100%.

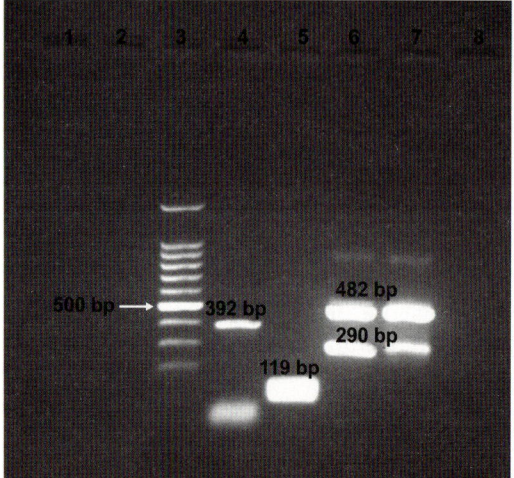

Lane 3: Mol. marker 500 bp;
Lane 4: Type 4 dengue virus (392 bp);
Lane 5: Type 2 dengue virus (119 bp);
Lane 6 & 7: Co-infection of type 1 (482 bp) and type 3 (290 bp);
Lane 8: Negative control

Fig. 10.5: Agarose gel analysis of the dengue type specific RT-PCR

Real-time RT-PCR for dengue 1–4 has been developed by CDC. The test has been found to be highly reproducible with high sensitivity and specificity and has been approved by FDA.

The advantages and disadvantages of molecular test are mentioned in Table 10.5.

LAMP assay: LAMP assay is a relatively new method of nucleic acid amplification based on the principle of strand displacement and stem loop structure that amplifies the target under isothermal condition. The assay generates large amount of target DNA production along with magnesium pyrophosphate as by-product causing turbidity.

The method has also been developed as multiplex assay for detection along with chikungunya infection. Table 10.6 describes the advantages of LAMP assay.

Figure 10.6 depicts the practical approach to dengue virus infection diagnosis in patient and Table 10.7 summarizes the methods used for dengue diagnosis.

Table 10.5: Advantages and disadvantages of molecular tests for dengue diagnosis

Advantages	Disadvantages
Covers the window period of IgM	Applicable only in early phase of illness
Can detect both in primary and secondary infections	Need of sophisticated equipment
	Expertise is required
Detection of virus sero-types and genotypes	Expensive

Table 10.6: Advantages of LAMP assay

- Thermocycler is not required
- Product can be visualized by naked eye/under UV light
- The turbidity can be measured by photometer and thus quantitation can be done
- No post PCR analysis is involved.

DENGUE VACCINE

Several vaccine strategies have been attempted for the development of dengue vaccine. These are live attenuated vaccine, inactivated vaccine, subunit vaccine, DNA vaccine, vector-based vaccine and virus-like particles. Amongst these, several live attenuated vaccines have reached phase III clinical trial. The only dengue vaccine which has got the licensure so far is also a variety of live attenuated vaccine.

Live Attenuated Vaccine

The methods used for attenuation are passage in cell culture, site directed mutagenesis and chimeric vaccine. Low replication capacity in humans and mosquitoes are desirable to decrease the risk of symptomatic infection and vector borne transmission of the vaccine virus.

Cell culture passage based live attenuated vaccine: Mahidole University, Bangkok, Thailand, was the first one to develop a tetravalent dengue LAV by attenuating DEN1, DEN2 and DEN4 strains in primary dog kidney (PDK) cells and DEN3 in primary African monkey kidney cells followed by fetal rhesus monkey lung cells. However, the

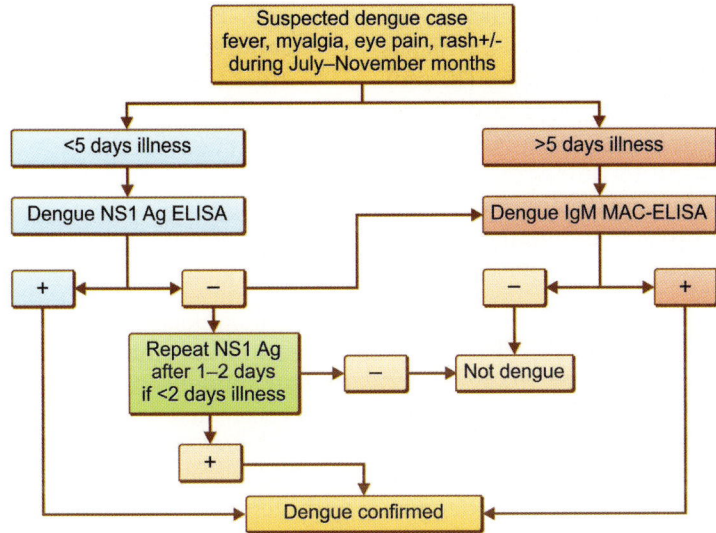

Fig. 10.6: Practical approach to lab diagnosis of dengue

Virus isolation	Table 10.7: Summary of methods of lab diagnosis		
	Serology		
	Conventional	Newer tests	Molecular tests
• Cell culture: C6/36	CFT	MAC-ELISA	RT-PCR
• Mice inoculation:	HAI	GAC-ELISA	Nested RT-PCR
Intracerebral	Neutralization	NS1 antigen	LAMP
• Mosquito inoculation:			
Intrathoracic			

vaccine failed to mount a balanced immune response to all the dengue serotypes and due to occurrence of increased frequency of adverse reactions like fever, rash, myalgia and retro-orbital pain, the further development of vaccine was stopped.

Walter Reed Army Institute of Research (WRAIR), Maryland, USA developed a tetravalent LAV with GlaxoSmithKline (GSK) where all the four dengue serotypes were attenuated by passaging in PDK cell line. Several formulations of this vaccine were tried with different concentration of four dengue serotypes. However, phase II clinical trials with different formulations of tetravalent vaccine showed inconsistent result regarding DEN-4 potency.

Target mutagenesis based live attenuated vaccine: Laboratory of Infectious Disease at National Institute of Allergy and Infectious Disease (NIAID) and National Institute of Health (NIH), Maryland, USA used this strategy of site-directed mutagenesis for attenuation of dengue virus strains. Dengue virus strains (DEN 1–4) were attenuated by targeted deletion of 30 nucleotides in 3' untranslated region (UTR) of full length complementary DNA clones. This candidate vaccine contains DENV1Δ30 and DENV4 Δ30 attenuated DEN 1 and DEN 4, respectively. DEN 2 and DEN 3 attenuated strains were prepared by using the backbone of DEN-4 Δ30 which is replaced with serotype specific prM and E gene. Various tetravalent preparations are under clinical trial. TV003 contains 10^3 pfu of each dengue virus serotype and TV005 contains 10^3 pfu of all except DEN 2 which is present at a 10-fold higher concentration (10^4 pfu). Various vaccine manufacturers in different countries have got license from NIH for its development (Instituto Butantan in Brazil, Vabiotech in Vietnam and Panacea Biotech and Serum Research Institute in India). Presently phase III clinical trial of this vaccine is ongoing in Brazil.

Chimeric dengue vaccine: Chimeric dengue vaccines are made of two varieties: (i) Chimera with another attenuated flavivirus, and (ii) inter-typic chimera with another dengue virus sero-type.

i. *Dengvaxia:* Chimeric yellow fever dengue-tetravalent dengue vaccine (CYD-TDV) is the only licensed vaccine available so far against dengue virus infection.

This is a recombinant chimeric live attenuated tetravalent vaccine. It contains the structural genes (pre-membrane and envelop proteins) of dengue virus 1, 2, 3 and 4 in the backbone of 17 D strain of yellow fever virus (genes of capsid and non-structural proteins). This is made with the concept that the structural proteins of dengue virus (prM and E gene) are responsible for production of neutralizing antibodies during natural infection, so chimeric vaccine containing these genes would mount the protective immunity in vaccine.

It induces neutralizing antibodies against the envelop protein of the virus and thus prevents the entry of virus into the host cell.

Two large scale phase III clinical trials have been conducted—one in five countries of Asia Pacific region in children of 2–14 years of age where CYD-TDV or placebo was given in 3 doses (0, 1, 6 months) and were followed up till 25 months, another phase III clinical trial was conducted in five dengue endemic Latin American countries in children of 9–16 years age.

- The overall efficacy was found to be 56–61% against dengue infection and 67–80% against hospitalization.
- Vaccine efficacy was 67.8% in children >9 years and 44.6% in children <9 years

age. Efficacy was further lower in children <9 years age in dengue naïve areas (14.4%).

- Increased risk of hospitalization was observed in children <9 years after 2 years of vaccination.
- Efficacy was maximum against dengue type 4, followed by type 3, type 1 and type 2.

It is manufactured by Sanofi-Pasteur and licensed in December, 2015. In 2017, Sanofi-Pasteur announced that the people who have been vaccinated with dengvaxia but not been infected previously with dengue virus may develop severe disease if infected naturally with dengue virus after vaccination. Considering this, WHO currently recommends dengvaxia only in persons who have had confirmed previous infection.

Considering the poor efficacy of vaccine in children <9 years and population in dengue naïve areas, vaccine has been recommended for the age group of 9–45 years individuals in the dengue endemic countries with three-dose schedule in 0–1–6 months.

ii. *Intertypic chimeric vaccine:* DENVax is a intertypic chimeric tetravalent vaccine. It contains DEN 2 strain which is attenuated by several PDK cell passage. This attenuated DEN 2 strain is used as the backbone for construction of other three dengue virus strains by replacing its prM and E genes with the corresponding serotype specific genes. Vaccine is manufactured by Inviragen Inc., Fort Collins Co, USA. Phase I study shows the vaccine as safe and immunogenic. Currently the vaccine is under phase III trial.

Tetravalent inactivated virus vaccine, recombinant dengue E protein subunit vaccine and tetravalent DNA vaccines with prM and E gene have also been developed and currently under phase I clinical trial.

Bibliography

1. Bhatt S, Gething PW, Brady OJ, et al. The global distribution and burden of dengue. Nature 2013; 496(7446): 504–7.

2. Chambers TJ, Nestorowicz A, Mason PW, et al. Yellow fever/Japanese encephalitis chimeric viruses: construction and biological properties. Journal of Virology1999; 73:3095–3101.

3. Dhanoa A, Hassan SS, Ngim CF, et al. Impact of dengue virus (DENV) co-infection on clinical manifestations, disease severity and laboratory parameters. BMC Infect Dis 2016 Aug 11;16(1): 406.

4. Gubler DJ. Dengue and dengue hemorrhagic fever. Clin Microbiol Reviews 1998; 480–96.

5. Gupta N, Srivastava S, Jain A, Chaturvedi UC. Dengue in India. Indian J Med Res 2012 Sep; 136(3):373–90.

6. Guzman MG, Halstead SB, Artsob H, Buchy P, Farrar J, Gubler DJ, Hunsperger E, Kroeger A, Margolis HS, Martínez E, Nathan MB, Pelegrino JL, Simmons C, Yoksan S, Peeling RW. Dengue: a continuing global threat. Nat Rev Microbiol 2010;8(12 Suppl):S7–16.

7. Halstead SB. Dengue. Lancet 2007 Nov 10; 370(9599):1644–52.

8. Halstead SB. Licensed dengue vaccine: Public health conundrum and scientific challenge. Am J Trop Med Hyg 2016 Jun 27. pii: 16-0222. [Epub ahead of print]

9. Halstead SB. Pathogenesis of Dengue: Dawn of a New Era. F1000Res. 2015 Nov 25;4. pii: F1000 Faculty Rev-1353. doi: 10.12688/f1000research. 7024.1. eCollection 2015.

10. Halstead SB. Pathogenesis of dengue: Challenges to molecular biology. Science 1988; 239(4839): 476–81.

11. http://www.sciencemag.org/news/2013/10/first-new-dengue-virus-type-50-years

12. http://www.who.int/immunization/research/ development/dengue_q_and_a/en/

13. Lu X1, Li X, Mo Z, Jin F, Wang B, Zhao H, Shan X, Shi L. Rapid identification of Chikungunya and Dengue virus by a real-time reverse transcription-loop-mediated isothermal amplification method. Am J Trop Med Hyg 2012;87(5):947–53.

14. Martina BE, Koraka P, Osterhaus AD. Dengue virus pathogenesis: An integrated view. Clin Microbiol Rev 2009;22(4):564–81.

15. Nagamine K, Hase T, Notomi T. Accelerated reaction by loop mediated isothermal amplification using loop primers. Mol Cell Probes 2002;16: 223–29.

16. Notomi T, H Okayama, H Masubuchi T. Yonekawa, K Watanabe, N Amino, T Hase.

2000. Loop-mediated isothermal amplification of DNA. Nucleic Acids Res 28:e63 (i-vii).

17. Ooi EE, Gubler DJ. Dengue in Southeast Asia: Epidemiological characteristics and strategic challenges in disease prevention. Cad Saude Publica 2009;25 Suppl 1:S115–24.

18. Parida M, Kouhei Horioke, Hiroyuki Ishida, et al. Rapid detection and differentiation of dengue virus serotypes by a real-time reverse transcription-loop-mediated isothermal amplification assay. Journal of Clinical Microbiology 2005; 43: 2895–2903.

19. Precioso AR, Palacios R, Thom´e B, et al. Clinical evaluation strategies for a live attenuated tetravalent dengue vaccine. Vaccine 2015; 33: 7121–5.

20. Bhatt S, Gething PW, Brady OJ, et al. The global distribution and burden of dengue. Nature 2013; 496: 504–7.

21. Sanofi Pasteur Media Release, http:// www.sanofipasteur.com/en/articles/First-Vaccinations-against-Dengue-Mark-Historic-Moment-in-Prevention-of-Infectious-Diseases.aspx.

22. Santiago GA, Vergne E, Quiles Y, Cosme J, Vazquez J, et al. (2013) Analytical and Clinical Performance of the CDC Real Time RT-PCR Assay for Detection and Typing of Dengue Virus. PLoS Negl Trop Dis 7(7): e2311. doi:10.1371/ journal.pntd.0002311

23. Soo KM, Khalid B, Ching SM, Chee HY. Meta-Analysis of Dengue Severity during Infection by Different Dengue Virus Serotypes in Primary and Secondary Infections. PLoS One 2016 May 23;11(5):e0154760.

24. Teoh BT1, Sam SS, Tan KK, Johari J, Danlami MB, Hooi PS, Md-Esa R, AbuBakar S. Detection of dengue viruses using reverse transcription-loop-mediated isothermal amplification. BMC Infect Dis 2013; 21;13:387.

25. Whitehead SS, Falgout B, Hanley KA, et al. A live-attenuated dengue virus type 1 vaccine candidate with a 30-nucleotide deletion in the 3′untranslated region is highly attenuated and immunogenic in monkeys. Journal of Virology 2003;77: 1653–57.

26. Wilder-Smith A, Ooi EE, Vasudevan SG, Gubler DJ. Update on dengue: Epidemiology, virus evolution, antiviral drugs, and vaccine development. Curr Infect Dis Rep 2010;12:157–64.

27. World Health Organization, Dengue: Guidelines for Diagnosis, Treatment, Prevention and Control,WHO,Geneva, Switzerland, 2009.

28. World Health Organization, Geneva, 2013. Dengue and severe dengue [factsheet no. 117, revised September 2013]. Available from: http://www.who.int/mediacentre/factsheets/fs117/en/

29. Yacoub S, Mongkolsapaya J, Screaton G. Recent advances in understanding dengue. F1000Res. 2016 Jan 19;5. pii: F1000 Faculty Rev-78.

B. JAPANESE ENCEPHALITIS VIRUS

Japanese encephalitis is the single most important cause of acute viral encephalitis globally, accounting for 30,000 to 50,000 cases and 10,000 to 15,000 annual deaths. The disease is more prevalent in eastern and southern Asia covering especially rural and suburban areas with a population of more than 3 billion.

VIRUS

The disease Japanese encephalitis is caused by the virus Japanese encephalitis virus (JEV), a member of genus _Flavivirus_ and family Flaviviridae. JEV is a single stranded positive sense, enveloped RNA virus. The genome of the virus is approximately 11 kb length, encodes for 3 structural proteins and 7 non-structural proteins. The structural proteins are capsid (C), premembrane (prM) and envelop (E) and non-structural proteins are NS1, NS2A, NS2B, NS3, NS4A, NS4B and NS5. Japanese encephalitis virus (JEV) is a member of the JEV serological complex. The complex consists of eight species and two strains/subtypes: Japanese encephalitis (JEV), MurrayValley encephalitis (MVEV), St. Louis encephalitis (SLEV), West Nile (WNV), Kunjin (KUNV), Alfuy, Cacipacore, Yaounde, Koutango, and Ustusu viruses.

JEV is classified into one **single serotype** with at least four distinct **genotypes (I–IV)**. Genotype V also has been described. Genotypes have been classified based on the sequence variation of the capsid, prM and E gene (C/prM, E). Genotypes I, II, and III are the most prevalent, having accounted for 98% of the isolated strains since 1935. Genotypes IV and V form the oldest JEV lineage, which probably have originated from an ancestral virus in the Indonesian-Malaysian region indicating the origin of JEV from this region. Genotypes I and III frequently occur in epidemic regions, whereas genotypes II and IV are mostly associated with endemic transmission.

JEV genotypes and their geographic distribution:

- **Genotypes I to III:** Australia and Asia (Cambodia, China, Thailand, Japan, Korea, Indonesia, Malaysia, Taiwan, and Vietnam, India)
- **Genotype IV:** Eastern Indonesia
- **Genotype V:** Muar region of Malaysia, China, Republic of Korea
- **Genotype III** is the most prevalent genotype worldwide.

In temperate epidemic areas, genotypes I and III are common, whereas in tropical endemic areas genotypes II and IV are also seen.

EPIDEMIOLOGY

The first documented JEV encephalitis outbreak was reported in 1871, whereas the virus was first isolated from the brain of an encephalitis patient during an outbreak in 1934 in Japan. Initially JEV was classified as group B arbovirus under the family Togaviridae. Originally the term "type B" encephalitis was used to differentiate from summer epidemic from von Economo's encephalitis lethergica or "type A" encephalitis. The term B was then abandoned. In 1985, JEV was included as a member of the genus Flavivirus and family Flaviviridae.

Global Scenario

Japanese encephalitis is mostly seen in south east Asian countries, such as China, Indonesia, Korea, Vietnam, Japan, Malaysia, Singapore, Philippines, Siberia, Burma and Southern countries of Asia such as India, Nepal, Bangladesh, Sri Lanka. In last two decades, the disease has spread to Pakistan, Nepal and Australia.

Several factors have been attributed for geographic expansion of JEV as described below:

- Irrigation projects facilitating rice cultivation and agricultural development leading to increase vector population have been associated with increase in JE cases.
- Migration of viremic birds and animal trade are thought to be the possible mechanism of import of JE virus to a naïve location.
- Wind-borne mosquitoes have been shown to be associated with the reintroduction of virus from the endemic southern parts of Asia. This has been observed as the

occurrence of JEV transmission after the onset of south west winds in the temperate parts of Asia.

Indian Scenario

In India, near 600,00,000 populations live in JE endemic region. Before 1970, JEV was mainly reported from southern parts of the country. Since 1970 onwards, the disease has been reported from northern and eastern states. In India, presently three states are severely affected with JEV; these are **Uttar Pradesh, Assam and West Bengal**. In 1973, major outbreak occurred in Bankura district of West Bengal with more than 40% mortality. Since then several outbreaks have been reported from several states and union territories covering almost all over the country. Figure 10.7 depicts the states with

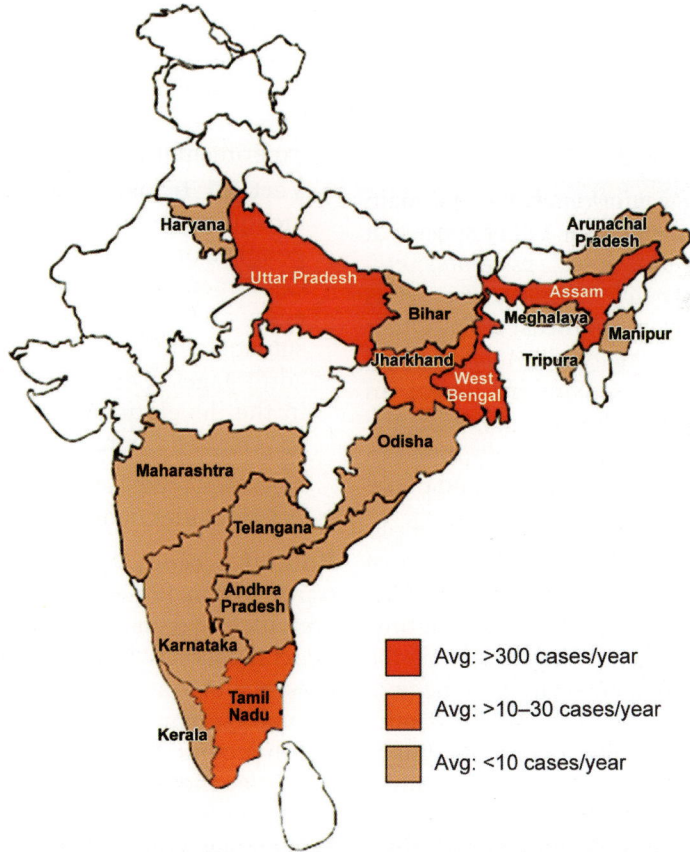

Fig. 10.7: Japanese encephalitis endemic states (schematic)

schematic representation of different grades of endemicity.

In Uttar Pradesh, Gorakhpur is amongst the worst affected area. Since 1988, a large number of cases are occurring almost every year affecting more than 1000–2000 cases with around 500 deaths. The largest outbreak occurred in 2005 affecting more than 6000 cases with 1500 deaths.

TRANSMISSION CYCLE

JEV is maintained in an **enzootic** cycle in nature. The cycle is maintained between **pig, bird and mosquito**. Wild and domestic birds and pigs act as the primary host and play an important role in maintenance and amplification of the virus.

More specifically birds such as pond herons, egrets and other ardeid birds act primarily as maintenance host and pig acts as amplifying host.

Humans get the infection accidentally by the bite of the infected mosquito. Human to human transmission through mosquito is not observed in JEV infection. Hence, humans are considered as the dead end host.

Vector: *Culex tritaeniorhynchus* is the main vector for JEV transmission. Other species of *Culex* are *Culex vishnui*, *C. pseudovishnui*, *C. gelidus*, *C. whitmorei*, *C. pipiens* also act as the vector mainly in Southeast Asia. Other mosquito species of *Mansonia*, *Anopheles* and *Aedes* also act as the secondary vector. In Northern Australia, *Culex annulirostris* is the main vector. These vectors mainly breed in stagnant water, paddy field, ditches and pools.

Amplifying host: **Pig** plays the role of amplifying host in JEV transmission. The titer of virus in infected pig becomes very high which lasts for 2–3 days. This provides ample time for the mosquito to get the infection from the pig. Infected pigs remain asymptomatic with rare instance of fetal loss. The close proximity of pig to human dwelling makes the amplifying host even more important for human transmission. High birth rate in swine also provides a large number of susceptible populations every year.

Maintenance host: Both bird and pig play the role of maintenance and amplifying hosts. However, pig is more known as amplifying host and bird as maintenance host. The titer of virus in bird becomes sufficient enough to infect mosquito.

Blocking host: The vector mosquito *Culex tritaeniorhynchus* prefers to feed on **cattle** than pig's blood. However, the titer of viremia in cattle is not sufficient enough to facilitate transmission of virus. Therefore, cattle play a negative or blocking role in JEV transmission.

Culex vectors are, however, opportunistic feeder and feed on pigs when the pig density is high in the locality.

Overwintering: This is thought to be the mechanisms of JE virus survival during the cold season in the temperate and subtropical endemic areas. The overwintering of JEV is evident by:

- Presence of sustained low level viremia in bats, snakes, frogs and lizards under experimentally simulated hibernation.
- Vertical transmission of JE virus in mosquito from one generation to next generation. This has been shown in experimentally infected mosquito through eggs at the time of fertilization and subsequently it passes onto larva, pupa and adult stages of mosquito. Virus can survive in the developmental stages during the adverse condition.

However, robust evidence of overwintering is lacking under natural condition.

Role of ecology and environment in JE transmission: JE is primarily a disease of rural area. In most of the southeast Asian countries, rice cultivation is the main agriculture practice, presence of water logged paddy fields is a common scenario in rural agricultural settings which provides a favorable breeding place for the *Culex* mosquito vector and attracts the ardeid birds where farmer has to work in the field for hours together. This ecological setting brings all the required components for transmission together, where mosquito gets the infection from the host ardeid bird and transmits to human. Time taken for mosquito to become infectious to other host is around 14 days and known as extrinsic incubation period.

In a poor rural domestic set-up, presence of open drainage (which provides favorable vector-breeding site) and pig rearing near the human dwelling (amplifying host) are favorable for JEV transmission. The presence of infected mosquito in the vicinity, when it bites the amplifying host pig, the virus gets multiplied and the pig develops viremia. When mosquito bites the viremic pig it gets the infection and transmits to humans present in the locality (after the extrinsic incubation period) (Fig. 10.8).

Epidemiological pattern: Two epidemiological patterns are observed.

1. *Epidemic pattern:* This is observed in temperate areas with distinct summer seasonality. This is seen in northern Vietnam, northern Thailand, and northern India, Nepal, Korea, China and Japan.

2. *Endemic pattern:* This is observed in southern tropical areas, such as southern Thailand, southern Vietnam, southern India, Sri Lanka, Malaysia, Indonesia and Philippines, where JE virus transmission is endemic. Sporadic JE cases occur throughout the year in these parts with a peak during the rainy season.

PATHOGENESIS AND PATHOLOGY

Infected mosquito bites the susceptible individual and releases the virus into the skin. Virus then multiplies in the dendritic cells and macrophages present in the skin. Subsequently the virus spreads to the lymphatic system. Replication of virus at these sites produces transient primary viremia. During this phase, virus is seeded in several non-neural tissues such as skeletal muscle, smooth muscle, connective tissue, lymphoreticular tissues and endocrine and exocrine glands. In majority of the cases, infection gets restricted at this stage without involving the central nervous system (CNS). In a few cases (1 in 25–1000) where virus enters the CNS, patients manifest with encephalitis.

The exact mechanism by which JEV crosses the blood–brain barrier to enter the CNS is not known. However, evidence of diffuse involvement of brain indicates hematogenous spread of virus. Replication of flaviviruses in the endothelial cells in experimental model also suggests the possible passage to CNS through antipodal transport of virus through vascular endothelial cells.

JEV generally affects the gray matter of thalamus, cerebral cortex, midbrain, cerebellum, basal ganglia and anterior horn cell. Pathological changes include congestion, degeneration, perivascular cuffing, neuronophagia and hemorrhages and thrombus formation in the affected areas. The changes are, however, not specific for JEV infection.

Bird–mosquito cycle   Pig–mosquito cycle

Fig. 10.8: Japanese encephalitis virus transmission cycle

CLINICAL FEATURES

JEV infection is mostly asymptomatic in nature. The average ratio between symptomatic to asymptomatic is 1 in 300 with a range of 1 in 25 to 1 in 1000. The reason for this variation is not known. It is thought to be due to various viral and host factors, such as age, immunity, host genetic makeup and viral factors like neuro-virulent strain or high viral load.

It affects mostly children of the endemic areas. However, adults can also be affected.

Incubation period ranges from 4 to 14 days. Symptoms started with sudden onset of fever with constitutional symptoms like malaise, myalgia, anorexia, nausea, vomiting and abdominal pain. These manifestations may improve without the involvement of CNS.

The involvement of CNS is manifested by gradual development of behavioral abnormality, altered sensorium, convulsion, tremor, ataxia, loss of memory, neurological deficit and muscle stiffness. In children, mask-like faces with wide unblinking eyes are commonly observed. Convulsion is seen more commonly in children as compared to that of adults.

The outcome of the symptomatic patients is variable. In general, recovery occurs in about one-third of the patients, mortality in one-third and remaining third develops sequela. Epileptic seizure, mental retardation, deformities are commonly seen as features of sequela.

LABORATORY DIAGNOSIS

The primary sample required for diagnosis is cerebrospinal fluid (CSF). Serum is the second important sample. In case of postmortem diagnosis, brain biopsy can be collected. Samples are to be sent in cold chain for virus isolation or RT-PCR and at room temperature for serology.

Virus isolation and RT-PCR are mainly done from CSF sample or from brain tissue in postmortem cases. Serum sample is not recommended for these purposes as by the time encephalitis develops, patient is no longer viremic and also because of low viremic titer.

Virus isolation: The methods of virus isolation are same as described for dengue virus such as cell culture, mosquito inoculation and intracerebral inoculation of suckling mice.

Isolation in *Toxorhynchites splendens* mosquito intrathoracic inoculation is highly sensitive.

JEV culture in Vero, LLCMK2 and porcine (PS) cell lines produce cytopathic effect. Like dengue virus, JEV can also grow in C6/36 or AP61 cell line.

Isolation or propagation of JEV in any of the host system needs to be confirmed by detection of viral antigen by immunofluorescence using specific antibody or detection of viral RNA by RT-PCR.

Serology: Due to difficulties with isolation and PCR, serology is the mainstay of diagnosis in JEV infection.

Detection of JEV-specific IgM by MAC ELISA or demonstration of fourfold rise of antibody titer by other serological test such as hemagglutination inhibition, complement fixation test, and neutralization test in serum sample is considered as the presumptive diagnosis for JEV in suspected patients. However, in clinical setting, positive JE IgM antibody in serum sample is considered as marker of acute infection in a patient with encephalitis from endemic region.

Detection of JEV-specific **IgM by MAC ELISA** or demonstration of fourfold rise of antibody titer by hemagglutination inhibition, complement fixation test, and neutralization test in CSF sample in a suspected JE case is considered as confirmed JE case. WHO criteria for JEV confirmation are given in Table 10.8.

JEV IgM antibody is generally positive by 4–5 days of illness in majority of the patients (70–75% cases) and almost 100% by 7–8 days of post-illness.

Several commercially available captures IgM ELISA are available such as Panbio, InBios JE detect kit and Xcyton. The sensitivity of these kits has been found to be low with the range

Table 10.8: WHO lab criteria for JEV confirmation

- Positive JEV IgM in CSF by MAC ELISA
- Positive JEV IgM in acute phase serum by MAC ELISA
- JEV antigen in tissue by immunohistochemistry (IHC)
- JEV RNA in CSF/tissue/blood by RT-PCR
- JEV isolation in CSF/tissue/blood
- Fourfold or more rise in JEV antibody in acute and convalescent sera by HAI/PRNT

of 17–57% and specificity has been reported high (>95%) in various studies. Inconsistent results have been reported regarding better sensitivity of any of these available tests. In India, JEV MAC ELISA has been developed by National Institute of Virology, Pune and is supplied for testing to various labs across the country through National Vector Borne Control Programme (NVBDCP), Ministry of Health and Family Welfare, Government of India.

JEV antigen detection: JE antigen detection can be done in tissue sample using the specific monoclonal antibodies by immunohisto-chemistry. Presently there is no JEV antigen detection test available for blood or CSF sample.

VACCINE

Several licensed vaccines are available for immunization against Japanese encephalitis.

Inactivated Vaccines

Mouse brain-derived inactivated vaccine: This is one of the oldest JE vaccines available for prevention of JEV infection. Vaccine is prepared either from **Nakayama strain** or **Beijing 1 strain**. First it was prepared in Japan, and then many countries started preparing it. In India, it was manufactured at Central Research Institute, Kasauli. Vaccine efficacy was more than 90%.

Vaccine was given in three doses at 0, 7, and 30 days, subcutaneously to children between age of 1 and 3 years with booster dose after 1 year and every three years till 10 years of age.

Vaccine preparation was expensive and due to adverse reactions, and availability of better cell culture derived vaccine, it is no more in use.

Disadvantages
- Requirement of multiple doses
- Potential for producing neurological side effects
- Hypersensitivity reaction
- Less stringent method of preparation.

Primary hamster kidney (PHK) cell culture inactivated vaccine: This is also one of the oldest JE vaccines used in China since 1967. It was prepared from **Beijing—P3 strain**. The side effects were relatively less and its efficacy was in the range of 76–90%.

Vero cell culture inactivated vaccine: Several Vero cell derived formalin inactivated JE vaccines have been developed and licensed in many countries including India. Three such vaccines have been described in Table 10.9. **SA14-14-2** is the master seed virus used in inactivated vaccines. The wild virus SA14 was isolated from pool of *Culex pipiens* larva following 11 passages in mice brain. After serial passage in primary hamster kidney (PHK) cells it was then named SA14-14-2 (SA14 clone 14–2). Virus was then adapted in primary dog kidney (PDK) cells followed by Vero cells and was used as the seed virus for inactivated vaccine. The seed virus was manufactured by Intercell AG, Austria and the vaccine has been licensed in various names in different countries. Table 10.9 lists out the Vero cell derived inactivated vaccine.

Live Attenuated Vaccines

Live attenuated SA14-14-2: This vaccine was made in China by passaging SA14 strain in PHK cells. Attenuation of the strain was associated with the changes in six amino acids in E protein and three amino acids in NS gene. Vaccine efficacy was >98% with two doses. Presently this vaccine is widely used in China as well as in several Asian countries including India. Vaccine is given subcutaneously at

Table 10.9: Vero cell culture derived inactivated vaccine

Vaccine	Strain	Age	Route	Dose	Country approved
IXIARO (Intercell)	SA 14-14-2	≥2 months	IM	2 m – <3 yr: 0.25 mL × 2 doses (4 weeks apart) ≥3 yr: 0.5 mL × 2 doses (4 weeks apart)	USA, EU
JEEV (Biological E. Ltd.)	SA 14-14-2 (Al. hydroxide adjuvanted IXIARO vaccine)	18–49 yrs	IM	0.5 mL (0.6 µg) × 2 doses (4 weeks apart)	India
JENVAC (Bharat Biotech)	Kolar strain 821564 XY	≥1yr	IM	0.5 mL (0.6 µg) × 2 doses (4 weeks apart)	India

8 months age. Booster is recommended at 2 years age.

Disadvantages: Attenuated strain is produced in primary cells. So, the method is slower as compared to that of continuous cell line.

JE Yellow fever chimeric vaccine (ChimeriVax-JE): This is a live attenuated JE vaccine in the backbone of 17 D yellow fever vaccine strain. Single dose is recommended. The vaccine has been licensed in Australia and Thailand. It is marketed in the name of IMOJEV, manufactured by Sanofi Pasteur.

Target Population for JE Vaccine

In endemic area, the ideal recommendation is universal immunization to all children below 15 years age. This strategy has been based on the evidence of seroconversion in majority by 15 years of age.

To start with vaccination is recommended in the epidemic foci as one time catch-up campaign for all children followed by introduction of routine immunisation programme.

Vaccine is also recommended to travellers to JE endemic countries particularly when the duration of stay is more than one month and during the expected JE season.

Bibliography

1. Erlanger TE, Weiss S, Keiser J, Utzinger J, Wiedenmayer K. Past, present, and future of Japanese encephalitis. Emerg Infect Dis 2009; 15(1):1–7.

2. Ghosh D, Basu A. Japanese encephalitis-A: Pathological and clinical perspective. PLoS Negl Trop Dis 2009; 3(9): e437. doi:10.1371/journal.pntd. 0000437

3. http://www.nvbdcp.gov.in/Doc/je-aes-cd-June16.pdf

4. http://www.cdc.gov/vaccines/hcp/vis/vis-statements/je-ixiaro.pdf.

5. Japanese encephalitis. National Vector Borne Disease Control Programme Ministry of Health and Family Welfare, Government of India.

6. Lewthwaite P, Shankar MV, Tio PH, et al. Evaluation of two commercially available ELISAs for the diagnosis of Japanese encephalitis applied to field samples. Trop Med Int Health 2010;15(7):811–18.

7. Misra UK, Kalita J. Overview: Japanese encephalitis. Progress in Neurobiology 2010; 91:108–20.

8. Ravi V, Robinson JS, Russell BJ, Desai A, Ramamurty N, Featherstone D, Johnson BW. Evaluation of IgM antibody capture enzyme-linked immunosorbent assay kits for detection of IgM against Japanese encephalitis virus in cerebrospinal fluid samples. Am J Trop Med Hyg 2009;(6): 1144–50.

9. Robinson JS, Featherstone D, Vasanthapuram R, et al. Evaluation of three commercially available Japanese encephalitis virus IgM enzyme-linked immunosorbent assays. Am J Trop Med Hyg 2010;83(5):1146–55.

10. Saxena V, Dhole TN. Preventive strategies for frequent outbreaks of Japanese encephalitis in Northern India. J Biosci 2008; 33: 505–14.

11. Singh M M, Gadey S, et al. A Japanese encephalitis vaccine from India induces durable and cross-protective immunity against temporally and spatially wide-ranging global field strains. The Journal of Infectious Diseases 2015; 212: 715–25.

12. Tiwari S, Singh RK, Tiwari R, Dhole TN. Japanese encephalitis: A review of the Indian perspective. Braz J Infect Dis 2012;16(6):564–73.

13. Vashishtha VM, Kalra A, Bose A, et al. Indian Academy of Pediatrics (IAP) recommended immunization schedule for children aged 0 through 18 years, India, 2013 and updates on Immunization. Indian Pediatrics 2013; 50: 1095–1108.

C. WEST NILE VIRUS

West Nile virus (WNV) is a mosquito-borne virus. It belongs to the family Flaviviridae, genus Flavivirus and Japanese encephalitis serocomplex along with Japanese encephalitis virus and St. Louis encephalitis virus.

WNV is a spherical enveloped virus with a diameter of around 50 nm. It contains single stranded RNA genome that encodes the capsid (C), envelop (E), pre-membrane (prM) and seven non-structural proteins.

The virus has **5 phylogenetic lineages**, of which lineages 1 and 2 are associated with human outbreaks. Lineage 1 has been subdivided into three sublineages: 1a, 1b and 1c consisting of virus isolates from different geographic areas as mentioned:

- **1a:** Isolates from Western hemisphere, Africa, Europe and the Middle East
- **1b:** Kunjin virus from Australasia
- **1c:** Indian isolates

TRANSMISSION CYCLE

WNV is maintained in the nature in an enzootic transmission cycle between **bird–mosquito–bird**. It can infect several vertebrates including human but many of them including humans suffer from severe disease and death. Those vertebrates also develop low level of viremia which is not sufficient enough to infect the mosquito and are considered as the dead end host.

Several bird species can be infected with the virus. WNV infection in some bird species like American crow leads to severe disease and death, in some species like house sparrows it leads to high viremia with less mortality. Several passerine birds are considered important for human transmission as they develop high viremia sufficient to infect mosquito.

Culex pipiens, *C. quinquefasciatus* and *C. tarsalis* species are the major vector of WNV transmission particularly between infected birds and humans.

The virus is primarily transmitted through mosquito bite. However, transmission of WNV has also been reported through transfusion of blood and its components, organ transplant and transplacental transmission.

EPIDEMIOLOGY

West Nile virus was first discovered in 1937 in the West Nile district of Uganda from a patient with a mild febrile illness. The virus is widely distributed throughout Africa, southern Europe, the Middle East, Eastern Russia, southwest Asia and Australia. Until 1990s, occasional human outbreaks were reported with mild febrile illness. Over the years, large human outbreaks were reported from several parts of Russia and Europe with more severe disease. In America, human cases started occurring since 1999, and then the virus spreads from New York to other states of USA and America. Since 1999 to 2015, near 44,000 WNV cases and 2000 deaths have been reported to CDC.

In India, the activity of WNV has been reported from birds, mosquitoes and humans as the virus has been isolated from all of them. Neutralizing antibodies to WNV has been detected in human sera collected from several eastern and western coastal states. Confirmed cases have been reported previously from Vellore and Kolar districts of Tamil Nadu.

Recently in 2010, WNV associated retinitis has been reported during a febrile outbreak in Tamil Nadu. In 2011, outbreak of WNV encephalitis was reported from Kerala.

PATHOGENESIS

WNV is inoculated onto the skin through bite of infected *Culex* mosquito. Then the virus replicates in keratinocytes and dendritic cells. These infected cells migrate to the draining lymph nodes and then to the bloodstream. Through blood, the virus reaches various organs and to the central nervous system. Systemic replication of WNV leads to high viral load in the blood. This induces the increased level of proinflammatory cytokines which facilitates the entry of virus by increasing the permeability as well as by disrupting the blood–brain barrier.

The gray matter of brain, brainstem and spinal cord are commonly affected by the virus. Pathologic changes in the brain can occur due to direct replication of the virus in the neuron or due to cytokine mediated. The risk of developing neuroinvasive disease occurs in 1 in 150–250 infected persons. Advanced age is one of the important risk factors for development of neuroinvasive disease.

CLINICAL MANIFESTATIONS

Incubation period ranges from 2 to 14 days. Majority of the patients (≈80%) remain asymptomatic. Around 20–25% of infected individuals develop clinical manifestations. Majority of them develop self-limiting West Nile fever with acute onset of fever, headache, malaise, weakness. Macular rash appears towards defervescence, over torso and extremities and usually sparing the palm and sole.

The neurologic manifestations can be meningitis, encephalitis or paralysis. WNV meningitis is manifested by abrupt onset of fever, headache, meningeal signs. Severity of encephalitis ranges from disorientation to severe encephalopathy, coma and death.

Paralysis in WNV infection occurs due to destruction of anterior horn cells of the spinal cord leading to poliomyelitis like acute, asymmetrical flaccid paralysis. Involvement of other organs may manifest as choroiditis, myocarditis, hepatitis, pancreatitis, etc.

LAB DIAGNOSIS

Serology: Detection of WNV-specific IgM antibody by IgM capture ELISA in CSF or serum is the mainstay of diagnosis. In patients with symptoms of neuroinvasive disease, IgM antibody in CSF is usually positive by the time symptoms appear, whereas in other patients it may take 8 days to develop. In serum, IgM may persist for 12–16 months after subsidence of symptoms. Therefore, IgM positivity in serum of asymptomatic patients may not be due to acute infection and can be due to past infection. However, demonstration of seroconversion is diagnostic.

WNV infection can also be diagnosed by demonstration of fourfold change of antibody titer by hemagglutination inhibition (HAI), complement fixation test (CFT) or plaque reduction neutralization test (PRNT). However, these tests are not useful for patient diagnosis as they require both acute and convalescent serum samples at 2–3 weeks interval.

Isolation: The virus can be isolated in suckling mice brain or cell culture system as applicable for other arboviruses.

Molecular test: Detection of WNV RNA by nucleic acid amplification tests such as RT-PCR, NASBA has been developed. Molecular detection tests are used mostly as an adjunct to MAC-ELISA and useful in early part of illness and in immunocompromised patients where antibody development is poor or in case of screening for blood transfusion.

TREATMENT AND PREVENTION

Treatment of WNV infection is mainly supportive like other acute encephalitic syndrome. Several immunotherapeutic agents and antiviral have been tried but no definitive advantages have been reported.

So far no human vaccine is available, and currently there is no progress in vaccine trial because of its doubtful marketing potential. Prevention mainly depends on the reduction in infected mosquito population.

Bibliography

1. Anukumar B, Sapkal GN, Tandale BV, Balasubramanian R, Gangale D. West Nile encephalitis outbreak in Kerala, India, 2011. J Clin Virol 2014;61(1):152–55.

2. cdc.gov/westnile/resources/pdfs/data/2-west-nile-virus-disease-cases-reported-to-cdc-by-state_1999-2015_07072016.

3. Chancey C, Grinev A, Volkova E, Rios M. The global ecology and epidemiology of West Nile virus. Biomed Res Int 2015;2015:376230. doi: 10.1155/2015/376230.

4. Dhiman RC. Emerging vector-borne zoonoses: Eco-epidemiology and public health implications in India. Front Public Health. 2014 Sep 30;2:168. doi:10.3389/fpubh.2014.00168. eCollection 2014.

5. Montgomery RR, Murray KO. Risk factors for West Nile virus infection and disease in populations and individuals. Expert Rev Anti Infect Ther 2015;13(3):317–25.

6. Paramasivan R, Mishra AC, Mourya DT. West Nile virus: the Indian scenario. Indian J Med Res 2003;118:101–8.

7. Patel H, Sander B, Nelder MP. Long-term sequelae of West Nile virus-related illness: a systematic review. Lancet Infect Dis 2015; 15(8):951–59.

8. Shukla J, Saxena D, Rathinam S, et al. Molecular detection and characterization of West Nile virus associated with multifocal retinitis in patients from southern India. Int J Infect Dis 2012 Jan;16(1):e53–59.

9. Suen WW, Prow NA, Hall RA, Bielefeldt-Ohmann H. Mechanism of West Nile virus neuroinvasion: a critical appraisal. Viruses 2014;18;6(7):2796–2825.

10. Ulbert S, Magnusson SE. Technologies for the development of West Nile virus vaccines. Future Microbiol 2014;9(10):1221–32.

11. Winkelmann ER, Luo H, Wang T. West Nile Virus Infection in the Central Nervous System. F1000Res. 2016 Jan 26;5. pii: F1000 Faculty Rev-105. doi: 10.12688/f1000research.7404.1.

12. Winston DJ, Vikram HR, Rabe IB, et al. West Nile Virus Transplant-Associated Transmission Investigation Team. Donor-derived West Nile virus infection in solid organ transplant recipients: report of four additional cases and review of clinical, diagnostic, and therapeutic features. Transplantation 2014 May 15; 97(9): 881–89.

13. Zeller HG, Schuffenecker I. West Nile virus: an overview of its spread in Europe and the Mediterranean basin in contrast to its spread in the Americas. Eur J Clin Microbiol Infect Dis 2004;23(3):147–56.

D. CHIKUNGUNYA

Chikungunya virus (CHIKV) is an enzootic virus found in tropical and subtropical regions of Africa, south and south east Asia, and Indian Ocean Islands. The name chikungunya is used to describe the virus as well as the disease caused by it and is derived from **Makonde or Swahili** word Kun quanla in reference to the **stooped posture** developed as a result of arthritic symptoms of the disease. The disease is characterized by fever, headache, myalgia, rash and joint pain, usually self-limiting in nature but can persist in some for years.

VIRUS

The CHIKV is a single-stranded, positive sense RNA virus. The virus is around 70 nm in size, with an icosahedral-like nucleocapsid surrounded by an envelop embedded with multiple copies of major glycoproteins E1 and E2. The molecular mass of these two proteins is about 50 kDa, anchored in the membrane by conventional membrane spanning anchors. These two polypeptides E2-E1 form a stable heterodimer which forms the spikes on the virus surface. The virus can be inactivated by acid pH, lipid solvents, detergents, bleach, phenol, 70% alcohol and formaldehyde.

CHIKV belongs to family Togaviridae, genus Alphavirus, which is composed of various serocomplexes that are grouped together on the basis of antigenic properties.

CHIKV belongs to the Semliki Forest Virus (SFV) antigenic complex along with other mosquito-borne alphaviruses like Ross River Virus, Mayarovirus, O'Nyong Nyong, Grtah, Be baru and SFV. All alphaviruses are antigenically related and have worldwide distribution.

CHIKV has three lineages with distinct antigenic and genotypic characteristics: (i) **West African**, (ii) **East** and **Central African** and (iii) **Asian**.

TRANSMISSION CYCLE

The maintenance and transmission cycle of CHIKV occurs differently in Africa and Asia.

In Africa, the virus is maintained in an enzootic sylvatic cycle involving non-human primates and small mammals, and forest dwelling *Aedes* mosquitoes like *Ae. furcifer*, *Ae. luteocephalus*, *Ae. taylori* or *Ae. africanus*. Outbreaks usually occur due to heavy rainfall leading to spillover of the virus from enzootic forest cycle. During epidemics, man–mosquito–man cycle occurs without animal reservoir. Increase in mosquito population in rural areas can also lead to outbreaks in non-immune population.

In Asia, CHIKV is maintained by man–mosquito–man cycle resulting in urban epidemics. *Ae. aegypti* and *Ae. albopictus* are the two main vectors which play a role in human transmission (Fig. 10.9).

Mother to fetus and blood transfusion have been reported as rare modes of transmission.

History of outbreaks: The first known outbreak of CHIKV was reported from Makonde plateau near Tanzania border, during 1952–53 when the virus was first isolated from blood of a patient. Since then, the virus has spread to the east, central and west Africa and Asia and Europe. Several outbreaks have been reported from various parts of Africa.

The first confirmed outbreak in Asia occurred in Philippines in 1954 with subsequent outbreaks in 1958 and 1968. During

Fig. 10.9: Sylvatic and epidemic chikungunya virus transmission cycle

1970, outbreaks occurred frequently in south and south east Asia. Thereafter, the activity of the virus became quiescent with only sporadic cases.

Global Re-emergence: CHIKV re-emerged both in Asia and Africa after remaining quiescent for several decades. Previous to 1999–2000, human infection of CHIKV in Africa was occurring at a very low level. The re-emergence started in the year 1999–2000 in Democratic Republic of Congo (DRC), and then it spreads to other parts of Africa. In 2005, Comoros island was affected severely where near two-thirds of population were infected. The virus then spreads to the neighbouring islands and Asian countries such as Indian Ocean islands, La Reunion, Malaysia, Indonesia, and India. Reunion experienced massive outbreak whereby April 2006, more than two lakhs cases (an estimated 2,44,000 cases) were infected with an attack rate of 35% and resulted in 203 deaths, involving almost one-third of its population. During the years 2006–2008, some of the SE Asian countries like Singapore and Maldives recorded the entry of virus for the first time. Imported cases were reported in Europe and USA. In 2007, a localized outbreak was reported in Italy affecting 197 cases. In US, since 2006, a median of 26 cases were reported each year, mostly among adult travellers returning from area with ongoing CHIKV epidemics.

Re-emergence in India: In India, CHIK virus was first isolated in 1963 from Kolkata and the activity was observed till 1971. Thereafter the virus got almost disappeared and was thought to lose its pathogenic potential. Re-emergence of CHIK virus occurred after more than three decades of quiescence during the end of 2005 when outbreak occurred in several coastal states of the country like Andhra Pradesh, Karnataka, Gujarat and Odisha. The virus then spreads to several other states during 2006–2007 sparing the northern part. The outbreak continued for >3 years resulting in millions of cases. The vast number of immunologically naïve population probably contributed to the persistence of the virus. After this outbreak, the virus has become endemic in several eastern and western coastal states, such as Kerala, Karnataka, Maharashtra, Gujarat, Rajasthan, Madhya Pradesh, West Bengal, Odisha, Andhra Pradesh, Tamil Nadu and Delhi (Fig. 10.10). In India, Asian genotype was circulating before re-emergence, whereas central/east African genotype was isolated during the latest epidemics. Strains with and without **A226V mutation** were reported from different parts of the country. Thus both *Ae. aegypti* and *Ae. albopictus* were considered as the main vector during the last decade outbreak.

Possible factors for re-emergence: The explosive outbreak in Indian Ocean islands have been attributed to the East African strain with mutation in envelop glycoprotein E1-A226V (Ala-Val). This mutation has been shown to increase the CHIKV infectivity of *Ae. albopictus* with more efficient dissemination to mosquito salivary gland and increase in transmissibility.

Fig. 10.10: Chikungunya active part of India

Decrease in herd immunity has also been considered as an important factor for re-emergence. The rapid transmission of virus during the early period of outbreak has been attributed to concurrent infection of CHIK virus with microfilaria which is endemic in coastal states.

CLINICAL FEATURES

The average incubation period is 2–4 days, which ranges from 1 to 12 days. Chikungunya is characterized by the **clinical triad of sudden onset of fever, rash and joint pain** followed by other constitutional symptoms for a period of 1–7 days. Fever is usually high grade (>40°C), associated with chill and rigor, nausea and vomiting. Fever may disappear and return in 1–2 days and known as **"saddle back fever"**.

Arthralgia is present in almost all patients with fever which commonly affects more than one joint with **symmetrical involvement** and swelling. **Finger, wrist, elbows, toes, ankle and knee** are commonly affected. The acute features of CHIKV infection usually resolves within 1–2 weeks, however, arthralgia may persist for months to years.

Cutaneous manifestations are normally present during acute infection and present as redness on trunk, face and limb, followed by maculopapular rash which gradually fades away or desquamate.

Neurological manifestations due to CHIKV infection have been reported both in children and adults, more commonly in patients with underlying illnesses like stroke, epilepsy and diabetes mellitus. Acute encephalitis, febrile seizure, acute flaccid paralysis, meningitis and Guillain-Barré syndrome are the common presentation. Optic neuritis, hearing loss and hypokalemic paralysis have also been reported.

Mother to child transmission of CHIKV was recorded for the first time during the last outbreak in reunion. Encephalopathy was the most common manifestation among the children born to CHIKV infected mother.

Prior to 2005, chikungunya was not considered as a fatal disease. However, the last decade CHIKV outbreak documented 1 in 1000 death. Heart failure, multiorgan failure, hepatitis and encephalitis have been attributed as the common causes of death.

PATHOGENESIS

Both viral and host immune responses have been attributed for CHIKV disease pathogenesis. Increased expression of cytokines both in animal model and in human studies has been shown to be associated with disease process.

Replication of CHIKV in muscle and joint tissue in high concentration, leading to recruitment of inflammatory cells like monocytes, macrophage and NK cells has been reported in non-human primate and mice models as well as in muscle tissue of CHIKV infected patients.

Increased plasma level of proinflammatory cytokines along with high viral load has been found during acute CHIKV infection. IL-6 in particular is described as the biomarker of disease severity, symptom persistence and has also been associated with genesis of joint pain.

Susceptibility of skeletal muscle progenitor cells has been hypothesized as the possible factor responsible for chronic and recurrence of myalgia.

DIAGNOSIS

The diagnosis of chikungunya virus infection is made on the basis of clinical, laboratory and epidemiological criteria. Three categories of cases, possible cases, probable cases and confirmed cases, are described.

- *Possible cases:* Defined on the basis of only clinical criteria like acute onset of fever and arthritis/arthralgia.
- *Probable cases:* Defined on the basis of both clinical and epidemiological criteria.
  - Epidemiological criteria are defined as residing or travelled to an area with active CHIKV infection.
- *Confirmed cases:* Confirmation of CHIKV infection by any of the following laboratory tests: Virus isolation, detection of viral RNA, specific IgM antibody or demonstration of fourfold rise in IgG antibody titer.

Table 10.10: Important points in common chikungunya diagnostic test

Sample: Serum
Transport in sterile container at room temperature
Test: Chikungunya IgM antibody detection by MAC ELISA

The virus kinetics is grossly similar to other arboviruses like dengue virus. The laboratory tests are essentially in the line of diagnosis of dengue viral infection; in terms of virus isolation in mice, mosquito cell line and intra-thoracic inoculation in mosquito; serological tests and molecular tests like IgM antibody detection and RT-PCR or LAMP assay respectively. However, presently antigen detection tests are not available for CHIKV infection (neither in ELISA nor immunochromatographic test format) and IgM antibody detection by **MAC-ELISA** is the most commonly employed test for diagnosis of acute infection (Table 10.10).

The details of laboratory diagnosis of arbovirus infection are described in dengue virus chapter.

TREATMENT

Acute infection is often treated with non-steroidal anti-inflammatory drug (NSAID) along with bedrest and fluid therapy.

Chloroquine:
- Chloroquine has been tried in chronic arthritis cases for its anti-inflammatory activity.
- Treatment with chloroquine has been shown to provide relief in chronic arthritis cases.
- However, several clinical trials have reported no beneficial role of chloroquine in acute CHIKV infection.

Steroid: The combination of low dose systemic steroids with NSAID has been reported to improve the pain and quality of life in patients with acute chikungunya arthritis as compared to NSAID alone.

Ribavirin:
- Ribavirin, a broad-spectrum antiviral, acts by inhibiting viral polymerase, IMP dehydrogenase. Moderately beneficial effect of ribavirin has been shown to reduce the arthralgia and swelling in patients with chronic chikungunya arthralgia.
- The antiviral effect of ribavirin has been shown against CHIKV in cell culture.
- The combination of ribavirin and interferon-α has shown synergistic antiviral effect suggesting their possible beneficial role in treatment of CHIKV infection.

siRNA and shRNA molecules have shown reduction in virus titer in various cell lines but still needs extensive studies in *in vivo* models before clinical trials.

VACCINE

The potential of explosive resurgence and ongoing CHIKV activity, lack of effective antiviral agents emphasize the need of vaccine to prevent CHIKV infection.

The possibility of developing an effective vaccine for chikungunya virus is evident from:
- Lifelong protective immunity induced by CHIKV infection.
- Relatively less antigenic variation.

Several vaccine strategies have been employed for the development of CHIK virus vaccine. However, no vaccine has so far been licensed for use.

Vaccine strategies are:
- Inactivated vaccine
- Live attenuated vaccine
- Virus-like particle
- Chimeric vaccine
- DNA vaccine.

Inactivated vaccine: Whole virus formalin inactivated vaccine prepared in various host system like embryonated egg, suckling mouse and cell culture was the first chikungunya vaccine to have developed during the period of 1960–1970. The inactivated vaccine prepared from mice brain and African green monkey kidney cell induced neutralizing antibody against CHIKV with no observed adverse effect in human volunteers. No further studies were carried out using inactivated vaccine for a long period. A recent study in 2009 using Vero cell derived formalin inactivated vaccine has shown good immunogenicity potential to neutralize the virus.

Live attenuated vaccine: A live attenuated CHIKV vaccine candidate (strain181/clone25) was developed by the US Army Reed Institute of Infectious Disease (USAMRIID) by serial passage of the parent strain 15561 in MRC 5 cells. The vaccine has shown to be safe and immunogenic but the phase 2 trials have reported transient arthralgia in some of the vaccines. A recent study has reported the possibilities of genetic instability because of attenuation only at two points.

Virus-like particle: Vaccine candidate using VRC-CHKVLP059-00-VP has been developed. The vaccine has been shown to be immunogenic in non-human primate. In a phase I clinical trial in healthy adults aged 18–50 years, the vaccine has been reported to be safe, immunogenic and well tolerated. The persistence of neutralizing antibodies for near five months in all participants also appears to be promising towards providing long-term protection against the virus.

Chimeric vaccine: Several chimeric vaccines have been developed containing non-structural protein of Venezuela equine encephalitis virus (VEEV), Eastern equine encephalitis virus (EEEV), Sin Nombre virus (SNV), vesicular stomatitis virus (VSV) and encephalomyocarditis virus (EMCV). Vaccines with VEEV, EEEV, SINV had the capability of infecting mosquito and thereby the possibility of transmission was the cause of concern.

To overcome this problem, chimeric vaccine was constructed containing internal ribosome entry sequence (IRES) of EMCV between CHIKV non-structural and structural genes. This vaccine was non-infectious for mosquito as it is unable to translate the IRES protein. However, the vaccine was found to be poorly immunogenic.

Vaccines containing EMCV IRES into the subgenomic promoter of CHIKV and vsvΔG-

CHIKV chimeric vaccine have been reported to be highly immunogenic along with reduced transmission capability in mosquitoes.

DNA vaccine: DNA vaccines using the capsid expressing DNA construct and envelop proteins have been developed. The latter vaccine has reported to produce neutralizing antibodies.

Bibliography

1. Chandrakant Lahariya SK Pradhan. Emergence of chikungunya virus in Indian subcontinent after 32 years: A review. J Vect Borne Dis 43, December 2006, pp. 151–60.

2. Chang LJ, Dowd KA, Mendoza FH, et al. VRC 311 Study Team. Safety and tolerability of chikungunya virus-like particle vaccine in healthy adults: a phase 1 dose-escalation trial. Lancet 2014 Dec 6; 384(9959):2046–52.

3. Chhabra M, Mittal V, Bhattacharya D, Rana U, Lal S. Chikungunya fever: A re-emerging viral infection. Indian J Med Microbiol 2008 Jan-Mar;26(1):5–12.

4. Felicity J Burt, Micheal S Rolph, Nestor E Rulli, Suresh Mahalingam, Mark T Heise. Chikungunya: A re-emerging virus. Lancet 2012;379:662–71.

5. Francesca Cavrini, Paolo Gaibani, Anna Maria Pierro, Giada Rossini, Maria Paola Landini and Vittorio Sambri. Chikungunya: An emerging and spreading arthropod-borne viral disease. J Infect Dev Ctries 2009; 3(10):744–52.

6. Mishra B, Ratho RK. Chikungunya re-emergence: Possible mechanisms. Lancet 2006;368(9539):918.

7. Schwartz O Albert ML. Biology and pathogenesis of chikungunya virus. Nature Reviews Microbiology 2010.

8. Staples JE, Breiman RF, Powers AM. Chikungunya fever: An epidemiological review of a re-emerging infectious disease. Clin Infect Dis 2009 Sep 15;49(6):942–48.

9. Tiwari M, Parida M, Santhosh SR, Khan M, Dash PK, Rao PV. Assessment of immunogenic potential of Vero adapted formalin inactivated vaccine derived from novel ECSA genotype of Chikungunya virus. Vaccine 2009 Apr 21; 27(18):2513–22.

10. Weaver SC, Osorio JE, Livengood JA, et al. Chikungunya virus and prospects for a vaccine. Expert Rev Vaccines 2012; 11(9): 1087–1101.

E. CRIMEAN-CONGO HEMORRHAGIC FEVER VIRUS

Crimean-Congo hemorrhagic fever (CCHF) is a tick-borne zoonotic viral disease. It is caused by the virus named Crimean-Congo hemorrhagic fever virus. The disease is mostly known for its highly fatal severe hemorrhagic manifestations.

The disease was first described in military personnel of Crimea, Soviet Union during the World War II and was named Crimea hemorrhagic fever. The viral agent was isolated from patients' blood and tissues. Later in 1969, it was identified as the same agent that is responsible for the Congo fever in Africa. Finally both the names were linked and the disease was termed Crimean-Congo hemorrhagic fever.

VIRUS

CCHF virus is a member of the genus Nairovirus and family Bunyaviridae which includes five genera and more than 350 species. The five genera are: Nairovirus, Hantavirus, Orthobunyavirus, Phlebovirus and Tospovirus. Within the genus Nairovirus, there are seven species of which CCHF virus is the most important human pathogen.

CCHF virus is a spherical enveloped virus containing single stranded, negative sense segmented RNA genome. The diameter of the virus is around 100 nm. Envelop contains 8–10 nm length glycoprotein surface spikes. The RNA genome has three segments—small (S), medium (M) and large (L), which encode for nucleoprotein, glycoprotein (Gn and Gc) and RNA dependent RNA polymerase (RdRp), respectively. Recently it has been found that the S segment also encodes for another protein called non-structural S (NS-S) protein in the opposite orientation to the nucleoprotein, i.e. positive sense which indicates that CCHF virus might be considered as an ambisense virus.

Based on the nucleotide sequence of S gene segment, CCHF virus has been divided into seven distinct clades, each of which is distributed in one genetic group (Table 10.11).

Table 10.11: Geographic distribution of clades and genogroups of CCHF virus

Clade	Genogroup	Distribution
I.	Africa 1	Senegal
II.	Africa 2	Uganda and South Africa
III.	Africa 3	South and West Africa
IV.	Asia 1	Pakistan, Middle East, Iran, Afghanistan, and India
V.	Europe 1	Russia, Turkey and Balkan region (Bulgaria, Kosovo)
VI.	Europe 2	Greek AP92 (single isolate)
VII.	Asia 2	Pakistan, China, Uzbekistan and Kazakhstan

EPIDEMIOLOGY

Geographic Distribution

CCHF virus has the widest geographic range among the **tick-borne** viruses of human importance and is the second most widespread among the medically important arboviruses. It is endemic in more than 50 countries of Africa, the Middle East, Eastern and Southern Europe and Asia. Occurrence of the virus is mostly correlated with the distribution of its principal vector, Hyalomma tick. Since the discovery of virus, more than 140 outbreaks have occurred affecting >5000 cases from all over the world. Over the past decade, the number of sporadic cases and focal outbreaks has been increasingly reported.

In India before the first human case reported in 2011, serological evidence was available indicating the presence of the virus in the country. Anti-CCHF antibody had been shown in the serum samples of human from Kerala and Pondicherry and in domestic animals from all over the country.

In 2011 January, first human cases were reported from Ahmedabad district of Gujarat during a nosocomial outbreak. A total of four deaths were reported. Subsequent to this, CCHF-specific IgG antibody was detected in the serum of domestic animals and isolation of virus and viral RNA detection were demonstrated in Hyalomma ticks from the village of the index case.

During 2012–2015, several sporadic cases and outbreaks have been reported from various districts of **Gujarat** (Ahmedabad, Amreli, Patan, Surendranagar, Kutch and Aravalli) and its neighboring state **Rajasthan** (Sirohi, Jodhpur, Jaisalmer).

Reservoir and Vector

CCHF virus circulates in nature between tick and vertebrate in an enzootic cycle. **Ixodid tick** acts as both the vector and reservoir, whereas **small vertebrates act** as reservoir only.

The virus has been found in more than 30 species of seven genera of Ixodidae (hard tick) family. *Hyalomma marginatum* and *Hyalomma anatolicum* species are considered as the important vector. Ixodid ticks from other genera Rhipicephalus, Boophilus, Dermacentor and Ixodes may also transmit the virus locally.

Several species of small vertebrates can transmit the virus to ticks during the viremic phase. Immature ticks (nymph) usually infest the small animals and acquire the infection from them. Animals such as hares and hedgehogs are the common source of infection for the ticks. Once the ticks are infected, they remain infected throughout their life by passing the virus from one developmental stage to other (transstadial transmission) and to the offspring by transovarial transmission. The mature ticks can transmit the virus to large vertebrates such as livestock animals (Fig. 10.11).

Modes of Transmission

Man gets the infection either through the skin, mucosa or by ingestion.

Transmission from tick to man occurs either through infected tick bite or crushing of infected tick on the skin.

Transmission from infected animals to man can occur through contact with blood or tissues.

Man-to-man transmission occurs by contact of skin or mucosa with infected blood, tissues or body fluids. As most of the CCHF cases get

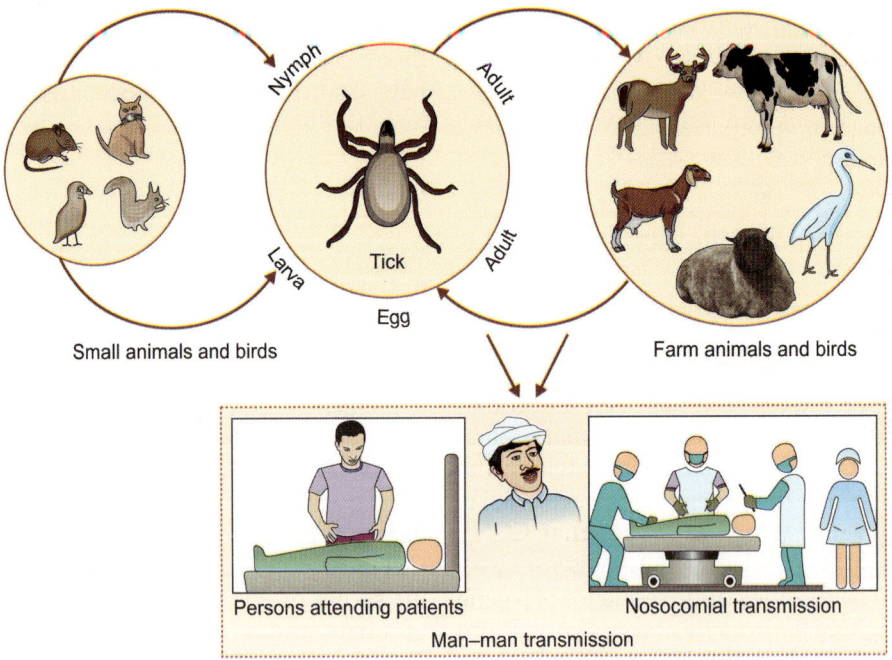

Fig. 10.11: Transmission cycle of CCHF virus

admitted to the hospital, nosocomial spread of infection is commonly reported. Possible aerosol transmission has been suspected in a Russian outbreak. Mother to fetus transmission has been reported too.

CLINICAL FEATURES

The average incubation period is 3–6 days. The length of incubation period may depend on the route of acquisition, viral load, source of infection. The minimum viral load required for transmission is 1–10 organisms.

The clinical course of disease passes through three phases: Pre-hemorrhagic phase, hemorrhagic phase and convalescent phase.

Pre-hemorrhagic phase: Disease starts with sudden onset of fever, headache, nausea, vomiting, abdominal pain, non-bloody diarrhea, myalgia, dizziness and photophobia. There may be relative bradycardia, tachycardia, hypotension and petechial rash. This phase generally lasts for 4–5 days.

Hemorrhagic phase: This usually shows a rapid progressive course with hemorrhagic manifestations, such as petechiae, hematemesis, hemoptysis, conjunctival hemorrahage. In severe cases, death can occur due to alveolar hemorrhage, multi-organ failure, disseminated intravascular coagulation, and circulatory shock. Death occurs in around 30–40% cases.

Convalescent phase: In general, this phase starts 10–20 days after the onset of symptoms. Patients at this stage usually have generalized weakness, weak pulse, tachycardia. Loss of hearing, poor vision, memory loss and temporary loss of hair have also been reported in some patients.

PATHOGENESIS

The exact pathogenesis of CCHF is not known. The pathological manifestations are multifactorial.

Release of high level of proinflammatory cytokines or cytokine storm by the infected cells is the key factor in disease pathogenesis.

Endothelial damage occurs either directly by replication of virus or through production of proinflammatory cytokines. This leads to dysregulated platelet aggregation leading to activation of coagulation cascade which causes intravascular coagulation leading to consumption of clotting factors and hemorrhage.

Vascular dysfunction leads to hemorrhage through leakage of plasma and erythrocyte into tissues. This could be due to the direct effect of the virus or due to the effect of endothelial damage.

Upregulation of vascular activation markers ICAM1 and VCAM1 by the virus has been demonstrated. Increased levels of cytokines have been found in severe fatal cases as compared to milder cases.

LAB DIAGNOSIS

Isolation of virus: This can be attempted in various cell lines, such as Vero, LLCMK2, SW13.4 or BHK21 or in newborn mice. Samples like blood, plasma or affected tissues can be subjected to virus isolation. Identification of virus can be done by immunofluorescence method using specific antibody or by RT-PCR. This technique has low sensitivity, applicable only within first five days of illness and requires BSL-4 facility.

Molecular tests: Presently CCHF is mostly confirmed by the molecular tests, such as reverse transcriptase polymerase chain reaction (RT-PCR) or real-time PCR from blood, plasma or tissue samples of the patients. Most of the PCR systems have been developed according to the locally prevalent strains. Quantitative real-time PCR system has been developed. Various other assay, such as loop mediated amplification assay (LAMP), low density macroarray and high density microarray have also been developed.

Antigen detection: ELISA for detection of viral antigen (antigen capture ELISA) is available and can detect during the early part of the illness. Viral antigen can be demonstrated in tissue by immunohistochemistry using specific antibody.

Serology: ELISA and immunofluorescence (IFA) tests are available for detection of IgM and IgG antibodies.

Detection of IgM antibody by capture ELISA from a single blood sample is commonly employed for confirmation of diagnosis. Fourfold rise in IgG titer can also be demonstrated. Recombinant nucleoprotein (rNP) has been used for development of IgM and IgG ELISA.

IgM and IgG usually appear towards 7–9 days of illness. IgM antibody persists for about four months, whereas IgG persists for a few years.

TREATMENT

CCHF patients are mainly managed with supportive therapy in the form of blood, platelet and plasma replacement. In severe cases, use of antiviral ribavirin has been recommended by WHO with a dose of 30 mg/kg followed by 15 mg/kg for four days and 7.5 mg/kg for six days. The beneficial role of ribavirin, however, is not conclusive. Role of immunoglobulin prepared from the convalescent serum of CCHF patients has been tried in the past but their role is doubtful.

PREVENTION

Prophylaxis

Postexposure prophylaxis is considered in case of direct mucosal exposure to blood or body fluids of confirmed or probable CCHF case. All such contacts should be kept on observation for two weeks and fever is monitored twice daily. In case temperature of >38.5°C is developed along with other constitutional symptoms, patient is admitted and it is recommended to start antiviral ribavirin. Though the role of ribavirin is not foolproof, based on its *in vitro* susceptibility activity, Center for Disease Control and Prevention (CDC) USA, recommends the use of ribavirin both for prophylaxis and for treatment of CCHF.

Individuals working in high-risk areas such as slaughter house or animal husbandry in endemic locality and health care personnel should follow proper infection control measures with barrier precautions.

All samples should be handled in lab with BSL-4 facility.

Isolation of patient must be done in case of suspected or confirmed CCHF case.

Individuals living in endemic area should take measures to prevent tick bite.

Vaccine

Mouse brain inactivated vaccine was developed by Soviet Union way back in 1970 and was licensed in Bulgaria in 1974. Vaccine was administered in 3 doses to military, medical and agricultural workers working in endemic localities. Bulgarian Ministry of Health had reported a fourfold decrease in number of cases over a period of 22 years after the administration of vaccine. However, no vaccine is available in any other country.

Bibliography

1. Akinci E, Bodur H, Leblebicioglu H. Pathogenesis of Crimean-Congo hemorrhagic fever. Vector Borne Zoonotic Dis 2013;13(7):429–37.

2. Appannanavar SB, Mishra B. An update on Crimean-Congo hemorrhagic fever. J Glob Infect Dis 2011;3(3):285–92.

3. Burt FJ, Swanepoel R. Molecular epidemiology of African and Asian Crimean-Congo haemorrhagic fever isolates. Epidemiol Infect 2005; 133(4):659–66.

4. Ergönül O. Crimean-Congo haemorrhagic fever. Lancet Infect Dis 2006;6:203–14.

5. Fajs L, Jakupi X, Ahmeti S, Humolli I, Dedushaj I, Avši?-•upanc T. Molecular epidemiology of Crimean-Congo hemorrhagic fever virus in Kosovo. PLoS Negl Trop Dis 2014 Jan 9;8(1): e2647.

6. http://www.searo.who.int/entity/emerging_diseases/links/CCHF_Fact_Sheet_SEARO.pdf?ua=1

7. Khurshid A, Hassan M, Alam MM, Aamir UB, Rehman L, Sharif S, et al. CCHF virus variants

in Pakistan and Afghanistan: Emerging diversity and epidemiology. J Clin Virol 2015;67:25–30.

8. Lahariya C, Goel MK, Kumar A, Puri M, Sodhi A. Emergence of viral hemorrhagic fevers: Is recent outbreak of Crimean-Congo Hemorrhagic Fever in India an indication? J Postgrad Med 2012;58(1):39–46.

9. Mardani M, Keshtkar-Jahromi M. Crimean-Congo hemorrhagic fever. Arch Iran Med 2007; 10(2):204–14.

10. Mourya DT, Yadav PD, Shete AM, Gurav YK, Raut CG, et al. Detection, isolation and confirmation of Crimean-Congo hemorrhagic fever virus in human, ticks and animals in Ahmedabad, India, 2010–2011. PLoS Negl Trop Dis 2012;6(5): e1653.

11. Papa A, Mirazimi A, Köksal I, Estrada-Pena A, Feldmann H. Recent advances in research on Crimean-Congo hemorrhagic fever. J Clin Virol 2015;64:137–43.

12. Sherifi K, Cadar D, Muji S, et al. Crimean-Congo hemorrhagic fever virus clades V and VI (Europe 1 and 2) in ticks in Kosovo, 2012. PLoS Negl Trop Dis 2014; Sep 25;8(9):e3168.

13. Spengler JR, Bergeron É, Rollin PE. Seroepide-miological studies of Crimean-Congo hemorrhagic fever virus in domestic and wild animals. PLoS Negl Trop Dis 2016 Jan 7;10(1):e0004210.

14. Swanepoel R, Shepherd AJ, Leman PA, Shepherd SP, McGillivray GM, Erasmus MJ, et al. Epidemiologic and clinical features of Crimean-Congo hemorrhagic fever in southern Africa. Am J Trop Med Hyg 1987;36:120–32.

15. Zehender G, Ebranati E, Shkjezi R, et al. Bayesian phylogeography of Crimean-Congo hemorrhagic fever virus in Europe. PLoS One 2013 Nov 4;8(11):e79663.

16. Zivcec M, Scholte FE, Spiropoulou CF, Spengler JR, Bergeron É. Molecular insights into Crimean-Congo hemorrhagic fever virus. Viruses Apr 21;8(4):106.

Respiratory Viruses

Acute respiratory tract infection is one of the leading causes of morbidity and mortality worldwide. Majority of the infections are due to viral agents. The important viral agents of respiratory infection are: SARS Coronavirus-2, Influenza A and B, parainfluenza type 1, 2 and 3, respiratory syncytial virus (RSV), and human metapneumovirus (hMPV). Other viruses, such as rhinovirus, enterovirus, adenovirus, bocavirus, human coronaviruses (NL63 and HKU1) also cause respiratory tract infection. Recently, parvovirus type 4, 5 and mimivirus have also been associated with respiratory infection, but their exact clinical importance is not known. Coronaviruses such as severe acute respiratory syndrome corona-virus 2 (SARS-CoV-2) , SARS virus, middle east respiratory syndrome (MERS) virus can cause severe respiratory syndrome and have been described separately. The present chapter will describe influenza virus, parainfluenza virus, respiratory syncytial virus, human metapneu-movirus and human bocavirus.

A. INFLUENZA VIRUS

Influenza (Inf) is a major cause of morbidity and mortality worldwide. It is caused by influenza virus. The uniqueness of this virus is its ability of emergence of new strains and thereby the potential of causing pandemic.

Influenza virus belongs to the family of Orthomyxoviridae which has seven genera or types; of which three are influenza A, influenza B and influenza C and other three are Thogotovirus, Isavirus and Quaranjavirus. **Influenza D has been proposed as a novel genus** in the family Orthomyxoviridae.

Influenza A: Influenza A viruses are spherical or pleomorphic virus of around 100–120 nm diameter. It contains single-stranded, segmented RNA genome which is surrounded by matrix protein and envelop (Fig. 11.1).

Envelop: Virus has lipid envelop that is derived from the host cell. Three surface projections are present in the envelop: **Hemagglutinin (HA) and neuraminidase (NA)** are two glycoproteins present with an average ratio of 4 to 1 between HA and NA and M2 protein acts as an ion channel. The length of the glycoprotein surface spikes is 10–14 nm.

Matrix protein: This layer (M1) is present under the envelop.

Viral genome consists of single stranded negative sense RNA of **eight segments** encoding 10 proteins: Hemagglutinin (HA), neuraminidase (NA), matrix protein 2 (M2) and matrix protein 1 (M1), non-structural proteins NS1 and NS2, and three polymerases; polymerase basic 1 (PB1), polymerase basic 2 (PB2) and polymerase acidic (PA).

Influenza B: The structure is similar to Influenza A. It has four surface proteins: HA, NA, NB and BM2.

Influenza C: It has hemagglutinin-esterase-fusion (HEF) and CM2 proteins on the surface. Unlike influenza A and B, it has **seven RNA segments**.

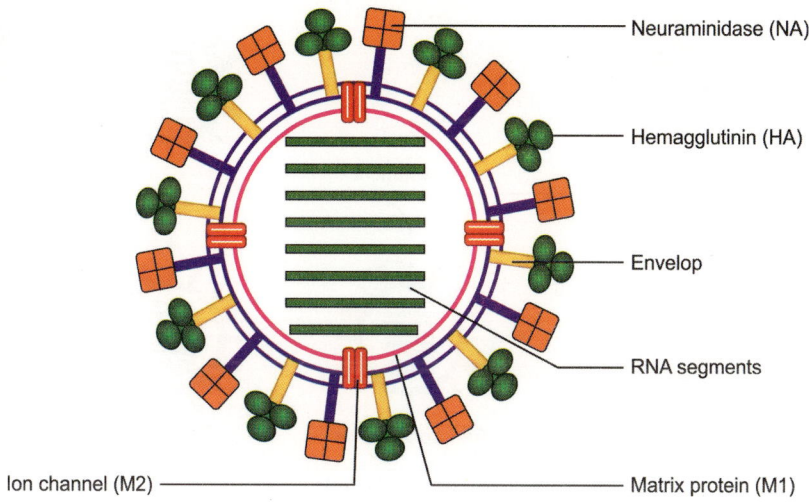

Fig. 11.1: Influenza virus (schematic)

Influenza A virus has been subdivided further based on HA and NA surface proteins. According to current information by Center for Disease Control and Prevention (CDC, September 2016), there are 18 HA types (H1–H18) and 11 NA types (N1–N11). Between two HA subtypes, the amino acid sequence difference is ≥30%.

Viral Proteins

Hemagglutinin (HA)

Structure: HA is a type I glycoprotein. It is present as a rod-shaped structure with a globular head, in the form of trimmer. Each monomer of the trimmer contains the receptor binding site. The carboxyl terminal end of the protein gets anchored into the envelop and hydrophilic N terminal projects away from the virus surface.

Function: It helps in attachment of virus with the host cell receptor sialyloligosaccharide. Receptor binding site is present within the globular head. On low pH, it changes its morphology and becomes susceptible to protease. This helps in fusion process and uncoating.

It is also responsible for the hemagglutinating property of the virus, acts as the major antigenic determinant and produces neutralizing antibody.

Neuraminidase (NA)

It is a type II glycoprotein. Its top part is mushroom shaped, present in the form of a tetramer. It has sialidase activity and prevents aggregation of virion. Antibody against NA protects from infection.

M2 Protein

It is present on the surface of the virion and acts as H2 channel. It gets activated in the endosome by low pH and acidify the inside of the virus which helps in uncoating of virion.

Influenza Virus Genetics

The unique property of influenza virus is its potential for emergence of new strains of epidemic and pandemic potential. This occurs due to antigenic variation in the new strain. Therefore, it is important to understand the genetic changes that occur in influenza virus.

Reassortment: The segmented genome of influenza virus makes this phenomenon possible. The process of reassortment can occur when infection with two different influenza virus strains occurs inside one cell. Ideally each virus should produce its own progeny with identical genetic content. But, during the process of assembly, there is a

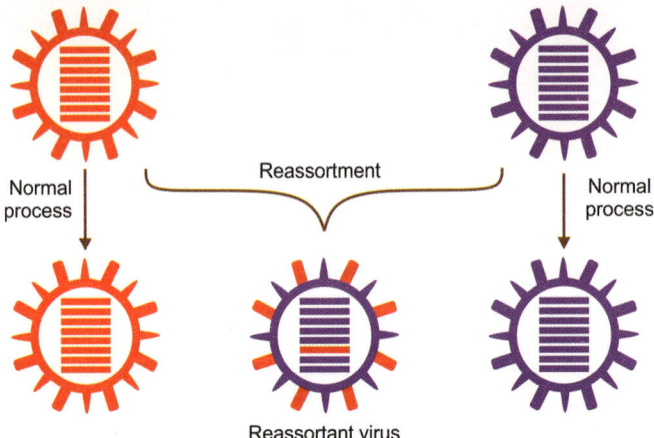

Fig. 11.2: Reassortment (schematic)

possibility that the genome segments of one virus may get assembled inside the other virus. This results in production of a new virus with genetic component that is different from the parent virus (Fig. 11.2). The reassortant virus is important because of its new genetic makeup. This is affected most when the genome segment encoding the surface proteins such as hemagglutinin or neuraminidase is different than the parent strain, as antibodies against these proteins are responsible for host immunity.

Reassortment occurs within influenza virus types (genus) but not between different genera or types. This means it occurs within the members of influenza A or B or C, but not between influenza A, B and C. Most commonly it is seen in influenza A.

Example: Influenza A (H1N1)pdm09 virus which was the latest pandemic influenza virus is a reassortant virus of 4 different strains consisted of North American avian strain, Eurasian swine strain, human H3N2 and classic swine strains. The reassortant virus of first three was already circulating amongst the North American swine which form the new reassortant virus with classic swine strain.

Recombination: In this, the genetic material of each segment comes from multiple origins.

Role of Antigenic Variation in Influenza Virus Epidemiology

Antigenic drift: This is caused by accumulation of point mutations leading to gradual change of antigenicity. The immunological pressure to surface proteins HA and NA is mainly responsible for point mutation. The resulting antigenic change in drift virus is minor, but sufficient enough to overcome the immune response produced by the vaccine strain. The appearance of drift strain thus makes the population susceptible. This in turn requires the change in vaccine strain in every few years.

Antigenic drift is more commonly observed in human influenza virus than avian strains, possibly because of constant immunological pressure from the humans than birds.

Antigenic shift: Sudden emergence of influenza virus with a major antigenic change than the previously circulating strains occurs due to antigenic shift. This can occur by various processes:

- Genetic reassortant between two different influenza viruses leading to emergence of a novel virus (1957, 1968 and 2009 pandemics).
- Antigenic change in avian/non-human strain that makes it capable of infecting humans with the ability to spread (For example, 1918 pandemic by avian influenza-like strain).

- Re-emergence of previously circulating virus in an immunologically naïve population (e.g. 1977 pandemic).

Maintenance of Influenza Virus in Nature

Influenza A: Of the 18 HA and 11 NA subtypes, aquatic birds are the natural reservoir of all types of influenza A viruses except H17N10 and H18N11 which have been detected only in bats. Aquatic birds do not manifest the disease. Ducks, shore birds and gulls are the main reservoir. Replication of virus occurs mainly in the intestinal tract of birds and virus gets excreted in the fecal matter. Transmission of virus amongst the aquatic birds occurs by ingestion of contaminated water.

Besides the water birds, several wild and domestic bird species also can get infected with influenza virus. These birds are chicken, turkey, ducks, geese, quail, sea birds, shore birds, gulls. Infection may be asymptomatic in some avian species while symptomatic in some others species. Water birds are usually resistant and do not manifest with symptoms.

Influenza A virus can also infect human, swine, horses, dogs and several other mammals.

Influenza A virus subtypes that have established themselves in various hosts (Table 11.1).

Influenza B virus infects mostly human host, also has been isolated in seals.

Influenza C virus infects human, swine and dogs.

Influenza D virus can infect swine, cattle, sheep and goat. Cattle have been shown as the natural reservoir.

Determinants of Host Range Specificity

All known HA and NA types of influenza A virus infect the avian species, only a few of them infect other animals including man. This indicates that influenza A viruses broadly maintain a host range restriction.

Hemagglutinin (HA) glycoprotein is one of the major determinants of the host range restriction because of its capability of recognizing the host cell receptor, sialyloligosaccharide.

Table 11.1: Influenza A virus subtypes in various hosts

Host	Influenza A subtypes
Water bird	HA: H1 to 17 NA: N1 to N9
Human	HA: H1, H2, H3 NA: N1, N2
Pig	HA: H1, H3 NA: N1, N2
Horse	H3N8, H7N7
Dog	H3N8

Human influenza virus binds preferentially to sialyloligosaccharide terminated by N-acetyl sialic acid linked to galactose by **α-2, 6 linkage**, the receptors that are predominantly present in humans. Until now, only a few HA types: H1, H2 and H3 have been found to infect human host.

Influenza virus infecting birds binds preferentially to the sialyloligosaccharide receptor having **α-2, 3 linkage** between sialic acid and galactose. These strains are called avian influenza virus. All HA types can bind to receptor having α-2, 3 linkage between sialic acid and galactose. So, avian species can be infected by all HA types of influenza A virus.

Influenza virus infecting pigs binds to the sialyloligosaccharide receptor having α-2, 3 linkage or α-2, 6 linkage between sialic acid and galactose. As pigs possess both types of receptors, they can be infected by both avian and human influenza viruses. Therefore, pigs are considered as **"mixing vessel"** and provide the scope for reassortant and emergence of novel strains.

Neuraminidase (NA) also shows host restriction by its preference for particular type of sialic acid linkage in specific host. Avian viruses cleave sialic acid of α-2, 3 linkage and human viruses cleave sialic acid of α-2, 6 linkage. Certain NA type of avian virus when infects human host, adapts to cleave sialic acid of α-2, 6 linkage in addition to α-2, 3 linkage.

PANDEMICS IN HUMANS

Spanish influenza 1918: Spanish influenza or influenza A pandemic of 1918 is considered as the largest pandemic of the previous century. It killed 25 million people in around 25 weeks, which is equal to the number of deaths caused by AIDS in 25 years.

The name "Spanish influenza" was given as the pandemic came to limelight when it occurred in large scale in Spain. The pandemic is believed to start in a military troop in Kansas City, USA, where within 3 weeks almost 1100 people got affected.

The virus strain responsible for the pandemic was long thought to be an avian strain. The genetic constituent of 1918 pandemic strain became clear only in 2007 when Taubenberger et al. published their research on autopsy material recovered from Armed Force Institute of Pathology. Tissue samples of 13 cases whose case record and histopathology indicated as possible influenza-related disease were subjected to reverse transcriptase polymerase chain reaction (RTPCR).

- Tissues were found to be positive for influenza A RNA fragment of <140 base pair length, H1N1 subtype and lack of a cleavage site mutation.
- The virus was not a reassortant virus.
- 1918 pandemic strain was finally confirmed to be an avian influenza-like virus, where all eight segments were from a new source as they were different from contemporary avian virus.

Spanish influenza pandemic had occurred in three waves during 1918–19. The first wave started in March, 1918 when the disease was more contagious but mortality was less. During second and third waves, pandemic became a cause for concern due to high mortality all over the world.

Most of the deaths were due to pneumonia and respiratory failure. The causative virus strain has been shown to be extremely virulent in mice and also evident from histopathology findings of massive pulmonary hemorrhage and edema in patients.

Asian influenza (H2N2): Pandemic started in 1957 from Southern China in Southeast Asia. The pandemic strain was a reassortant virus containing three genes (HA of H2 subtype, NA of N2 subtype, PB) from avian virus and remaining genes from circulating human H1N1 virus.

The virulence was less as compared to Spanish influenza virus with a mortality of >1 million worldwide.

Hong Kong influenza (H3N2): Hong Kong pandemic started in 1968 in Southern China. Pandemic virus was first identified in Hong Kong.

Pandemic strain was a reassortant virus containing two genes from avian virus (HA of H3 subtype, PB1) and remaining six genes from circulating human H2N2 virus.

Virulence of pandemic virus was less as compared to Spanish influenza virus which is probably due to pre-existing antibody to N2 protein.

Russian influenza (H1N1): Russian influenza pandemic started in 1977 in China and United Sates of Soviet Union (USSR).

Pandemic strain was influenza A H1N1 strain that is closely similar to the H1N1 strain circulating before 1950. Initially it was thought to be due to the reemergence of the previously circulating H1N1 strain, but presently it is believed that the virus appeared due to accidental release from the laboratory.

Virulence of pandemic virus was less as compared to Spanish influenza virus. This is probably due to pre-existing antibody to N2 protein.

Influenza A (H1N1) pdm09: In late March 2009, the first case of a novel influenza A virus strain (initially named 2009 Inf A H1N1) was reported in a child in Mexico. The novel virus then spreads rapidly in USA and to more than 70 countries all over the globe affecting near 30,000 cases by **June 11, 2009**. World Health Organization (WHO) on June 11 declared the first influenza virus pandemic of the century. Pandemic was declared to be over on 10th August 2010 when the total deaths reported were 18449 worldwide. The pandemic strain was later renamed by WHO as influenza A (H1N1) pdm09.

Indian scenario: India reported its first case in May 2009 and by the end of the year 2010, more than 20,000 cases and 1763 deaths were reported.

2009 pandemic occurred in three waves; first wave occurred in September followed by second wave in December, and third wave during August 2010.

Resurgence of pandemic influenza cases occurred during the year 2012–13 and again during 2015. The number of cases were more than 5000 each in 2012 and 2013 with deaths more than 400 and 600, respectively, whereas number of cases crossed 30000 with more than 2000 deaths in 2015.

The Inf A (H1N1) pdm09 strain is a **quadruple reassortant** virus which emerged due to the **reassortment between Eurasian avian like swine virus with triple reassortant North American swine virus**. The later is a triple reassortant of classical swine, North American avian and human H3N2 strains, which was circulating amongst the North American swine since 2000.

The origin of different genes of Inf A (H1N1) pdm09 is shown in Fig. 11.3.

Severity of H1N109 pandemic virus: The pandemic strain does not contain any amino acids associated with high virulence.

- Severity of pdm09 virus has been found to be higher as compared to seasonal H1N1. The virus has the ability to bind sialic acid of α-2, 6 linkage as well as sialic acid of α-2, 3 linkage. In human trachea sialic acid of α-2, 6 linkage receptor is present, whereas in lungs sialic acid of α-2, 3 linkage is present. It also infects both type I and type II pneumocytes.

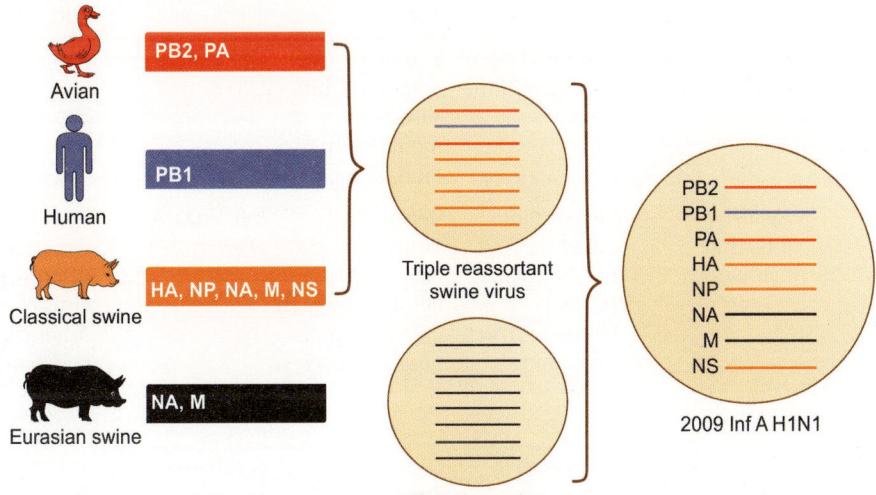

Fig. 11.3: Reassortant influenza A (H1N1) pdm09

- The Inf A (H1N1) pdm09 has been shown to attach both trachea as well as lungs in ferret in contrast to seasonal H1N1 which binds predominantly to trachea only.

The 2009 pandemic affected all age groups but death rate was maximum in young adults. The infection rate in elderly was observed to be less but death was more.

Clinical features of H1N109 pandemic virus:

- The symptoms caused by A (H1N1) pdm09 are similar to other influenza or upper respiratory tract infection. Fever, sore throat, headache were common symptoms.
- However, involvement of lungs leading to pneumonia was more common during pandemic period.
- Gastrointestinal symptoms are associated with a few cases.

In severe cases, respiratory failure and multiorgan failure can occur. The comparative feature of influenza pandemics is shown in Table 11.2.

Avian Influenza (H5N1) Epidemic in Human

In 1997, a total of 18 cases of H5N1 were reported in Hong Kong of which six died. The virus was found to be a pure avian strain (all eight RNA segments were of avian origin) of highly pathogenic avian influenza type (HPAI). The source of this highly virulent H5N1 was found to be reassortant of two different avian influenza viruses. Source of HA gene (H5) was from geese and other genes were from H9N2 and H6N1 virus prevalent in quail. The unique feature of this strain was deletion in the stalk region of NA which was responsible for adaptation of this strain to terrestrial poultry such as chicken.

It was found that the source of virus was infected poultry in Hong Kong. Around 1.5 million birds were culled in the affected poultry market which restricted further spread of infection in human.

In 2003, H5N1 virus reemerged in humans in People Republic of China. Subsequent cases were also reported from several other south eastern Asian countries, such as Vietnam,

Table 11.2: Comparative features of influenza pandemics

Features	Spanish influenza (H1N1)	Asian influenza (H2N2)	Hong Kong influenza (H3N2)	Russian influenza (H1N1)	Influenza A (H1N1) pdm09
Pandemic year	1918	1957	1968	1977	2009
Place of origin	Kansas City, USA	Southeast Asia: Southern China	Southeast Asia: Southern China. Virus was identified in Hong Kong	China and USSR	Mexico
Pandemic strain	Avian influenza-like virus	Reassortant virus: H2, N2, virus: H2, N2, PB Human H1N1: Remaining genes	Reassortant virus: Avian virus: H3, PB1 Human H2N2: Remaining six genes	Influenza A H1N1	Reassortant virus: Eurasian avian-like swine virus North American triple reassortant swine virus
Virulence	Virulent (so far considered to be most virulent)	Less as compared to Spanish influenza	Less as compared to Spanish influenza	Less as compared to Spanish influenza	Virulent but less as compared to Spanish influenza

Indonesia, Cambodia, Thailand, etc. amounting to 850 cases with 449 deaths till 9th May 2016.

After the bird culling in poultry markets in 1997, the virus was no more found in chicken, but found in geese. From 2000 onwards, series of reassortant virus emerged, but by 2003, genotype Z became the predominant one amongst the terrestrial poultry in southern China. The Z genotype of HPAI has been found to be pathogenic even for aquatic wild birds.

Transmission of H5N1 virus: The source of H5N1 virus for human was the infected live poultry markets. As multiple organs were involved in birds, direct handling of infected poultry, slaughtering, close contact acted as various modes of transmission. Respiratory route was the major mode of entry. Transmission through gastrointestinal tract and conjunctiva has also been reported.

Clinical features of H5N1 virus: The average incubation period is 2–4 days. All age group can be affected. Maximum numbers of cases reported were below 20 years possibly due to more exposure to poultry.

Fever, cough, and breathlessness are the common manifestations. Radiological finding of lungs shows bilateral pneumonia with diffuse interstitial infiltrate and segmental or lobar consolidation. The disease often progresses rapidly towards development of pneumonia and acute respiratory distress syndrome leading to respiratory failure and death. Multiorgan dysfunctions involving liver, kidney and heart are other severe complications responsible for death. Mortality rate more than 50% has been reported reflecting the high severity in humans.

Pathogenesis of H5N1 virus: Several viral and host factors contribute to the severity of disease due to H5N1 virus.

- *High viral load:* H5N1 virus has the capacity of high viral replication leading to high viral load. The lack of immunity in humans also permits high viral replication.
- *High level of cytokines:* High viral replication leading to hyperinduction of proinflamma-

tory cytokines such as tumor necrosis factor-α (TNF-α), CCL2, CL3, CCL4, CCL5, interferons- α and β, CXCL10, IL6, IL8. These high levels of cytokines are responsible for severe disease pathology in H5N1 viral infection.

- *Immune evasion:* Viral protein NS1 helps the virus to evade the antiviral activity of interferon.

EPIDEMIOLOGY

Incubation period in influenza A is around 1–3 days and slightly more (1–4 days) in influenza B.

Modes of Transmission

Person-to-person transmission is the major mode of transmission for human influenza. Infected persons are the main source of infection. In a few percentages of cases, infection can be asymptomatic or subclinical; they also can act as the source of transmission.

Transmission of avian virus from bird to human can also occur. In general, influenza viruses are host specific. Restriction of host range is maintained by receptor specificity. Occasionally there can be crossing of the species barrier and avian strains have infected humans and other animals.

Transmission of avian strain usually occurs due to close contact with infected birds through aerosol/droplet/fomites from contaminated bird excreta or infected tissues or blood. Infection in human in such cases is usually mild and self-limiting as the avian strain is not able to replicate efficiently in human host. However, in 1997, the transmission of H5N1 avian strain infection led to severe infection causing several deaths and later in 2003 again re-emerged creating a pandemic threat.

Circulating strains: Till the last pandemic in 2009, seasonal influenza A H1N1, H3N2 and influenza B were the three circulating strains worldwide. After emergence of Inf A (H1N1) pdm09, it replaced the seasonal influenza A

H1N1 and presently influenza A (H1N1) pdm09, H3N2 and influenza B are the three circulating strains.

Seasonality: Influenza virus generally shows a seasonal activity. The seasonality of virus has been observed to be different in temperate and tropical climate.

- In **temperate regions**, influenza virus shows a seasonal peak during winter months in both northern and southern hemispheres.
- In **tropical climate**, a year round influenza activity is observed.
- In **India:** In major parts of the country with tropical climate usually biannual peak is observed, one during monsoon and second peak in winter months. In extreme northern parts with temperate climate, peak of influenza activity is seen during winter.

Shedding of virus: In human host, influenza virus replicates in the respiratory tract. Because of preponderance of sialic acid of α-2, 6 linkage receptor in the upper respiratory tract, replication of virus is more seen in URT as compared to lower respiratory tract. However, influenza virus can also infiltrate into the lower respiratory tract and cause pneumonia.

Inf A (H1N1) pdm09 can bind to both sialic acid of α-2, 6 linkage receptor which is present predominantly in URT and sialic acid of α-2, 3 linkage receptor present in the lower respiratory tract of humans.

Replication of virus takes its peak after 48 hours then declines and continues for a week. This goes hand in hand with the appearance and persistence of symptoms.

Shedding of virus starts occurring from one day before the appearance of symptom and continues for a week thereafter. In children, and immunocompromised individuals, shedding may prolong for another 5–7 days.

CLINICAL FEATURES

Infection with influenza virus can be asymptomatic or subclinical. The clinical manifestations are usually restricted to symptoms of upper respiratory tract. Onset of illness is abrupt with, headache, cough, malaise and fever. Subsequently dry cough may be productive with expectoration. These symptoms are accompanied with sneezing, rhinorrhea, nasal obstruction, pharyngitis, conjunctival congestion.

Influenza H7N9 is more commonly associated with conjunctivitis. Gastrointestinal symptoms like vomiting, diarrhea are seen in influenza A (H1N1) pdm09 virus.

COMPLICATIONS

Acute otitis media, croup and pneumonia are common complications in children. Pneumonia in seasonal influenza strains is commonly seen in elderly or children. Individuals with underlying cardiac diseases are more prone to severe diseases. Influenza virus can cause primary viral pneumonia or there can be secondary bacterial pneumonia.

Primary viral pneumonia usually starts abruptly after onset of influenza-like illness. It progresses rapidly within a few hours, with increase in shortness of breath, tachycardia, high fever, hypotension and hypoxemia. Chest radiograph shows bilateral involvement with features of acute respiratory distress syndrome without consolidation. Bilateral lungs involvement is seen in chest radiograph. The disease does not respond to antibiotics and mortality is high. Figures 11.4 and 11.5a and b

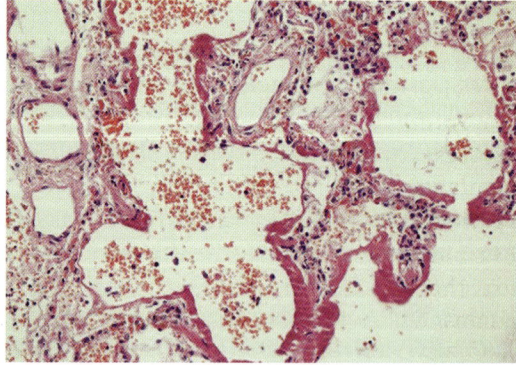

Fig. 11.4: Histopathology of lungs with H1N1 virus pneumonia showing hyaline membrane formation in alveolar duct and space. *Image Courtesy:* Dr Amanjit Bal, Professor, Histopathology, PGIMER, Chandigarh

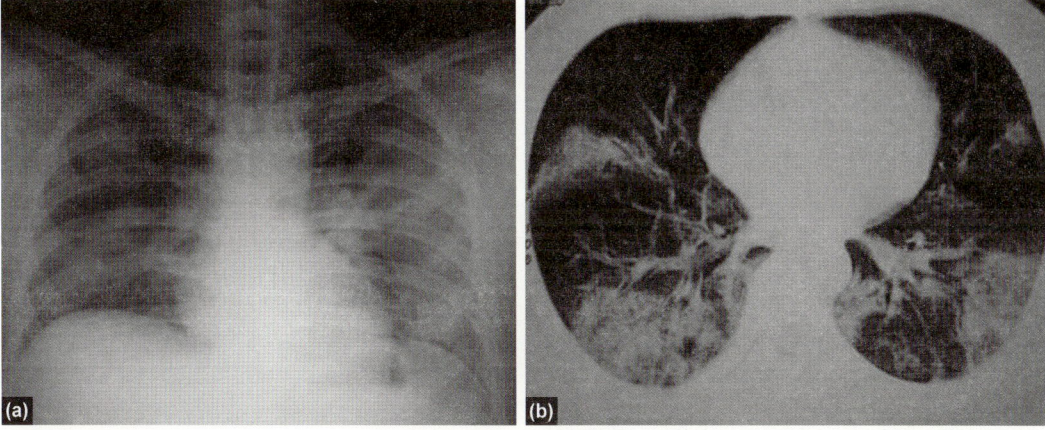

Fig. 11.5a and b: (a) Chest radiograph showing ill-defined bilateral lower and midzone air-space patchy opacities due to H1N1 influenza pneumonia; (b) Lung windows of CT thorax showing bilateral multi-focal patchy, rounded consolidation without any lobar distribution

show the histopathology and radiological picture of H1N1 pneumonia, respectively.

Secondary bacterial pneumonia commonly occurs in elderly or individuals with underlying cardiopulmonary disorders and typically occurs with a short phase of improvement from influenza-like illness. *Streptococcus pneumoniae*, *Staphylococcus aureus* and *Haemophilus influenzae* are the common bacteria involved. Symptoms start with fever, cough with sputum production and progress for consolidation.

Myositis, myocarditis, pericarditis and Guillain-Barré syndrome have also been reported as complications of influenza.

ANTIVIRALS

Antiviral agents against influenza are most commonly used amongst the respiratory viruses. Currently two groups of drugs are available for influenza virus. These are M2 blockers and neuraminidase inhibitors.

M2 blockers: Amantadine and rimantadine are the two chemically related drugs of adamantine group. Rimantadine is the α-methyl derivative of amantadine.

Mechanism of action: Both act by inhibiting the M2 ion channel. M2 protein allows the entry of proton inside the virion during the uncoating process. This induces an acidic environment in the intracellular compartment, which helps in fusion of virus envelop with the plasma membrane of the host cell and release of the virus nucleic acid into the cytoplasm. Blocking of M2 protein inhibits the process of virus replication.

Uses:
- Acts against influenza A virus.
- Used as prophylactic agent against seasonal influenza A virus.
- Decreases the severity and duration of illness when given within 48 hours of onset of clinical symptoms.

Resistance: Resistance to adamantine group of drugs occurs due to single point mutation in the M2 protein at positions 26, 27, 30, 31, 34, 38 amino acids of which mutation at position 31 is the commonest one. There occurs complete cross-resistance between two drugs.

- During the year 2003–2004, resistance to adamantines (amantadine, rimantadine) started appearing in influenza A H3N2 and by 2005, all most all H3N2 strains are considered as resistant. Because of this reason, adamantines are no more in use against seasonal influenza.
- Influenza A H1N1 2009 pdm strains are inherently resistant to adamantines.

- Strains of H5N1 that emerged in 1997 were sensitive to amantadines. 2003 H5N1 strains contained mutations at Sr31Asn and leu27Ile and showed high level resistance.

Neuraminidase inhibitors (NI): This group of drug first came to market in 1999 and presently there are four NI compounds available: **Oseltamivir (Tamiflu), zanamivir (Relenza), peramivir (Rapivab) and laninamivir (Inavir)**. These four drugs differ in their chemical structure, bioavailability and mode of administration. Amongst these only oseltamivir and zanamivir are the two approved drugs and oseltamivir is most widely used.

- **Oseltamivir** is approved for use in >1 year age patient and is available in oral form.
- **Zanamivir** is approved for >7 years age patient. It has no oral bioavailability and available in dry powder form to be administered locally by oral inhalation.
- **Peramivir** is approved for intravenous use only during pandemic. Presently it is also approved in Japan, China and Republic of Korea.
- **Laninamivir** is approved and used in Japan.

Mechanism of action: During the virus replication process, neuraminidase is synthesized and comes to the surface of the virus and destroys the receptor by removing the sialic acid. This helps the virus in getting out from the infected cell and thereby available to infect the adjacent cells and helps in spreading the infection. Neuraminidase inhibitor inhibits the spread of virus and reduces the severity of infection by inhibiting neuraminidase.

Uses: Used for treatment of seasonal influenza A H1N1, Inf A (H1N1) 09pdm, influenza A H3N2 and HPAI H5N1. The drug is better effective when given early phase of illness.

- Effective against both influenza A and B.
- To decrease the severity of infection.
- To shorten the duration of infection.

The dose of NIs is given in Table 11.3.

Resistance: Common mutation sites responsible for H3N2 virus resistance are: R 292K, E119V and for Inf A (H1N1) pdm09 virus are: H275Y, I223R and I223V.

Oseltamivir resistant strains are generally resistant to peramivir but sensitive to zanamivir. This is due to the requirement of conformational change in oseltamivir for binding to the active site of neuraminidase which is not required in zanamivir.

NI resistant strains usually isolated in patients during or after treatment/prophylaxis with NI. However, resistant strains have also been reported without previous exposure

	Table 11.3: Treatment and prophylaxis for influenza		
Agent, group	*Treatment*		*Chemoprophylaxis*
Oseltamivir			
Adults		75 mg capsule twice per day for 5 days	75 mg capsule once per day
Children ≥12 months	15 kg or less	30 mg BD	30 mg once per day
	16–23 kg	45 mg BD	45 mg once per day
	24–40 kg	60 mg BD	60 mg once per day
	>40 kg	75 mg BD	75 mg once per day
Zanamivir			
Adults		Two 5 mg inhalations (10 mg total) twice per day	Two 5 mg inhalations (10 mg total) once per day
Children		Two 5 mg inhalations (10 mg total) twice per day (age, 7 years or older)	Two 5 mg inhalations (10 mg total) once per day (age, 5 years or older)

to drug. More commonly resistance is observed in immunocompromised patients. This occurs because of delayed viral clearance in these patients which leads to prolong duration of treatment and selection of resistant strains.

The global status of **neuraminidase inhibitor** (NAI) susceptibility data by WHO is as follows:

- 2012–13: 0.5% of influenza virus is resistant (showed reduced inhibition or high reduced inhibition).
- 2013–14: 1.9% of influenza virus is resistant (showed reduced inhibition or high reduced inhibition).
- 2014–15: 0.6% of influenza virus is resistant (showed reduced inhibition or high reduced inhibition).

This indicates >99% of all influenza strains (considering all circulating strains in human) are susceptible to neuraminidase inhibitors, thus these drugs are still the **choice of treatment and prophylaxis for influenza virus infection**.

However, clusters of NI resistant strains of influenza A (H1N1) 09pdm have been reported in various countries including USA, Japan and China.

VACCINE

Seasonal influenza vaccine is usually given in multivalent form consisting of the circulating strains. Presently trivalent influenza vaccine is given which contains components of influenza A H1N1, H3N2 and influenza B.

Aim of influenza vaccination: To protect against infection and disease and to increase the herd immunity to prevent the spread of infection.

Recommendation: The recommendation for influenza vaccination in majority of the countries worldwide is for the elderly and high-risk individuals.

In the USA, national influenza vaccination in 2010 included 6 months to 18 years old.

Timing of vaccination: Annual vaccination scheme is recommended for influenza vaccination. This is based on the rationality that the antigenic variation in the circulating strains

becomes sufficient enough to decrease the efficacy of the vaccine in the subsequent year. Therefore, every year the vaccine strains are selected based on the circulating strains worldwide.

- Vaccine is recommended prior to the start of the influenza season.
- In western countries where influenza season occurs in the winter, it is given prior to winter season.

Types of vaccine: Available in inactivated and live attenuated forms.

Preparation of seed virus: Seed virus contains the HA and NA surface antigens from the vaccine virus which are selected from the circulating influenza virus strain for that particular year for vaccine preparation and remaining six genes from the donor virus.

Seed virus is prepared either by co-infecting these two types of virus strain in egg or by the process of reverse genetics.

Inactivated vaccine: Inactivated vaccines are available in three types of formulations; whole virus vaccine, split virus and subunit vaccine.

- Whole virus vaccines are prepared from the harvested allantoic fluid which has been infected with the seed virus. The harvested fluid is chemically inactivated using formalin or beta-propiolactone, then concentrated and purified.
- Split virus vaccine is prepared in the similar manner with an additional step of detergent treatment which removes the virus envelop exposing the virus nucleic acids.
- Subunit vaccine has an additional step of purification.

Dose: One dose of inactivated vaccine contains 15 µg of HA per strain.

- For individuals >9 years: One dose is given.
- For 6 months to 8 years: Two doses at 4 weeks interval.

Immunogenicity: Antibody production against HA antigen is the main mechanism of protection. Wild virus vaccine is more immunogenic than split or subunit vaccine particularly in children.

Table 11.4: Differences between inactivated and live attenuated influenza vaccine

	Inactivated vaccine	Live attenuated vaccine
Preparation	Chemical inactivation	Attenuation
Immunogenicity	Induces serum Ab against HA antigen	Induces mucosal IgA, serum IgG and cellular immunity
Route of administration	Intramuscular	Intranasal
Dose schedule	6 months to 8 years: Two doses at 4 weeks interval >9 years: One dose	2–9 years: 0.2 ml × 2 dose at 4 weeks interval 10 years: 0.2 ml
Adverse reaction	Local reactions: Pain, induration, redness Systemic reactions: Fever, malaise, headache.	Running nose, fever, sore throat

Adverse reaction:
- Local reactions: Pain, induration redness at the local injection site.
- Systemic reactions such as fever, malaise and headache.
- This is mainly due to the presence of residual egg protein.
- More commonly seen with wild virus vaccine than split or subunit vaccine.

Live attenuated vaccine (LAIV): This is prepared using temperature sensitive, cold adapted strain attenuated by serial passage.

It induces mucosal IgA , serum IgG as well as cellular immunity and considered to be more immunogenic than inactivated vaccine.

LAIV is recommended in above 2 years age group.

Contraindication:
- Pregnant and immunocompromised individuals.
- Individuals with asthma.

Table 11.4 describes the differences between inactivated and live attenuated influenza vaccines.

Inf A (H1N1) 09pdm virus vaccine: Both inactivated and live attenuated vaccines are available against Inf A (H1N1) 09pdm.

During the pandemic, monovalent vaccine of Inf A (H1N1) 09pdm virus was given. Presently, the virus has become a component

of seasonal vaccine. The differences between inactivated and live attenuated vaccines are principally same as that for seasonal vaccine.

H5N1 vaccine: Several vaccine candidates including high pathogenic HA, low pathogenic HA, DNA vaccines and virus-like particles (VLP) are under clinical trials.

B. HUMAN PARAINFLUENZA VIRUS (HPIV)

Human parainfluenza viruses (HPIVs) are the second most common viral causes of severe respiratory tract infection in infants and children. Human parainfluenza viruses belong to the family Paramyxoviridae. The recent classification of Paramyxoviridae is given in Table 12.1. HPIV has four serotypes—1 to 4. HPIV 1, 2, and 3 are first isolated from children with lower respiratory tract infection and HPIV 4 was isolated from children with mild respiratory tract infection.

The name "Parainfluenza" was given because it produces influenza-like symptoms and the virus particle also resembles influenza virus as it possesses lipid envelop and hemagglutinin and neuraminidase surface glycoproteins.

VIRUS

HPIVs are spherical or pleomorphic, enveloped viruses with single stranded, non-segmented, negative sense RNA genome. The nucleocapsid is of helical symmetry. Hemagglutinin neuraminidase (HN) and fusion (F) protein are the two surface glycoproteins present on the envelop.

Classification: The virus has four serotypes—HPIV 1 to 4 which is based on the less than 50% antigenic similarities for F and HN proteins. According to the recent ICTV classi-fication, HPIV 1 and 3 belong to the genus Respirovirus (renamed human respirovirus 1 and 3, respectively) and HPIV 2 and 4 belong to the genus Rubulavirus (renamed human rubulavirus 2 and 4, respectively). Both the genera presently classified under the family Paramyxoviridae.

VIRAL SURFACE PROTEINS

Hemagglutinin Neuraminidase (HN) Protein

These proteins are present on the virus envelop as surface projections.

Structure

It is a type II glycoprotein, present in the form of a tetramer. It consists of a stem and a globular head. Globular head bears the antigenic sites and HA and NA activities.

Function

- Possess the activities of both hemagglutinin and neuraminidase.
- In the initial part of infection, it mediates the attachment to host cell surface receptor sialic acid by hemagglutination activity.
- In the later part of infection, it breaks the sialic acid residue by neuraminidase to release the progeny virions.
- Both hemagglutination and neuraminidase activities are modulated by pH.

Fusion (F) Protein

Structure

It is a type I glycoprotein. Present in the form of a trimmer. It is synthesized as inactive F0 form which is cleaved by endopeptidase to active form which consists of F1 and F2 that are linked by disulphide bond. The N terminal of F1 is responsible for fusion.

Function

- Responsible for fusion of viral envelop with the plasma membrane of the host cell. This in turn leads to entry of the virus and release of viral nucleic acid to host cell cytoplasm.
- Responsible for fusion with the neighboring cell producing syncytia (cytopathic effect in *in vitro*).

EPIDEMIOLOGY

HPIVs are ubiquitous virus and are present worldwide. These viruses are responsible for causing acute respiratory tract infections in all age groups but more commonly in infants and children less than 5 years. The peak rate of infection of different HPIV types is:

- *HPIV-1 and 2:* Children between 1 and 3 years age.
- *HPIV-3:* Infants less than six months.
- *HPIV-4:* No specific age group predilection.

Modes of Transmission

- Through person-to-person contact.
- Infected large droplet.

Seasonality

HPIV-1: Biennial epidemics during fall months (September to December). Mostly associated with croup (laryngotracheobronchitis)

HPIV-2: Biennial fall epidemics along with HPIV-1 or biennially alternate with HPIV-1 or annual epidemic

HPIV-3: Summer and spring months.

CLINICAL FEATURES

HPIVs cause acute respiratory tract infections (ARTI) affecting either upper or lower respiratory tract or both. Primary infection by HPIV is usually symptomatic in nature which manifests with respiratory illness. The most common symptoms are pharyngitis, rhinitis, laryngitis leading to fever, cough, and hoarse-ness. Cervical lymphadenopathy is uncommon, whereas otitis media occurs frequently.

Croup (acute laryngotracheobronchitis): As the name suggests, croup is inflammation of larynx, trachea and bronchus, which leads to hoarse barking cough, inspiratory stridor, fever and laryngeal obstruction. It is most commonly seen in children between 1 and 2 years age.

Croup is most commonly associated with HPIV-1 (more than 50%) followed by HPIV-2 and HPIV-3.

Bronchiolitis and pneumonia: These are the main symptoms of LRTI in children. It manifests with fever, tachypnea, expiratory wheeze, rales, retraction and trapping of air. Young infants and children up to 2–3 years age are most commonly affected. HPIV is the second common viral cause of LRTI to respiratory syncytial virus (RSV). All four serotypes of HPIVs can cause bronchiolitis or pneumonia but HPIV-1 and HPIV-3 are most frequently associated. HPIV-3 is more frequently associated in hospitalized patients. Interstitial or perihilar infiltrates are the common radiologic findings of chest.

In immunocompromised patients, more commonly hematopoietic transplant and severe combined immunodeficiency syndrome (SCID) patients and elderly individuals are more prone to develop severe respiratory illness caused by HPIV. HPIV-3 is most frequently associated with infection in this patient group though all four HPIV can cause infection.

C. HUMAN RESPIRATORY SYNCYTIAL VIRUS (HRSV)

Human respiratory syncytial virus (HRSV) is the most common viral agent of lower respiratory tract infection in children. The virus was first discovered in 1956 from a chimpanzee suffering from coryza and was known as chimpanzee coryza agent. Later it was discovered from infants with respiratory illness and was named respiratory syncytial virus (RSV).

According to the recent ICTV classification, HRSV has been renamed "human orthopneumovirus" and belongs to the genus Orthopneumovirus of family Pneumoviridae.

Morphology of HRSV and HMPV (human metapneumovirus) is similar to human respiroviruses and human rubulaviruses with some important differences in the surface proteins.

RSV and HMPV have three surface glycoproteins projected from the virus envelop—G, F and SH.

Attachment Protein (G)

- G protein is responsible for attachment of the virus with the host cell.
- There is no structural similarity or sequence homology with corresponding H or HN proteins of other paramyxoviruses.
- G protein is most variable of all the proteins.
- The variability of G protein is the basis of phylogenetic classification within the species.

Fusion (F) Protein

F protein is structurally similar to F protein of other members of Paramyxoviridae but with limited amino acid similarity. It is synthesized in inactive F0 form which gets cleaved by peptidase to F1 and F2.

It is responsible for fusion of virus membrane with plasma membrane of host cell as described for HPIVs.

Short Hydrophobic (SH) Protein

This is the third transmembrane protein present on the virus surface. The exact function of SH protein is still not known. It has been implicated in impairing the Th1 mediated host antiviral response.

Genetic Diversity

HRSV has single serotype and two antigenic groups: A and B. The division of two antigenic groups is based on the neutralization with hyperimmune serum and monoclonal antibodies. The two groups can also be differentiated on the basis of nucleotide sequence identity. The difference between the two groups is mainly due to the difference in their G protein which is around 47% amino acid identity difference between the prototype of each group.

F protein has 50% antigenic relatedness between subgroups as compared to 1–7% for G protein.

F protein shows equal protectiveness against both the subgroups, whereas G protein is 13-fold less effective against the heterologous antigenic group.

Genotypes of RSV are classified according to the phylogenetic clustering of G protein gene sequences. Presently there are 8 genotypes within the subgroup A and 11 genotypes within the subgroup B. However, with time, the G protein undergoes so much of sequence variation that the existing classification system gets changed with new genotypes being added.

EPIDEMIOLOGY

Disease Burden

The global RSV disease burden has been estimated as 64 million cases and 160,000 deaths every year. In a worldwide estimate, it was estimated that 33.8 million new episodes of RSV associated acute lower respiratory tract infection (ALRI) occur worldwide in children below 5 years of age and 66000–199000 RSV associated deaths in children <5 years. In a hospitalized study in India, RSV has been detected in around 17% children.

Primary infection of RSV occurs most commonly during the first year of life (near 70%) and by 24 months age >95% children have been found to be infected with RSV.

Modes of Transmission

- RSV is mostly transmitted by large infected droplets than aerosols.
- The infectivity of the virus remains for a long time in the inanimate environment. This facilitates the transmission of infection through infected fomites.

Seasonality

- In temperate climate seasonal peak of RSV is seen during winter months.

- In tropical and subtropical countries north to equator, RSV peaks are observed during rainy season.
- In Hong Kong, peak activity has been reported during spring and summer months.
- In South America and South Africa, during cold dry months.
- In India, RSV season is mostly observed during fall and winter.
- Thus in general, HRSV activity is observed mainly in two different temperature range—2–4°C and 24–30°C.

Reinfection

- Reinfection with RSV occurs throughout life. Reinfection can occur even during the same epidemic season.
- Reinfection occurs more commonly due to heterologous antigenic group but can also occur due to the same antigenic group.
- Repeated infection indicates the ability of RSV to infect even in the presence of pre-existing antibodies.
- With repeated infection, the severity becomes less with less risk of developing lower respiratory illness.

Host factors for severe HRSV disease: HRSV is known to cause severe lower respiratory tract illness in the following high-risk group of individuals leading to reduction of pulmonary function, more number of hospitalization and higher mortality.

- Individuals with underlying cardiac or lung disease.
- Elderly individuals with >65 years age.
- Immunocompromised individuals: Patients with T cell deficiency, severe combined immunodeficiency disease (SCID), hematological malignancies, hematological stem cell transplant recipients.
- Children with cystic fibrosis.

Molecular Epidemiology

Circulation of multiple strains is commonly seen in any given year with predominance of a few genotypes with predominance of either

A or B antigenic group. The predominant strain pattern is different in different geographic locations. Replacement of predominant strain usually occurs in every 1–2 years with different strains. Though replacement of strains reflects the advantage of heterologous strains, reinfection with the same genotype is a common phenomenon with HRSV in contrast to influenza virus.

Introduction of new genotype from other geographic location often leads to epidemic.

Unlike influenza, global spread of infection due to new strain is not a common scenario with HRSV. Global spread of HRSV has been observed with a new genotype known as BA of antigenic group B. The unique feature of BA genotype is 60 nucleotide duplication in the G gene, which is thought to be responsible for the spread of the strain across the continents.

PATHOGENESIS

Humans are the only natural host for HRSV. The incubation period is around 3–5 days. HRSV initially infects the nasopharynx, then spreads to the lower respiratory tract. This occurs most likely due to aspiration of virus in children. The peak virus titer in nasal secretion is around 10^4–10^6 infectious virus unit/mL. It apically infects the ciliated cells, inhibits the transport of Na^+, resulting in accumulation of fluid in the apical surface. Secretion of mucus in the airways, sloughed epithelial cell debris and accumulation of neutrophils leads to blockage of bronchioles and alveoli. **Airway mucus is the hallmark of RSV bronchiolitis**. Both type I and type II alveolar cells can be infected by HRSV. In case of pneumonia, there is also infiltration of mononuclear cells leading to interalveolar wall thickening in addition to the accumulation of fluid in the alveolar space.

CLINICAL FEATURES

Rhinorrhea, cough and mild fever are the initial manifestations. This may progress to lower respiratory tract illness. Bronchiolitis and pneumonia are the two manifestations of HRSV infection. Cough, wheeze, poor feeding and breathlessness are the predominant symptoms. Tachypnea, inspiratory crackles and expiratory wheeze are the important findings on examination.

Symptoms are usually mild when infection occurs in the newborns, this could be due to the protective effect of maternal antibody.

ANTIVIRALS

Ribavirin: This is a nucleoside analog, acts as a broad-spectrum antiviral agent mostly against RNA viruses. Ribavirin has been used in severe respiratory viral infections, but the efficacy of the drug is inconclusive. The drug is presently recommended for HRSV infection in immunocompromised and children with significant cardiopulmonary abnormalities.

Palivizumab: This is a humanized mouse monoclonal antibody against the protective epitope of F protein of HRSV. It does not inhibit RSV infection, but reduces the clinical severity. It has been found 50–100% more effective than RSV intravenous immunoglobulin and safe in high-risk children. The drug is recommended in high-risk children like premature babies, children with underlying cardiopulmonary disease.

Motavizumab: It is a humanized monoclonal antibody, derived from palivizumab. The drug is more effective than palivizumab.

D. HUMAN METAPNEUMOVIRUS (HMPV)

Human metapneumovirus (HMPV) was first **discovered in 2001** while isolating the possible viral pathogen from respiratory samples of infected children on monkey kidney cell line which produced cytopathic effect similar to respiratory syncytial virus (RSV) but nucleotide sequence similarity was found to be close to avian metapneumovirus (AMPV).

According to the recent ICTV, HMPV is a member of the **genus** Metapneumovirus in the family Pneumoviridae.

GENETIC DIVERSITY

HMPV has been divided into two major groups A and B based on the sequence variability of M, N, P, F and L genes. Each of the A and B groups or genetic lineages is further divided into sublineages A1, A2 and B1 and B2 based on the sequence identity of F and G proteins. Co-circulation of all the four lineages has been reported worldwide.

EPIDEMIOLOGY

Approximately 10% of all respiratory viral infections in children requiring hospitalizations are due to HMPV, whereas in adults it is responsible for 4–5% of severe respiratory disease.

HMPV infection can occur in all age groups but more common in children below 5 years age. First infection usually occurs around 6 months of age, after which there may be several repeated infections. As compared to RSV, **older children** are infected with HMPV. **Elderly sick patients** are the second common group of patients in whom HMPV can cause severe infection.

Seasonal peaks are observed with HMPV infection with a peak during cold winter months from December to February. The peak season of HMPV usually occurs after 1–2 months after RSV season.

CLINICAL FEATURES

Incubation period of HMPV infection is 4–6 days. In children, HMPV usually affects upper respiratory tract manifesting as fever, cough and rhinorrhea. Lower respiratory tract illness due to HMPV usually manifests as that of RSV with bronchiolitis, pneumonia, croup and acute exacerbation of asthma.

In healthy adults, HMPV generally produces mild symptoms, such as cough, sore throat, congestion, rhinorrhea and constitutional symptoms.

Severe lower respiratory tract illness due to HMPV can be seen in individuals with underlying chest diseases or immunocompromised

patients and old adults with comorbid conditions. Up to 40% pneumonia has been reported in HMPV infection.

Coinfections of HMPV with RSV have been reported by some groups to cause more severe infection as compared to single infection. However, there is no consensus so far regarding this information.

E. HUMAN BOCAVIRUS

Human bocavirus (HBoV) was **discovered in 2005** from pool of nasopharyngeal aspirate collected from children with suspected acute respiratory infection. The virus has been found in respiratory samples as well as from blood and stool. The role of HBoV as a respiratory pathogen is still not clear and has become complex with more intricate findings.

Bocavirus is a member of family Parvoviridae and genus Bocaparvovirus. The name "Boca" has been derived from "**Bovine parvo**" and "**Canine minute virus**" due to their similarities in genome structure and nucleotide sequence. It contains a single linear positive or negative sense single-stranded DNA genome of 4000–6000 nucleotide length. HBoV has been classified into 4 species: HBoV1 through HBoV4. HBoV1 has been found in respiratory tract, whereas HBoV 2–4 are found in stool.

HBoV as respiratory pathogen: Ever since the discovery of HBoV from nasopharyngeal sample of ARTI patients, several studies have been carried out to find out the association of HBoV in respiratory infection.

All most all the studies have shown statistically high HBoV DNA positivity in ARTI patients as compared to the asymptomatic controls. In one of the largest study on HBoV by Fry et al in association with Center for Disease Control and Prevention (CDC, Atlanta) that included 1168 patients with community acquired pneumonia, 512 patients with influenza-like illness and 280 asymptomatic individuals, HBoV DNA was detected in 1% asymptomatic individuals, 3% in patients with influenza-like illness and 4.5%

of pneumonia patients. It was found to be the third most common viral agent in children <5 years age with pneumonia.

High viral load (>10⁴ copies/mL) has been shown to be associated with symptomatic infection as compared to low and moderate viral load and also has been found more commonly as the sole viral agent without co-infection. High viral load in respiratory samples has been correlated with detection of DNA in blood sample of the patient. HBoV DNAmia has been found to decline with subsidence of symptoms.

The association of HBoV with respiratory infection has been shown by detection of HBoV DNA in respiratory samples. Detection of DNA in clinical samples is not sufficient to indicate the pathogenic association as the DNA has also been detected in asymptomatic individuals and the positivity could be due to bystander than pathogen. HBoV being a member of Parvovirus family can persist for a long time and can be detected in sample during its persistence. HBoV has also been detected more as a coinfecting agent with known respiratory pathogens than as a sole agent. The causative role has not been proved by modified Koch's postulate.

However, considering the association of high viral load in respiratory sample, DNA in blood and presence of virus as a sole pathogen during acute illness and decline of DNA in blood during resolution, it is presently accepted that the association of HBoV is of clinical relevance in presence of high viral load (>10⁴ copies/mL) or with the evidence of primary infection. The presence of at least two of the following—HBoV IgM antibody and fourfold rise in IgG titer and low avidity IgG during symptomatic phase, medium to high viral load are the markers for primary infection.

HBoV as a gastrointestinal pathogen: HBoV has been detected in stool samples of patients with acute gastrointestinal illness as well as in asymptomatic individuals. In most of the cases, it has been detected along with other agents. The exact role as a gastrointestinal pathogen has not been established.

DIAGNOSIS OF RESPIRATORY VIRUS INFECTION

Samples: In respiratory virus infection, samples from the affected organ system (respiratory tract and other organs) are collected for detection of virus or its components such as virus, viral antigen, or viral RNA.

- Nasal/throat swab, nasal/throat washing, nasopharyngeal swab/aspirate are the samples collected for virus isolation, antigen detection or genome detection in patients with acute upper or lower respiratory tract infection.
- Tracheal aspirate, bronchoalveolar lavage (BAL) or lung biopsy can be tested in case of lower respiratory tract infection (LRTI). However, BAL and lung biopsy involves invasive procedure, and recommended when samples of URTI comes negative.
- Sputum can be collected for testing when produced.
- Blood or serum sample is collected for detection of antibody against the virus.

Virus isolation: Respiratory viruses can be isolated in various host systems.

Embryonated egg: **Amniotic cavity** of the embryonated egg is a suitable system for the primary isolation of influenza virus. Previously this system was used commonly for isolation of influenza virus from the clinical sample. However, the system suffers from low sensitivity and due to less availability of suitable pathogen free egg and development of resistance, this method is not commonly used presently.

Embryonated egg host is, however, not suitable host for primary isolation of HPIV.

Cell culture: Various cell lines are commonly used for primary isolation as well as for propagation of respiratory viruses.

Kidney cell lines from various animals such as Madin darby canine kidney (MDCK) cell lines and rhesus monkey kidney cell line LLCMK2 are commonly used for isolation of influenza, parainfluenza viruses, and other respiratory viruses. LLCMK2 and African green monkey kidney cell lines (Vero cells) are commonly used for HMPV isolation. Hep2 and HepG2 cells also support the growth of HMPV. Mixtures of two or more cell lines are available commercially for culture of respiratory viruses.

Provisional identification of virus is done by cytopathic effect which is confirmed by viral antigen detection in the infected cells by using monoclonal antibodies or virus RNA detection by RT-PCR.

Isolation is still considered as the gold standard. However, the method is time consuming and less sensitive. Because of these limitations, the method is no more used for routine diagnosis from clinical sample and has largely been replaced by molecular technique. However, it is required for virus characterization, antigen and vaccine preparation.

Shell vial technique: Also known as **centrifugation enhanced technique**. In this method, sample is centrifuged over the cell line during the period of incubation. This facilitates the rapid entry of virus into the cell and early expression of viral antigen on the surface of the infected cell line. The virus antigen is then detected by staining the infected cell with specific antibody by 24–48 hours. This technique is **commonly used for respiratory syncytial virus (RSV)**.

Rapid Antigen Detection

Several tests are available for rapid antigen detection directly from respiratory sample. Immunofluorescence or enzyme immunoassay (EIA) methods are used for antigen detection. Of these, EIA format offers a point of care test system which can be done bedside without using any special equipment for testing and reading. These are mainly available for influenza virus and RSV from clinical sample.

The rapidity, simplicity and potential of point of care, makes the format useful.

For influenza virus: Sensitivity of these tests is poor and widely variable—20–90%. High sensitivity is achieved in the early part of illness due to high viral shedding.

Specificity in general is considered as good, but some recent report has shown near 40% false positivity.

Types of sample, transport media, transport temperature, processing of samples have influenced the sensitivity and specificity.

The test is based on the detection of conserved antigens like nucleoprotein and matrix protein. The method is not suitable to differentiate between types or subtypes of influenza viruses.

For RSV: Immunofluorescence assay on nasal or throat samples using monoclonal antibody demonstrates **intracytoplasmic fluorescence** (Fig. 11.6). However, this is expensive, requires fluorescence microscope and needs individual sample processing.

Molecular methods: Detection of viral RNA from clinical samples is done by various molecular tests. The advantages of molecular tests are:
- High sensitivity, specificity.
- Multiplexing to detect various types and subtypes of influenza viruses. This helps in detection of seasonal strains along with pandemic strains, also other respiratory viruses.

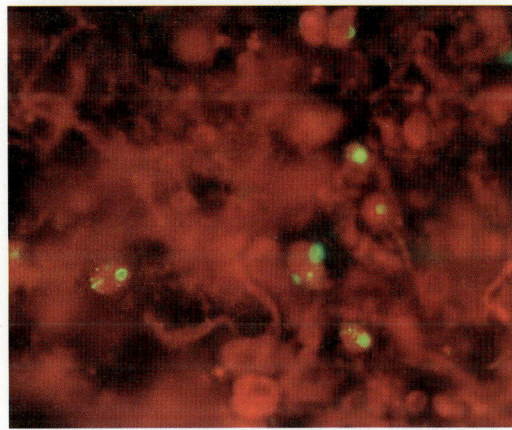

Fig. 11.6: RSV antigen showing cytoplasmic fluorescence in nasopharyngeal aspirate by monoclonal antibody

- Rapidity helps for short turnaround time.
- Ability to be integrated with automation platform and high throughput system.

Polymerase chain reaction (PCR) is the most commonly used molecular method for diagnosis of influenza and other respiratory viruses and almost has achieved the place of gold standard.

Reverse transcriptase real-time PCR (rRT-PCR): As most of the respiratory viruses are RNA viruses, RT-PCR is commonly used for detection of viral genome.

This method was the recommended test by CDC, Atlanta, for the detection of influenza A (H1N1) pdm09 and was the method followed for diagnosis almost worldwide including India. The test uses detection of four genes—influenza A, swine A, swine H with an in-house RNAseP gene as control.

Multiplex RT-PCR: Various multiplex PCR systems are available commercially for detection of respiratory virus panel which primarily includes influenza A, influenza B, influenza A (H1N1) pdm09, HPIV 1–4, HMPV, RSV, adenovirus, human bocavirus and coronaviruses.

Loop mediated amplification (LAMP) and nucleic acid sequence based assay (NASBA) has been used for detection of influenza virus genome.

LAMP assay has reported similar sensitivity, specificity as rRT-PCR. The test uses 3 sets of primer and does not require any sophisticated equipment. Amplification occurs in isothermal temperature at 60–65°C.

Serology: Antibody starts appearing after 2 weeks of infection and takes its peak in 4–7 weeks. Antibody detection from blood is done for various purposes:

- To look for seroconversion in acute and convalescent samples for diagnosis.
- For seroprevalence study in the community.
- To test for neutralizing antibody for immunity status.
- To check immune response to vaccine.
- Identification of virus strain using specific antisera.

Serological Methods Used

- Hemagglutination inhibition.
- Complement fixation.
- Neutralization.
- Enzyme immunoassay.

Demonstration of >4-fold rise in titer of antibody between acute and convalescent sera (at 2–3 weeks interval) diagnose recent infection. The time taken and requirement of two samples is the main disadvantage of the test and because of this limitation the test is not used for routine patient diagnosis. However, it helps in diagnosis retrospectively.

Antibody detection helps in studying the seroepidemiology of a recently emerged strain. This is helpful in estimating the exposure to the new strain or immune status in the population which is important to take policy decision on vaccination.

A titer of >20 by HAI and >40 by microneutralization (MN) test of a single convalescent sample is considered significant for seroprevalence study.

Immunogenicity of vaccine is estimated by determining the neutralizing or HAI antibody.

Bibliography

1. Broccolo F, Falcone V, Esposito S, Toniolo A. Human bocaviruses: Possible etiologic role in respiratory infection. J Clin Virol 2015;72:75–81.

2. Broor S, Bharaj P. Avian and human metapneumovirus. Ann N Y Acad Sci 2007;1102:66–85.

3. Broor S, Parveen S, Bharaj P, et al. A prospective three-year cohort study of the epidemiology and virology of acute respiratory infections of children in rural India. PLoS One 2007;6;2(6): e491.

4. Cumulative number of confirmed human cases for avian influenza A(H5N1) reported to WHO, 2003–2016. WHO/GIP, data in HQ as of 09 May 2016. Accessed from http://www.who.int/influenza/human_animal_interface/EN_GIP_20160509cumulativenumberH5N1 cases.pdf?ua=1

5. de Graaf M, Fouchier RA. Role of receptor binding specificity in influenza A virus transmission and pathogenesis. EMBO J 2014;33(8): 823–41.

6. Falsey AR, Hennessey PA, Formica MA, Cox C, Walsh EE. Respiratory syncytial virus infection in elderly and high-risk adults. N Engl J Med 2005; 352(17):1749–59.

7. Falsey AR, Walsh EE. Respiratory syncytial virus infection in adults. Clin Microbiol Rev 2000;13(3):371–84.

8. Ferguson L, Olivier AK, Genova S, et al. Pathogenesis of Influenza D Virus in Cattle. J Virol 2016;90(12):5636–42.

9. Feuillet F, Lina B, Rosa-Calatrava M, Boivin G. Ten years of human metapneumovirus research. J Clin Virol 2012; 53(2):97–105.

10. Fry AM, X Lu, Chittaganpitch M, Peret T, Fischer J, Dowell SF, Anderson LJ, Erdman D, and Olsen SJ. Human bocavirus: A novel parvovirus epidemiologically associated with pneumonia requiring hospitalization in Thailand. J Infect Dis 2007;195:1038–45.

11. Fukuyama S, Kawaoka Y. The pathogenesis of influenza virus infections: The contributions of virus and host factors. Curr Opin Immunol 2011;23(4):481–86.

12. Fukuyama S, Kawaoka Y. The pathogenesis of influenza virus infections: the contributions of virus and host factors. Curr Opin Immunol 2011;23(4):481–86.

13. Henrickson KJ. Parainfluenza viruses. Clin Microbiol Rev 2003;16(2):242–64.

14. Horimoto T, Kawaoka Y. Pandemic threat posed by avian influenza A viruses. Clinical Microbiology Reviews 2001;14:129–49.

15. http://www.who.int/influenza/human_animal_interface/EN_GIP_20160404cumulative numberH5N1cases.pdf

16. Hurt AC, Besselaar TG, Daniels RS, et al. Global update on the susceptibility of human influenza viruses to neuraminidase inhibitors, 2014–2015. Antiviral Res 2016;132:178–85.

17. Imai M, Herfst S, Sorrell EM, Schrauwen EJ, Linster M, De Graaf M, Fouchier RA, Kawaoka Y. Transmission of influenza A/H5N1 viruses in mammals. Virus Res 2013;178(1):15–20.

18. Jiang W, Yin F, Zhou W, Yan Y, Ji W. Clinical significance of different virus load of human bocavirus in patients with lower respiratory tract infection. Sci Rep 2016 Feb 1;6:20246. doi: 10.1038/srep20246.

19. Kahn JS. Epidemiology of human metapneumovirus. Clin Microbiol Rev 2006;19(3):546–57.

20. Kumara S, Henricksona KJ. Update on Influenza Diagnostics: Lessons from the Novel H1N1 Influenza A Pandemic. Clinical Microbiology Reviews 2012; 25: 344–61.

21. Mahony JB. Detection of Respiratory Viruses by Molecular Methods. Clinical Microbiology Reviews 2008; 21: 716–47.

22. Neumann G, Kawaoka Y. Transmission of influenza A viruses. Virology. 2015; 479–80:234–46. doi: 10.1016/j.virol.2015.03.009.

23. Novel Swine-Origin Influenza A (H1N1) Virus Investigation Team. Emergence of a Novel Swine-Origin Influenza A (H1N1) Virus in Humans. N Engl J Med 2009; 360: 2605–15.

24. Ong DS, Schuurman R, Heikens E. Human bocavirus in stool: A true pathogen or an innocent bystander? J Clin Virol 2016;74:45–9.

25. Peiris JS, de Jong MD, Guan Y. Avian influenza virus (H5N1): a threat to human health. Clin Microbiol Rev 2007;20(2):243–67.

26. Peiris JS, Poon LL, Guan Y. Emergence of a novel swine-origin influenza A virus (S-OIV) H1N1 virus in humans. J Clin Virol 2009;45(3):169–73.

27. Principi N, Piralla A, Zampiero A, et al. Bocavirus Infection in Otherwise Healthy Children with Respiratory Disease. PLoS One 2015 Aug 12;10(8):e0135640.

28. Principi N, Bosis S, Esposito S. Human metapneumovirus in paediatric patients. Clin Microbiol Infect 2006;12(4):301–8.

29. Schildgen O, Schildgen V. Respiratory Infections with Human Bocavirus. Clin Infect Dis 2016; 62(1):134. doi: 10.1093/cid/civ759.

30. Schildgen V, van den Hoogen B, Fouchier R, Tripp RA, Alvarez R, Manoha C, Williams J, Schildgen O. Human Metapneumovirus:

Lessons learned over the first decade. Clin Microbiol Rev 2011;24(4):734–54.

31. Tregoning JS, Schwarze J. Respiratory viral infections in infants: causes, clinical symptoms, virology, and immunology. Clin Microbiol Rev 2010;23(1):74–98.

32. Types of Influenza Viruses. https://www.cdc.gov/flu/about/viruses/types.htm

33. Watanabe T, Watanabe S, Maher EA, Neumann G, Kawaoka Y. Pandemic potential of avian influenza A (H7N9) viruses. Trends Microbiol 2014; 22(11):623–31.

34. Writing Committee of the WHO Consultation on Clinical Aspects of Pandemic (H1N1) 2009 Influenza. Clinical Aspects of Pandemic 2009 Influenza A (H1N1) Virus Infection. N Engl J Med 2010;362:1708–19.

Measles Virus

Measles virus is believed to have evolved in an environment where both humans and animals coexisted. The virus is closest to the rinderpest virus (RPV), an animal morbillivirus.

The characteristic skin eruption of measles was first described by an Arab physician Abu Becr as *"hasbah"* who described the disease as a modification of smallpox with "anxiety of mind, skin qualm and heaviness of heart oppress more in measles than smallpox".

The term "morbilli" means "little disease" in Italian. Sanvages first described morbillias "measles". It was Peter Panuma Danish physician who described the characteristic features of measles like respiratory route of transmission, highly contagious nature, 14 days incubation period and lifelong immunity after natural infection.

VIRUS

Measles virus belongs to the family Paramyxoviridae, Morbillivirus genus and species *Measles morbillivirus*. The classification of Paramyxoviridae is shown in Table 12.1.

Measles virus is a pleomorphic virus but mostly round in shape and 100–300 nm in diameter. It is an enveloped virus containing single stranded, non-segmented, negative sense RNA genome. The lipid envelop surrounds the inner helical nucleocapsid.

Envelop contains two transmembrane surface glycoproteins: Hemagglutinin (H) and fusion (F) protein. On the innerside of the envelop lies the layer of matrix (M) protein.

Table 12.1: Recent classification of Paramyxoviridae

Paramyxoviridae: Seven genera (4 of human importance)
1. *Human respirovirus* 1 and human respirovirus 3 (previous name: PIV1&3)
2. *Rubulavirus:*
   i. Human rubulavirus 2 and human rubulavirus 4 (previous name: PIV2&4)
   ii. Mumps rubulavirus (mumps virus)
3. *Henipavirus:* Hendra henipavirus, nipah henipavirus
4. *Morbillivirus:* Measles morbillivirus

Pneumoviridae: Two genera
1. *Metapneumovirus*
   • Human metapneumovirus (HMPV)
2. *Orthopneumovirus*
   • Human orthopneumovirus
     (previous name respiratory syncytial virus RSV)

The ribonucleocapsid is composed of RNA genome and protein complex and is of helical symmetry. The RNA genome is of 16 kilo base that is complexed with three of the viral proteins—nucleocapsid (N), phosphoprotein (P) and large polymerase protein (L). N protein wraps the RNA genome and L and P proteins are attached to the helix of ribonucleocapsid (Fig. 12.1).

VIRUS PROTEINS

Measles virus RNA genome encodes six structural and two non-structural proteins.

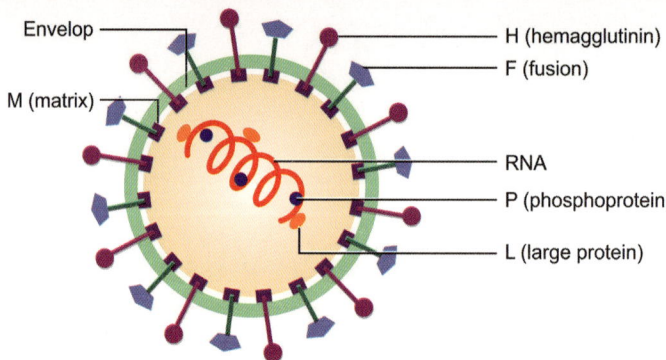

Fig. 12.1: Measles virus morphology

Amongst the six structural proteins, three are associated with envelop—H, F and M proteins, and three are complexed with RNA genome—N, P and L. The remaining two are nonstructural C and V proteins.

Hemagglutinin or H protein is a surface transmembrane protein, a type II glycoprotein responsible for hemagglutination (HA) and binding of virus to the cell surface receptor, and thus is the determinant of cellular tropism.

Fusion or F protein is another transmembrane surface protein, a type I glycoprotein. It is synthesized as F0, its inactive precursor which gets proteolytically cleaved to active F1 and F2 heterodimer. The F1 subunit interacts with H protein for receptor binding and responsible for cell-cell fusion and budding of virion.

Matrix (M) protein: A non-glycosylated protein, present below the bilayer of lipid envelop. It modulates the fusion by F/H complex and release of virion occurs by envelop glycoproteins. Absence or deletion of M protein leads to increase in fusion leading to spread of infection but decrease or inefficient production of infectious virion. Persistent measles virus infection often has mutation in M protein.

Nucleocapsid protein (N): Most abundant protein that completely covers the genomic RNA. Each monomer of N protein binds to six nucleotides. This rule of six has been put forth as the reason of efficient replication of virus genome containing nucleotide as multiple of six. It condenses the viral genome into a more stable and smaller form which gives the genome a "herringbone pattern" appearance in the electron microscope.

Large or L protein is the RNA specific RNA polymerase producing mRNA. It is multifunctional. It is associated with nucleocapsid. It works in association with P protein in viral replication.

C and V proteins interact with the cellular proteins and regulate the virus replication and response to infection. In addition, C protein has been implicated in suppression of innate immune response by inhibiting the induction of interferon.

MEASLES VIRUS GENOTYPES

According to the current status by World Health Organization (WHO), measles virus strains are divided into **8 clades** (designated as **A through H**) and **24 genotypes**. Genotyping is based on the sequence variation of COOH-terminal 450 nucleotide coding for 150 amino acids of nucleoprotein (N-150) or entire coding region of H. Difference in nucleotide by >2.5% in N protein or 2% in H from the closest reference strain is considered as a separate genotype. Within a genotype, there may be multiple lineages. Each lineage includes a

Table 12.2: MeV genotypes and their geographic distribution

Clade	Genotype	Present status	Geographic distribution
A	Single genotype	No endemic strain Sporadic infections	All vaccine strains
B	B1, B2, B3	B3 in circulation	African regions
C	C1, C2	Not in circulation since >10 years	
D	D1 to D11	D4, D8 and D9 are presently in circulation	Southeast Asian, Western pacific regions, Europe
G	G1, G2, G3	G3 was the last one to circulate till 2015	Western pacific regions
H	H1, H2	H1 is presently in circulation	Western pacific and southeast Asian regions

group of viruses that have identical N-450 nucleotide sequence. Viruses of same lineage indicates single chain transmission.

Clade A includes all the measles vaccine strains: Moraten, Edmonston, Zagreb. It has not been found till date in any endemic transmission and the wild virus types of clade A are considered to be extinct. Clades A, E and F consist of single genotype each. During the period 2005 to 2015, 11 wild type genotypes were found. In 2017, 5 wild type genotypes belonging to 3 clades—B3, D4, D8, D9 and H1 were detected. **In India** during 2011 to 2015, **D8 and D4** were the predominantly circulating genotypes along with a few B3 from Kerala state. Table 12.2 describes the genotypes and their geographic distribution.

Importance of Genotyping

1. It helps in tracking the transmission pathways of measles virus.
2. To monitor the status of endemic virus transmission as absence of endemic virus is a criterion for verifying the measles elimination in a particular region.
3. Genotyping helps to differentiate between natural infections or vaccine-related adverse events (as all measles virus vaccine strains belong to genotype A).

Pattern of Genotype Circulation

Three patterns of genotype circulation have been described.

1. *Regions with endemic circulation:* In countries where endemic transmission still persists, majority of measles cases occur due to the circulating endemic strains. Multiple co-circulating lineages are detected.
2. *Countries where measles is eliminated:* Small number of cases due to imported viruses with various genotypes.
3. *Countries with good measles control but increase in number of susceptible individuals due to poor maintenance of high vaccination coverage:* Reintroduction of virus usually leads to outbreak due to single genotype.

EPIDEMIOLOGY

Disease burden: Measles is still a leading cause of vaccine preventable infant mortality. According to WHO 2018 data, 189392 measles cases have been reported globally till the month of November.

Amongst the six WHO regions, maximum number of measles cases are reported from Southeast Asian region followed by European region, African region, Eastern Mediterranean region (EMR), Western pacific region (WPR) and American region.

India is among the top ten measles reported countries with an annual incidence rate of 50.44 per one lakh population and about 67,000 cases during 2018. Every year nearly 2.5 million people are affected with about 49,000 deaths. **India accounts for around 37% of global measles deaths.**

During the period 2000 to 2016, the reported annual measles incidence decreased by 87%, (from 145 to 19 cases per million persons), and annual estimated measles deaths decreased by 84% (from 550,100 to 89,780), amounting to an estimated 20.4 million deaths prevented due to measles vaccination.

Mode of Infection

Measles virus is transmitted mainly through respiratory droplets. However, it can be transmitted by aerosols of infected particles which may remain suspended in the air for a few hours. As there is no subclinical, latent or persistent virus infection, the maintenance of virus transmission in human population depends on the unbroken chain of acute infection. The transmission is common amongst school children and household contacts.

The average incubation period of measles, i.e. from infection to onset of fever is considered to be 10 days and to onset of rash is 14 days though longer incubation period has been reported (Fig. 12.2).

The period of infectivity ranges from several days before to several days after the appearance of rash during the period of respiratory manifestations like cough, sneezing, coryza which facilitates the spread of virus and the titer of virus in blood and body fluids is maximum.

Measles is highly contagious in nature with the secondary attack rate **(SAR) of >90%** among the susceptible population.

In majority of the measles high burden countries, the disease is most common in 1–4 years age group children. Shifting of age group towards older children occurs in settings with high vaccine coverage and low population density.

Epidemiology of measles virus is unique in several aspects. The key determinants of measles epidemiology are: high contagiousness of the virus, respiratory mode of transmission and lifelong immunity by natural infection and vaccination. Man is the only natural reservoir of measles virus. There is no animal reservoir. Due to the high contagiousness and high secondary attack rate of measles virus,

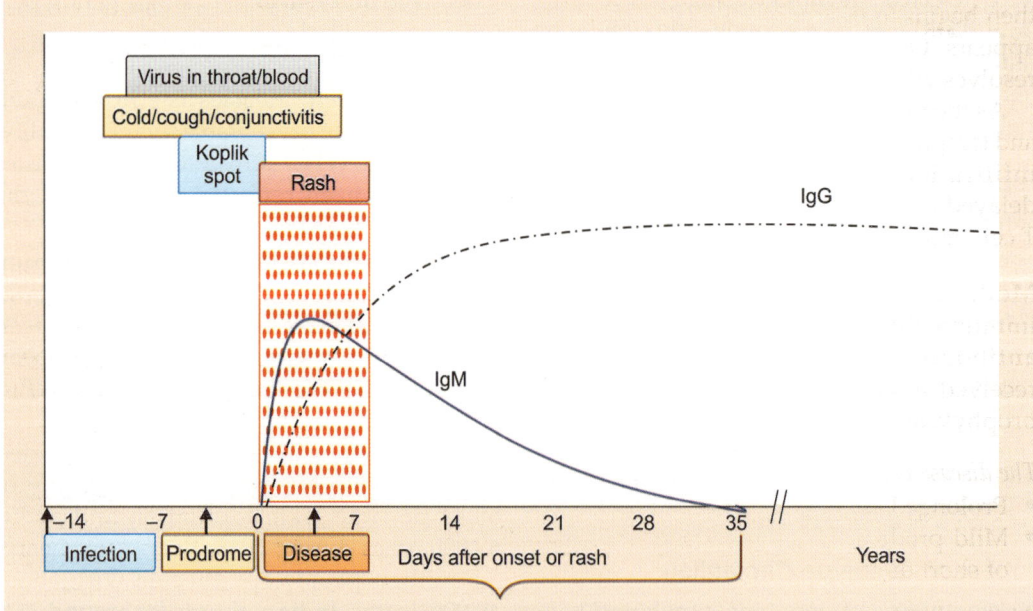

Fig. 12.2: Time course of measles virus infection

out breaks can occur even in populations with <10% susceptible individuals.

CLINICAL FEATURES

Classical measles: Disease manifestation starts after 10 days of exposure. Prodromal symptoms begin with fever followed by cough, coryza and conjunctivitis (typically, at least **1 of 3Cs: C**ough, **C**oryza and **C**onjunctivitis).

Appearance of **Koplik's spots** occurs after 2 days of prodromal symptoms and 2 days before the onset of rash. To start with, it appears first behind the lower molar then affects the entire buccal mucosa. Koplik's spots are bluish white papular lesions, about 1 mm in diameter, with white center and surrounded by an erythematous base. These are considered pathognomonic of measles.

Rash in measles is characteristic in nature. It starts 3–4 days after the onset of fever. The prodromal symptoms are maximum at the time of onset of rash. These are maculopapular, start from behind the ear, hair line, face and neck. Then spreads to trunk and extremities. Rash persists for 3–4 days and then begins to fade in the same manner as it appears. Overall an uncomplicated measles resolves within a week.

As the rash occurs due to cellular response and is represented by perivascular lymphocytic infiltration, the manifestation of rash is delayed or absent in children with impaired T cell immunity with high disease severity.

Modified measles: It occurs in partially immune children, with residual maternal antibodies or in individuals who have received immune globulin as postexposure prophylaxis.

The disease is characterized by:
- Prolonged incubation period
- Mild prodrome, and sparse, discrete rash of short duration and mild illness.

Atypical measles: Observed in individuals who were vaccinated with killed measles vaccine and were subsequently exposed to natural measles virus infection within years.

- It has been mostly observed in individuals who received killed measles vaccine during the period of 1963–1967 in US.
- It occurs due to abnormally intense cellular immune response due to sensitization to measles virus antigen by the killed measles vaccine which do not provide protection.
- Lack of production of antibody to fusion protein (F) allowing the virus to spread from cell-to-cell in presence of neutralizing Ab.
- Manifest with maculopapular or petechial rash, but may presents with hemorrhagic, urticarial, vesicular or combination. Unlike classical measles, rash first appears on wrist or ankle.
  - High fever
  - Interstitial pulmonary infiltrates, edema of extremities.

COMPLICATIONS

Measles associated complications are seen in about 30% of measles cases and more common in children below 5 years of age and adults more than 20 years. It presents either as severe measles infection or secondary bacterial infection.

Diarrhea, otitis media, pneumonia and encephalitis are amongst the common complications associated with measles. Diarrhea is the most common measles complication, reported in 8% of measles cases.

Immunocompromised individuals particularly with defective cellular immune response are prone for severe measles which manifest mainly as giant cell pneumonia **(Hecht pneumonia)** consisting of **multinucleated giant cells in the lungs**. These patients may present with or without rash.

Respiratory tract complications are amongst the **most frequent** complications due to measles. Otitis media and bronchopneumonia are the commonest complications. Otitis media is seen in majority of children with measles amounting to 7% of all complications.

Pneumonia is seen in 6% of measles cases and the most common cause of measles related deaths. Besides measles associated giant cell pneumonia, pneumonia can occur due to secondary viral or bacterial infections. Most often the bacterial pneumonia is due to *Streptococcus pneumoniae* or *Haemophilus influenzae*.

Central nervous system (CNS) complications are rare but severe. It can occur primarily in three different manners (Fig. 12.3).
1. Acute disseminated encephalomyelitis (ADEM)
2. Measles inclusion body encephalitis (MIBE)
3. Subacute sclerosing encephalomyelitis (SSPE)

Acute disseminated encephalomyelitis (ADEM):
It is an acute demyelinating autoimmune disease due to measles virus induced autoimmunity to myelin basic protein of brain. It is seen within days to weeks of acute measles. It occurs in 1 in 1000 measles cases. The disease is characterized by fever, seizure and other neurological deficits. Mortality rate reported is 10–20%. No measles virus antigen, RNA or measles specific antibody can be detected in brain or CSF.

Measles inclusion body encephalitis (MIBE):
It occurs in individuals with impaired cellular immunity. Progressive measles virus replication in brain results in neurological deterioration and death. It occurs after months of acute measles infection.

Subacute sclerosing panencephalitis (SSPE):
Slowly progressive disease, occurs after years (5–10 years) of acute measles. The disease is more common in children who had measles infection before 2 years of age. Manifestation is commonly seen in boys <20 years of age. It occurs in 1 **in 10000 to 1 in 100,000 cases.**

The disease process is thought to be due to host response to the mutated measles virions and occurs because of defective assembly and budding.

Disease is characterized by four phases:
- I: Progressive mental deterioration and memory defect often present as change in school performance
- II: Myoclonous, seizures
- III: Neurological deterioration
- IV: Optic atrophy, coma

Death occurs within months to years of onset.

Diagnosis of SSPE can be made by detection of **high level of measles antibody in CSF.** The CSF: serum ratio of hemagglutinating antibody of **≤1:68** by hemagglutination inhibition assay (HAI) is considered as significant.

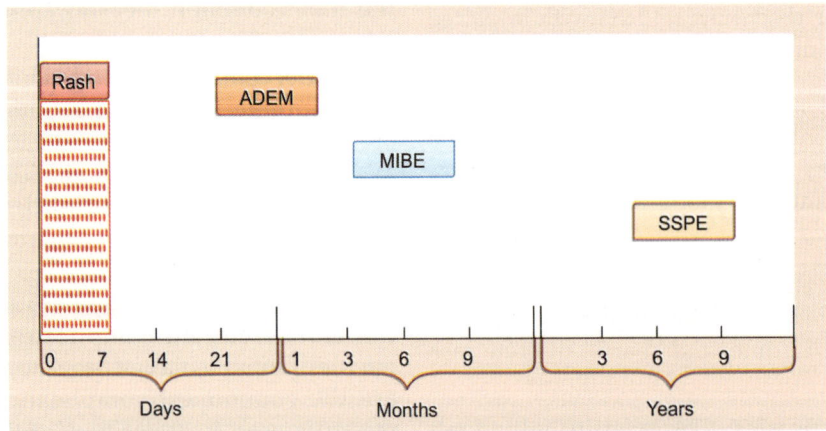

Fig. 12.3: Timing of neurologic complications of measles

PATHOGENESIS AND PATHOPHYSIOLOGY

Measles virus enters through infected respiratory droplets and primarily infects the oropharynx, tracheal and bronchial epithelial cells. The initial replication for 2–4 days occurs in the respiratory mucosa after which the virus spreads to draining lymph nodes. Replication continues in the lymph nodes and then the virus spreads to the blood leading to primary viremia and through blood disseminates to reticuloendothelial system. Extensive replication of virus continues in the lymph nodes and leads to over flow of virus to blood leading to secondary viremia. Through infected lymphocytes and monocytes in the blood, the infection then spreads to the various organs including spleen, lymph node, kidney, skin, brain, etc. as well as to the respiratory tract. Replication in these organs primarily occurs in the endothelial cells, epithelial cells and monocytes or macrophage.

The thin epithelial layer of the respiratory tract and conjunctiva gets destroyed due to the process of virus infection and starts breaking down and along with the inflammatory process it leads to the development of prodromal symptoms, such as cough, coryza, congestion and conjunctivitis.

During the prodromal period, **Koplik's spot** appears. This is the pathognomonic feature of measles. It appears as a raised spot with white center and erythematous base on buccal mucosa. This marks the beginning of the delayed type hypersensitivity.

The **maculopapular rashes** in measles are characterized by vascular congestion, oedema, epithelial necrosis and round cell infiltrates. Measles virus antigen is absent in the skin lesions unlike other sites. However, the viral antigens are concentrated in the blood vessel and in the endothelial cells of the dermal capillaries. Thus, virus is not present in the skin lesion and there is **no shedding of virus from the skin surface**. The containment of skin infection occurs due to development of cytotoxic T cells which is thought to be responsible for damage to the infected tissue and production of interferon causing cellular resistance to infection.

IMMUNE RESPONSE

Host immune response is essential for viral clearance, development of protective immunity and clinical recovery. Innate immune response to measles virus infection leads to activation of NK cells and increase in non-structural viral proteins that inhibits the interferon production and thereby facilitates the viral replication.

The adaptive immune response consists of both humoral response and cellular immune response. Humoral response is responsible for establishment of lifelong protective immunity, whereas cellular immune response is responsible for recovery.

The appearance of measles specific IgM antibody initiates the humoral response. It arises at the time of rash and persists for 6–8 weeks. The presence of IgM antibody during the rash phase makes it the marker of diagnosis. The rise of IgM antibody occurs only during the primary infection and does not occur due to reinfection or vaccination. Rise of IgM is followed by IgG antibody. The role of humoral immunity in measles infection is evident from the immunity in infants conferred by passively transferred maternal antibody and immunity provided by anti-measles immunoglobulin.

Cellular immune response in measles is first marked by Th1 response which is characterized by interferon γ. This is important for viral clearance and clinical recovery. During the convalescence phase, Th2 response occurs which facilitates the production of anti-measles antibodies such as interleukin-4, interleukin-10 and interleukin-13. The role of cellular immune response is evident from the fact that, children with agammaglobulinemia recover completely from measles whereas children with T cell deficiency suffer from severe disease.

The duration of protection after measles infection is usually lifelong. This is due to continuous production of anti-measles antibody and measles virus specific CD4 and CD8 T lymphocytes.

Contrary to the intense immune response in measles virus infection, measles virus is the first virus to be described as a cause of immune suppression. Transient lymphopenia and Th1 inhibition are mainly the reason of immune suppression which in turn leads to paradoxical depressed response to non-measles pathogens making the individual more prone to secondary bacterial and viral infections.

LABORATORY DIAGNOSIS

The clinical presentations of measles are usually classical in nature. The laboratory confirmation is required in cases that are clinically challenging such as immuno-compromised or individuals with pre-existing antibodies, or to confirm a suspected outbreak particularly in low or non-endemic countries (Flowchart 12.1).

Serology: Measles specific IgM antibody detection is the mainstay of confirmation of disease. Measles IgM appears around 4 days after the onset of rash. Therefore, it may be negative during the first few days of illness. Two-thirds (75%) of cases become positive by 3rd day of rash and almost all by 4th day of rash. It persists for 3–4 weeks and starts to decline by 4th week of illness and becomes non-detectable by 8th week.

Demonstration of four-fold rise in titer can be employed by detection of antibody in acute and convalescent serum collected at a gap of 10–30 days. This can be done by: IgG antibody by ELISA, hemagglutinating antibody by HAI assay.

Detection of measles specific IgG antibody or HAI antibody in a single serum sample indicates past infection and is used for sero-surveillance study.

Virus isolation: Measles virus can be isolated from several clinical samples using susceptible cell line. Respiratory samples, such as nasal or throat swabs, nasopharyngeal aspirate,

Flowchart 12.1: Lab diagnosis of measles

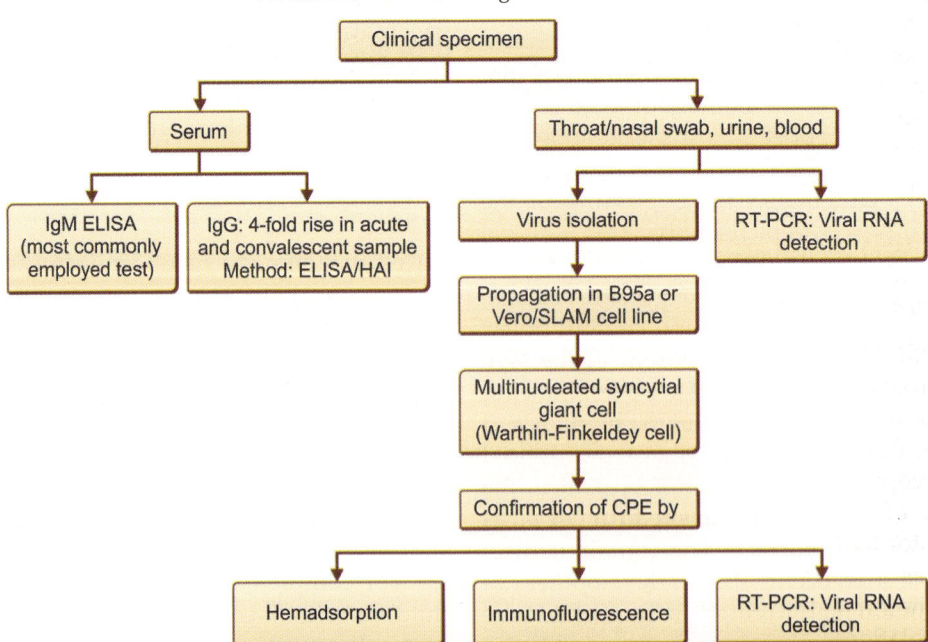

urine and blood samples can be collected for isolation of virus and for detection of RNA. Swabs should be collected in viral transport medium and transported in cold chain (4°C) to the lab. It can be stored at 4°C for a few hours then should be stored at –70°C. Samples for virus isolation should be collected as soon as possible after the onset of rash.

Urine collection: First morning urine is ideal to collect. Preferably around 50 mL should be collected. Immediately sample should be centrifuged at 4°C at 1500 rpm for 5–10 minutes. The deposit is transferred to 1–2 mL of VTM and then stored at –70°C.

Blood: Lymphocytes act as the source of measles virus. Blood is collected in heparinized vial, lymphocytes are separated using Ficoll hypaque gradient centrifugation technique. The purified lymphocytes are stored at –70°C.

Cell lines: B95a and Vero/SLAM are the two cell lines recommended for isolation of measles virus from clinical samples.

B95a: This is an Epstein-Barr virus transformed marmoset B lymphoblastoid cell line. These cells are 1000 times more sensitive than commonly used cell lines. Hence, one of the recommended cell lines for primary isolation of measles virus. The cell line is easy to maintain and cytopathic effect produced by measles virus is readily observed. However, as the cells are Epstein-Barr virus infected, these cells should be handled as infectious material. Hence, not preferred currently in measles virus laboratory network.

Vero/SLAM: This cell line is made up of Vero cells transfected with plasmid encoding genes of human signalling lymphocyte activation molecule (SLAM) also called CDw150. This is a glycoprotein present on the membrane of some of the T and B cells and act as the receptor for binding of measles virus. SLAM acts as a receptor for both wild type and vaccine strains. The sensitivity of Vero/SLAM for isolation of measles virus is same as that of B95a cells. These cells are also sensitive to laboratory adapted measles virus strains, vaccine strains and rubella virus. They possess less biological hazard than B95a cells. The disadvantage of Vero/SLAM is the requirement of Geneticin for expression of SLAM. This cell line is currently used for isolation by the WHO measles virus laboratory network.

Cytopathic effect: Measles virus produces two types of cytopathic effect (CPE). First, multinucleated syncytial giant cell **(Warthin-Finkeldey cell)** due to fusion of the infected cells induced by the virus. Multinucleated giant cells are formed by the wild virus. Second, **spindle cells or stellate cells** are produced by the adapted measles virus (strain that has been given repeated passage in the cell line).

Identification from infected cell can be made by (i) hemadsorption using 0.5% monkey RBC, (ii) immunofluorescence using measles virus specific antibody or (iii) by reverse transcriptase PCR using specific primers. Infected cells can also be stained for demonstration of intranuclear and intracytoplasmic inclusion bodies.

Reverse transcriptase polymerase chain reaction (RT-PCR): This is used for genetic characterization of the isolated strains rather than as a diagnostic procedure.

VACCINE

Live attenuated measles vaccines are available either in single or in combination with other viral vaccines. In 1963, the first licensed live attenuated measles vaccine became available. The first measles vaccine candidate was Edmonston strain isolated by Enders and Peebles in 1954 in cell culture from blood sample of David Edmonston a child with measles. The Edmonston B vaccine strain was the first candidate vaccine strain. It was prepared by serial passage in human and monkey kidney cells and chicken embryo fibroblast cell line. This strain was found to be inadequately attenuated. This led to the development of second-generation vaccine strains with further attenuation such as; **Schwarz, Moraten and Edmonston-Zagreb** which are currently in use. All these strains are the Edmonston lineage. The non-Edmonston live attenuated vaccine strains are: Shanghai-191, Leningrad-16 and CAM-70. The nucleotide sequence among the various vaccine strains are minimal (<0.6%).

Presence of maternal measles antibody is inhibitory for development of protective antibody after vaccination. Approximately 85% of vaccinated children develop protective antibody when vaccinated at 9 months of age and >90% when given at 12 months of age.

Measles vaccine induces both cellular and humoral immune responses as it happens with natural infection.

Protective antibody appears during second week of vaccination and concentration rises to its peak during 3rd to 4th week. The duration of protection after vaccination is usually considered shorter than the duration of protection conferred after natural infection.

Current recommendation: Two doses of measles vaccination are recommended for all countries.

Countries with high level measles transmission with high-risk of mortality (Ex. India):
- MCV1: 1st dose—9 months of age.
- MCV2: 2nd dose—16–24 months.

Countries with low level measles transmission:
- MCV1: 1st dose at 12 months of age
- MCV2: 2nd dose preferably at the time of school enrollment.

Recommendation of supplementary dose of measles vaccine to infants from 6 months age in following situations (MCV0: supplementary dose):
- During measles outbreak
- During campaign in endemic countries
- Refugee population

- Children with high-risk of contracting measles cases
- Children born to HIV positive mother.

Recommendation for children who have received MCV0: These children should receive both MCV1 and 2.

Disadvantages of presently available measles vaccination:
- Vaccine gets inactivated on exposure to light and heat.
- Once reconstituted, loses its potency within few hours of storage.
- Maintenance of cold chain is essential.

Presence of passively acquired maternal antibodies and immunological immaturity reduces the vaccine efficacy.

Integrated Measles-Rubella Control: Measles-Rubella (MR) vaccination: In 2012, Measles and Rubella Initiative (M&RI), WHO prepared planned document to achieve the goal of measles and rubella elimination in at least five WHO regions by end of 2020. In 2013, WHO regional committee of Southeast Asia (SEA) has decided to adopt the goal of measles elimination and rubella/CRS (congenital rubella syndrome) control in SEA regions by 2020. The core strategic objectives were articulated to achieve this goal:

a. To achieve and maintain high level of population immunity (at least 95%) with two doses of measles and rubella containing vaccines.

b. Develop and sustain a sensitive and timely case-based measles, rubella and CRS surveillance system.

c. Develop and maintain an accredited measles and rubella laboratory network that supports every country in the region.

d. Strengthen support and linkage to achieve the above three strategic objectives.

Measles-rubella (MR) campaign: India and ten other WHO Southeast Asian countries have resolved to achieve the goal to eliminate measles and to control rubella/congenital rubella syndrome by year 2020.

The inclusion of these two viral vaccines was decided by WHO as measles is still a killer disease in several countries like India and rubella causes severe irreversible congenital birth defects like deafness and blindness in near 40,000 children every year.

MR campaign has two phases: Campaign phase and implementation in routine vaccination phase (Table 12.3).

In the campaign phase, it targets the children of 9 months age to <15 years age irrespective of their vaccination status, i.e. even if they have received measles vaccine or MMR vaccine (measles, mumps, rubella) or history of measles or rubella disease.

In the implementation phase, MR vaccine has been implemented in the routine National Immunisation Programme.

First dose is given at 9–12 months of age and second dose at 16–24 months.

Table 12.3: Measles-rubella vaccine
Type: Live attenuated vaccine
Strain:
• Measles: Edmonston
• Rubella: RA27/3
Route: Subcutaneous
Dose:
• 1st: 9–12 months age
• 2nd: 16–24 months
Efficacy:
• Measles: 85–90%
• Rubella: 95–100%

ERADICATION

Measles is a potentially eradicable disease due to the following reasons:
• It has no animal reservoir.
• Single serotype
• Safe and effective vaccine is available.

• Chain of transmission is interrupted by vaccination.
• Establishment of surveillance system.

The criteria for measles elimination have been suggested as: (i) Absence of an endemic measles genotype for at least 12 months and (ii) one dose of measles coverage of at least 95% with scope for second dose.

Bibliography

1. Beaty SM and Lee B. Constraints on the genetic and antigenic variability of measles virus. Viruses 2016, 8, 109; doi:10.3390/v8040109

2. Kelly H, Riddell M, Heywood A, Lambert S. WHO criteria for measles elimination: a critique with reference to criteria for polio elimination. Euro Surveill. 2009 Dec 17;14(50):19445.

3. https://mohfw.gov.in/sites/default/files/Measles%20rubella%20vaccine%20operational%20guidelines.pdf

4. https://www.who.int/ihr/elibrary/manual_diagn_lab_mea_rub_en.pdf

5. https://www.who.int/immunization/monitoring_surveillance/burden/vpd/surveillance_type/active/measles/en/

6. https://www.who.int/immunization/policy/Immunization_routine_table1.pdf?ua=1

7. https://www.who.int/immunization/policy/Immunization_routine_table2.pdf?ua=1

8. Kelly H, Riddell M, Heywood A, Lambert S. WHO criteria for measles elimination: A critique with reference to criteria for polio elimination.

9. Moss WJ. Measles. Lancet 2017;390:2490–2502.

10. Progress Toward Regional Measles Elimination — Worldwide, 2000–2016. *Weekly*/October 27, 2017/66(42);1148–1153. https://www.cdc.gov/mmwr/volumes/66/wr/mm6642a6.htm.

11. Vaidya SR, Chowdhury DT. Measles virus genotypes circulating in India, 2011-2015. J Med Virol 2017 May;89(5):753–58.

Mumps Virus

Mumps is an acute viral infection caused by mumps rubulavirus (previously known as mumps virus) that mostly occurs in children. The disease is commonly manifested as swelling and tenderness of salivary glands. It is usually benign and self-limiting in nature but may be associated with severe complications like epididymo-orchitis and meningitis.

The first epidemic of mumps in history has been described by Hippocrates in fifth century BC as a disease in children with swelling in front of ear and sometimes accompanied by painful enlargement of testes.

Mumps disease was caused by a filterable agent was first suggested by Granata in 1908. In 1934, Johnson and Goodpasture first reproduced the disease in experimental animals by inoculating the saliva from acute mumps cases. Major progress occurred when the virus was cultured in chick embryo by Habel.

The etymology of the word Mumps is not clear. The possible sources are; English noun mump meaning **lump**; English verb **to sulk** a characteristic facial expression; or from Scottish verb **to speak indistinctly** which has been ascribed to the mumbling speech pattern of the affected person.

VIRUS

According to the recent classification of virus by ICTV, the previously called mumps virus is now named as "Mumps rubula virus". It is a member of family Paramyxoviridae which contain seven genera of which Respirovirus, Rubulavirus, Henipavirus and Morbillivirus are of human importance. Mumps rubulavirus belongs to the genus "Rubulavirus" along with human rubulavirus 2 and human rubulavirus 4 which were previously called parainfluenza virus 2 and parainfluenza virus 4, respectively. The classification of Paramyxoviridae is given in Table 12.1.

The virus is a single stranded, negative sense RNA virus of 100–200 nm diameter of spherical shape. The nucleocapsid is of helical symmetry and is enclosed by bilayer of lipid envelop. Hemagglutinin-neuraminidase (HN) and fusion (F) proteins are the two glycoproteins present as surface projections from envelop. The inner layer of the envelop consists of the non-glycosylated membrane protein. The SH or small hydrophobic protein is a putative membrane protein. The genome of the virus is a single stranded, negative sense RNA consists of 15384 nucleotides and encodes for seven proteins.

i. Proteins associated with nucleocapsid are: Nucleocapsid associated protein (NP), phosphoprotein (P) and large proteins (L)

ii. Envelop proteins: The hemagglutinin-neuraminidase (HN), fusion (F) and matrix proteins (M) are and

iii. Membrane associated protein: Small hydrophobic or SH protein.

The proteins that are associated with nucleocapsid are responsible for replication of RNA.

The surface proteins HN and F are the major antigenic determinants. Antibodies produced against these proteins are primarily responsible for neutralization of the infectious virus providing protection. The hemagglutinin-neuraminidase (HN) protein is responsible for attachment with the host cell receptor and along with fusion protein (F) fusion of plasma membrane with the virus envelop. The hemagglutination property of HN protein is used for identification of virus. F protein is mainly responsible for the fusion of the cell membranes and thereby for spreading of the virus from cell-to-cell within the host.

The SH protein is the most variable protein. The sequence of SH protein shows highest diversity. Therefore, genotyping of mumps virus is based on the sequence diversity of SH gene.

EPIDEMIOLOGY

Mumps is endemic all over the world and is a disease of childhood. School going age of 5–15 years is the most commonly affected age group. The disease is not commonly seen in infants below 1 year of age possibly because of existing maternal antibody. The secondary attack rate is relatively considered to be much less as compared to measles and varicella. During the pre-vaccination years, epidemics used to occur every 2–3 years. In temperate region, the disease is mostly reported during January to May (winter and spring) months with a peak in February–March. However, in tropical countries, no such seasonal variation has been reported.

Change in epidemiology has been observed in countries where mumps vaccine has been included in the routine immunization program. In post-vaccination years, seasonal variation is no more observed even in temperate regions. Shift in age group affected has been observed from 5–15 years to 18–24 years, possibly because of waning immunity.

Outbreaks of mumps have been reported both from low vaccine coverage as well as from high vaccine coverage areas. In later case, it has been attributed mostly to waning immunity and has been reported in student communities residing or studying in close environment facilitating the spread of virus.

In India, mumps is prevalent in all parts of the country. Outbreaks of mumps with different clinical presentations like classical parotitis, atypical mumps and aseptic meningitis have been reported from different parts of the country occurring mostly in children age group.

Molecular Epidemiology

Presently mumps virus has **12 genotypes** based on the sequence diversity of SH gene. The genotypes are named as **A to N**. Genotypes E and M have been removed as the members belonging to these genotypes were shifted to other genotypes.

Genotypes A and B were previously seen predominantly in Europe and Japan, respectively. Since 1990, wild virus strains of these genotypes have not been reported. Strains of genotype A and B are currently used for production of vaccine. Jeryl-Lynn strain a member of genotype A and Urabe AM9 of genotype B are the two most commonly used vaccine strains.

Till 2012, genotypes C, D, H and J were common in Western hemisphere and genotypes B, F, G and I in Asian countries. In India, genotypes C and G have been reported from several outbreaks across the country.

PATHOGENESIS AND PATHOLOGY

Humans are the only natural host of MuV.

Mumps virus is transmitted by respiratory route or through direct contact of oral mucosa with the infected respiratory droplets, fomites or saliva. This mode of transmission is evident from (i) experimental transmission of disease by Henle and colleague to children through both oral and nasal inoculation, (ii) isolation of virus from various respiratory samples and saliva from both asymptomatic and children with parotitis and (iii) transmission among close contacts.

Mumps virus (MuV) enters the respiratory epithelial cells by binding to its receptor sialic acid. The entry of MuV to the epithelial cells occurs through both apical and basolateral sides as sialic acid is present on both these sides. This is presumed to facilitate the spreads of infection to the neighboring cells and also causes secondary viremia. The release of virus from the infected cells mostly occurs through apical side which is believed to restrict the infection to the respiratory mucosa. In case of MuV, apical release results in glandular involvement. Parotid gland is most commonly affected. Other salivary glands like sub-mandibular and sublingual glands can also be affected.

After infecting the respiratory epithelial cells, virus spreads to the regional lymph nodes leading to viremia. This further spreads the virus to various extra-respiratory sites. Most of the organs of the body can be infected during viremia, though MuV has predilection for glandular organs. The organs that are commonly affected include testes, pancreas kidney, ovary, epididymis and central nervous system (CNS) (Fig. 13.1).

During the process of infection, MuV is present in the blood for 1–2 days. Isolation of MuV from blood has been reported but is extremely rare. The virus has been isolated from saliva and throat from 1 week prior to 1 week after the swelling of the gland and from urine till 14 days after disease onset.

Parotitis

Mumps virus infects the ductal epithelium of parotid gland. Replication of virus in parotid gland leads to perivascular and interstitial infiltration of lymphocytes, hemorrhage, edema and necrosis of acinar and periductal cells. The infected cells desquamate into the lumen of the duct. This along with lymphocytic infiltration and edema leads to blockage of salivary gland ducts and swelling of the gland.

Orchitis

Both direct infection by virus and indirect immune-mediated damage have been postu-lated for development of orchitis. Both Leydig and germ cells are involved. Blockage of the tubule occurs due to epithelial cell debris along with mononuclear cell infiltration and edema. This reduces the production level of testosterone. Atrophy of testicle can lead to oligospermia and reduced fertility.

CNS

MuV is a neurotropic virus. It can enter the CSF via choroid plexus or through infected mononuclear cells. This can finally lead to infection of the ependymal cells lining the ventricles and brain parenchyma. Lymphocytic infiltration of meninges occurs along with edema, gliosis, perivascular cuffing and neuronal degeneration.

CLINICAL FEATURES

The incubation period ranges from 12 to 24 days. Nearly one-third of infected persons remain asymptomatic. Amongst the sympto-matic patients, enlargement of parotid gland is the most common manifestation and seen in around 95% of cases. The disease starts with prodromal symptoms that include fever, anorexia, and malaise. Specific symptoms are evident from earache and tenderness. Swelling of parotid gland is visible within 2–3 days. To start with, the swelling is unilateral which

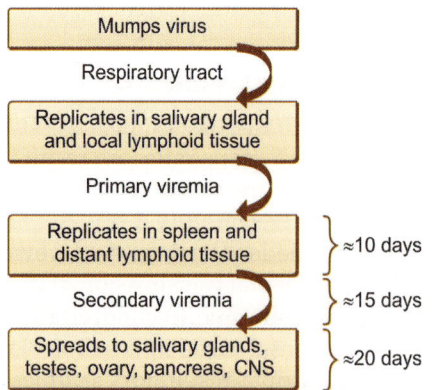

Fig. 13.1: Pathogenesis of mumps

becomes bilateral within 1–2 days in majority of the cases. Involvement of other salivary glands may occur, however, usually occur with parotitis.

Extra-salivary Manifestations

Meningitis: Involvement of CNS is the most common extra-salivary manifestation. Cerebrospinal fluid pleocytosis is seen in near 50% of mumps cases. Clinical manifestation of meningitis occurs only in 1–10% of parotitis cases whereas only half of mumps meningitis cases have parotitis. Mumps meningitis presents with the typical features of viral meningitis like headache, fever and nuchal rigidity. CSF shows pleocytosis which is usually <500 lymphocytes/mL with slightly decreased sugar and normal protein level.

Encephalitis: Encephalitis due to mumps virus infection has been reported in around 1 in 6000 mumps cases. The onset of encephalitis has been observed to occur in two patterns: (i) Occurrence along with parotitis, and (ii) 7–10 days after development of parotitis. The former is thought to be due to the direct viral infection whereas the latter is due to post-infection demylination process thought to be host immune response mediated.

Orchitis: Orchitis has been reported in 10–20% of postpubertal men. The symptoms are acute onset with fever, pain, swelling and tenderness. In majority of cases, it is associated with epididymitis. Bilateral testicular involvement occurs in around 20% of cases. Unilateral orchitis is not associated with infertility. Patients with bilateral orchitis may have hypofertility and sterility is rare.

Other manifestations like pancreatitis, oophoritis, myocarditis, nephritis, thyroiditis, prostatitis, etc. have also been associated due to mumps virus infection.

LABORATORY DIAGNOSIS

Mumps virus infection is often diagnosed on the basis of its classical clinical presentation. Laboratory confirmation is important in outbreak situation, non-classical clinical manifestation, in vaccinated persons and to find the genotype distribution.

Mumps virus infection can be diagnosed by serology, virus isolation or detection of viral RNA.

Isolation of virus and RNA detection can be done from saliva, throat swab, urine samples and CSF in case of meningitis. Samples should be collected during the acute stage of infection. MuV is present in the saliva 2–3 days before to up to 5 days of parotid swelling. Transport of samples to the laboratory for isolation and RNA detection should be done maintaining the cold chain. Storage for a few days can be done at 4°C and at –70°C when required for a prolonged period. Virus isolation can be done by using the Vero cell lines. Confirmation of infected cells is made by detection of antigen using immuno-fluorescence and RNA detection.

Serology is the mainstay of diagnosis. MuV specific IgM antibody detection in acute serum sample by ELISA is the most commonly employed test for confirmation. In a previously unexposed or unvaccinated person, MuV IgM is usually positive during the initial part of illness and reaches its peak during seventh day and remains positive for a few weeks to a few months. In previously exposed or vaccinated persons, IgM detection may be negative due to transient rise in titer and thus negative IgM in this case should not rule out Mumps virus infection. Demonstration of four-fold rise of antibody titer in acute and convalescent samples by complement fixation test, hemagglutination inhibition test or IgG seroconversion by ELISA also confirms the diagnosis.

VACCINE

Vaccine against mumps virus is available only in combinations as measles, mumps, rubella (MMR) and measles, mumps, rubella and varicella (MMRV).

Jeryl Lynn (JL), Leningrad-3 and Leningrad-Zagreb strains are the currently recommended vaccine strains for mumps by World Health Organization (WHO). Jeryl Lynn strain belong to the genotype A. Use of Urabe and Rubini strains for vaccine preparation has been stopped due to association of high rate of aseptic meningitis and inadequate vaccine efficacy, respectively. In India, mumps vaccine preparations are manufactured by Serum Institute of **India** using **Leningrad-Zagreb strain (Tresivac)** and by **GSK using Jeryl Lynn strain (Priorix)**. Mumps vaccine is prepared in chick embryo fibroblast tissue culture. The overall efficacy of mumps vaccine is nearly 80% after first dose and about 90% after second dose.

Dose schedule: The first dose of mumps vaccine is recommended at the age of 1 year or after. Second dose of vaccine should be given between the ages of 4 to 6 years.

In India, mumps vaccine is not included in the National Immunisation Programme. Indian Academy of Pediatrics (IAP) recommends the first dose of mumps vaccine as MMR at 9 months of age in place of single measles vaccine and second dose at 15 months of age.

Rubella Virus

Rubella is a mild exanthematous disease, caused by the rubella virus. The disease is mostly a self-limiting disease presenting with mild morbilliform rash with fever and lymphadenopathy. The major concern regarding this virus is its ability to cause severe irreversible teratogenic effects in fetus when the infection occurs in pregnant women.

It is also known as "German measles" when it was described by two German physicians De Bergen and Orlow in 1752 and 1758 respectively as a milder form of measles. The term "rubella" was given by a Scottish physician Henry Veale in 1866, which has been derived from the Latin word "*rubellus*" means "reddish" or "little red". The virus gained its importance when an Australian ophthalmologist Norman Gregg linked the cataract and congenital cardiac defects in children with occurrence of rubella in mothers during pregnancy. This observation was noticed due to spurt in the number of cataracts in newborns that was proceeded by large rubella outbreaks.

VIRUS

Rubella virus, now called Rubivirus rubellae is the only member of the genus Rubivirus, belong to the family Matonaviridae.

It is a single stranded, positive sense RNA virus. The shape is roughly spherical with a diameter of 50–85 nm. The genome is enclosed within the icosahedral capsid. The nucleocapsid is surrounded by an envelop. The genome consists of 9762 nucleotides and encodes three structural proteins (C, E1, E2) and two non-structural proteins (P90, P150) (Fig. 14.1).

E1 and E2 are the two glycosylated lipoproteins present as transmembrane spikes on the envelop. E1 which is the major antigenic

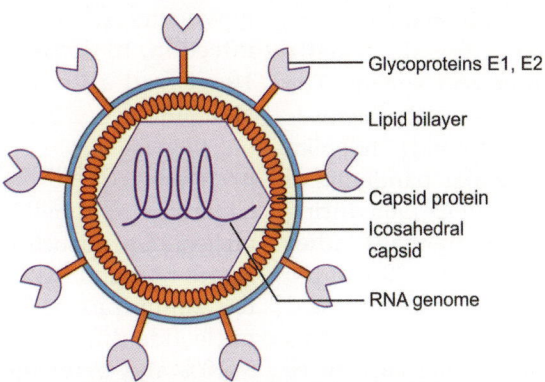

Fig. 14.1: Rubella virus (schematic)

Glycoproteins E1, E2

Lipid bilayer

Capsid protein

Icosahedral capsid

RNA genome

determinant, induces humoral antibody response. The antibody titer against this protein correlates with the level of protection. E1 protein is also responsible for binding to the host cell receptor myelin oligodendrocytes glycoprotein (MOG) and receptor-mediated endocytosis. E2 protein is embedded to the envelop and forms connections between the rows of E1 proteins.

The two non-structural proteins are responsible for viral replication.

EPIDEMIOLOGY

Humans are the only natural reservoir of rubella virus. There is no animal reservoir. The infection is mainly transmitted through infected respiratory droplets either by respiratory route or by direct contact with respiratory mucosa. The virus can also be transmitted transplacentally from infected mother to fetus.

The period of infectivity is from one week before to one week after the onset of rash. However, the congenitally infected infants shed high concentration of virus for several months and serves as a potential source for maintenance of virus in the community.

In endemic countries, the maximum number of cases occur in the age group of 5–9 years. With the implementation of universal rubella immunization, the affected age has shifted from children to young adults.

It has been observed that in children, both sexes, are affected equally whereas in adults infection is more in women than men.

The annual seasonal outbreaks usually occur during late winter and spring. The accumulation of susceptible population leads to epidemics in every 6–9 years and pandemics in 20–30 years. The last major pandemic of rubella had occurred from 1962–1965 during which approximately 10% pregnant ladies were infected with the virus and 30% of the infants born to infected mothers manifested with congenital rubella syndrome. During this pandemic, 12.5 million rubella case were reported only from US which resulted in more than 13,000 fetal deaths and 20,000 cases of congenital rubella syndrome (CRS). This pandemic triggered the development of rubella vaccine and a live attenuated rubella vaccine was implemented in 1969. After the vaccination, both the cases of rubella and CRS decreased by more than 95% in two decades. In 2005, Center for Disease Control and Prevention (CDC) has declared elimination of endemic CRS cases from US.

According to WHO data, the number of reported rubella cases from six WHO regions during 2018 was about 10,000 of which maximum were contributed by Africa, Southeast Asia and Western Pacific WHO regions.

In 2011, WHO recommended the rubella vaccination campaign, targeting the children of 9 months to 14 years. The percentage of WHO member states to include rubella containing vaccine in their routine immunization program has increased from 43% in 1996 to 78% in 2016. However, despite this, intermittent rubella outbreaks continue to occur. Protective immunity in at least 85% of population is required to prevent outbreaks.

In India, the number of confirmed rubella cases during the year 2015 to 2018 were 3265, 9084, 2854 and 1066, respectively. The maximum number of cases occurs in the age group of 5–9 years which is still consistent with the pattern of endemic country. In a study among pregnant women, 15.2% were found to be susceptible to rubella. The laboratory evidence of CRS has been found in 1–15% of newborns with suspected intrauterine infection, 10–15% of pediatric cataract and 10–50% of all children with congenital anomalies.

Molecular Epidemiology

Genotype of rubella virus is based on the nucleotide sequence variation of E1 gene. Two clades RG1 and RG2 consisting of more than ten genotypes have been identified. The nucleotide difference between two clades is 8–10% and between genotypes <5%. The strains from America, Europe and Japan belong to RG1 and strains from China and India belong to RG2.

Clade 1 (RG1) has 10 genotypes (A to J) of which 9 are confirmed and one is provisional (1A: Provisional; 1B to 1J: Confirmed).

Clade 2 or RG2 has three genotypes; 2A, 2B and 2C.

The knowledge of molecular epidemiology helps in tracking the source of infection. Genotypes 1E and 2B are most widely circulated.

CLINICAL FEATURES

Postnatal rubella infection is mostly mild, self-limiting or subclinical in nature. The incubation period ranges from 14–21 days. When symptomatic, it starts with constitutional manifestations, such as fever, anorexia, malaise and lymphadenopathy. Constitutional symptoms are more pronounced in older children and adults. After around 2–5 days of prodromal symptoms, rash appears and prodromal symptoms disappear with the onset of rash. In 20% cases, enanthem has been observed in buccal mucosa called **Forchheimer spot**.

Lymphadenopathy is typically prominent in suboccipital, posterior auricular and cervical lymph nodes and most pronounced and tender at the time of rash onset.

Exanthema is pinpoint, erythematous and maculopapular. Typically appears from face and spreads centripetally to all over body. Rash persists for about 3 days and fades in the same direction as it appears. The most probable mechanism of rash is immunological, but rubella virus has been isolated from skin biopsy taken from the area of rash, from area without rash and also from skin of patients with subclinical infection.

COMPLICATIONS

Arthralgia is the most common complication which occurs in 30 to 60% of teenagers and adult females. Joint involvement appears almost one week after rash and commonly persists for a week. It can affect any joint but fingers and knees are commonly involved.

Encephalitis and thrombocytopenia are the other complications of rubella.

Congenital Rubella Syndrome

Maternal immunity is protective against intrauterine infection. Rubella virus can cause severe congenital defects when acquired during *in utero* through transplacental transmission from infected mother. The infection when occurs in the pregnant women during the early part of gestational period up to 8–10 weeks the transplacental transmission occurs in up to 90% of cases, which decreases to 25% during second trimester, rises again to 35% at 27–30 weeks and near 100% at 36 weeks of gestation. However, the severity and risk of fetal malformation decreases with the acquisition of infection in mother during advanced gestational period. The risk is almost nil when infection in mother occurs after 16 weeks of gestation.

In susceptible pregnant women, the infection spreads to placenta then crosses the placenta and spreads to various organs of fetus through the vascular system. The possible mechanisms of fetal damage are epithelial necrosis of chorionic villi, apoptosis and inhibition of mitosis by direct viral effect on infected cells, restricted precursor cell development and damage to endothelial cell of blood vessels resulting in ischemia in developing organs.

The **classical triad** of congenital rubella syndrome consists of **cataract**, **cardiac defects** and **sensory neural hearing loss** which typically occurs when the neonate gets infected during first 8 weeks of gestation. Sensorineural hearing loss occurs in maximum (80%) cases and cataract and cardiac defects in 50–60% of cases.

Sensory neural hearing loss is usually bilateral. In some cases, it may be the sole manifestations whereas in some cases it may not be apparent till second year of life.

Patent ductus arteriosus and pulmonary artery and valvular stenosis are the common cardiac defects.

Amongst the ophthalmic abnormalities, cataract, pigmentary retinopathy and congenital glaucoma are the common ones. Pigmentary retinopathy is characterized by salt and pepper appearance of fundus.

The common manifestations of congenital rubella syndrome (CRS) are given in Table 14.1.

Diagnosis of CRS is often based on the detection of rubella specific IgM antibody in the cord blood or in the serum of infant within first six months of life.

LABORATORY DIAGNOSIS

As the clinical manifestation in rubella is not specific and can mimic many other exanthematous diseases, clinically suspected rubella cases should always be confirmed by laboratory test.

Detection of rubella infection can be made by isolation of virus and detection of viral genome from various clinical samples and detection of rubella specific antibodies in blood.

Serology

Detection of IgM antibody: Rubella specific IgM antibody detection in serum is the mainstay of diagnosis of acute infection. Presence of IgM in single sample indicates acute or recent infection except in individuals vaccinated with rubella vaccine within 8 days to 8 weeks of sample collection and in absence of rubella transmission in the community.

Positive IgM antibody in infants suggests congenital rubella syndrome or congenital rubella infection.

IgM antibody in rubella infection appears at the time of rash and persists for 4–8 weeks. Around 50% cases are positive for IgM at the time of rash, whereas nearly all cases are positive by 5 days of rash. ELISA is the most commonly used serological test for detection of IgM antibody. Most of the commercially available ELISA tests are 80% specific. False positive rubella IgM can occur due to presence of rheumatoid factor, Parvovirus B19 or CMV infection.

Detection of IgG antibody: IgG antibody starts appearing just after IgM but persists lifelong. Acute rubella infection can also be diagnosed by demonstration of fourfold or more rise in rubella specific IgG antibody in acute and convalescent serum samples collected at least at 10 days interval and tested in parallel. The detection of IgG can be done by quantitative ELISA. Detection of IgG antibody in single serum sample indicates past infection and in infants can be due to passive transfer of maternal antibody.

IgG avidity test: During the early phase of primary infection, IgG antibody is of low avidity in nature which becomes stronger with passage of time. Therefore, detection of low avidity IgG antibody is indicative of primary infection and detection of high avidity IgG

Table 14.1: Manifestations of congenital rubella syndrome	
Major manifestations	*Other manifestations*
Ophthalmic manifestations	• Microcephaly
• Cataract	• Hepatosplenomegaly
• Retinitis	• Jaundice
• Congenital glaucoma	• Thrombocytopenia
• Cloudy cornea	• Purpura
• Chorioretinitis	• Blue berry muffin spots
Cardiac defects	• Radiolucency of long bones
• Patent ductus arteriosus	**Delayed manifestations**
• Pulmonary artery stenosis	• Mental retardation
Hearing defects	• Psychomotor retardation
• Bilateral sensorineural hearing loss	• Attention deficit, autism

indicates past infection or previous rubella vaccination. Avidity test is useful in differentiating between primary and secondary infection.

Isolation of Virus

Rubella virus can be isolated from nasopharyngeal secretion, oral fluid, blood, CSF, urine in rubella and CRS cases, of which nasopharyngeal and throat samples are more commonly used. Tissue samples can also be subjected for virus isolation in CRS cases. Virus can be isolated on the day of rash and can be detected up to 10 days.

Vero/hSLAM (human signalling lymphocyte activation molecule) cell line is recommended for isolation of rubella virus by Global Laboratory Network, WHO. This cell line is made up of Vero cells transfected with plasmid encoding genes of human signalling lymphocyte activation molecule (SLAM) also called **CDw150**. The virus does not produce any cytopathic effect. Identification of virus is done by detection of E1 protein in the infected cell by immunofluorescence method, immunocolorimetric assay or by RT-PCR.

Isolation of rubella virus, however, is not used for routine patient diagnosis. It is useful for epidemiological and research purpose.

RT-PCR

The viral RNA can be detected from various clinical samples as described for isolation of virus. RT-PCR positivity remain up to 10 days. Sequencing of the PCR product is important for molecular typing, helps in differentiating between wild type virus and vaccine virus and also to track the source of infection. Table 14.2 gives the salient points of lab diagnosis of rubella virus infection.

VACCINE

After the successful isolation of rubella virus in cell culture, the progress on rubella vaccine led to the development of an effective live attenuated vaccine. Licensed rubella vaccine is available since 1970. Globally, rubella vaccine

Table 14.2: Lab diagnosis of rubella

Serology
- Rubella IgM antibody detection: Mainstay of diagnosis
- Rubella IgG avidity test: Differentiate between primary and secondary rubella infections

Isolation of virus
- Nasopharyngeal secretion, oral fluid, blood, CSF, urine
- Cell line: Vero/hSLAM

Reverse transcriptase PCR
- Nasopharyngeal secretion, oral fluid, blood, CSF, urine
- Detection of viral RNA
- Genotyping
- Differentiate between wild and vaccine strain (sequencing)

containing **RA 27/3 strain** is the most commonly used. The other live attenuated rubella vaccine contains Takahashi, Matsuura, TO-336 and BRD-2 strains. RA 27/3 strain was derived from kidney of a rubella virus infected fetus. The isolate has been attenuated by passing 4 times in human diploid fibroblast cell line followed by 17–25 times in WI-38 fibroblast cells. In North America, presently RA 27/3 is the only vaccine strain used. In Japan, Matsuura and TO-336 strains are in use, whereas in China, BRD-36 is used.

Rubella vaccine is available as single preparation, but more commonly as combination with measles (MR) or measles, mumps (MMR) and measles, mumps and varicella (MMRV). The combination does not affect the efficacy of rubella vaccine.

The first dose of vaccine is recommended at the age of 12–15 months and second dose before entering into school.

Vaccine is administered subcutaneously. Protective antibody response occurs after around 4 weeks of first dose of vaccination in nearly 95% of the vaccinated individuals and around 100% after the second dose. The duration of protection by vaccination is around 20 years or more. Vaccination causes mild rubella symptoms and arthritis in up to 30%

of adult women. Vaccine is contraindicated during pregnancy, febrile illness and in immunocompromised individuals. However, in an incident where more than 100 pregnant women were inadvertently vaccinated, no fetal or maternal complications were reported.

Most of the Western countries have rubella vaccination in their national immunization programe. With successful vaccination and surveillance program, rubella has been eliminated from regions of America while European and Western pacific regions have started accelerated rubella control and CRS prevention.

Integrated measles-rubella control: Measles-rubella (MR) vaccination: Refer to "Measles" chapter.

Congenital rubella syndrome surveillance in India:

Rubella infection during first trimester of pregnancy can seriously affect the foetus either in the form of spontaneous abortion or serious birth defects causing congenital rubella syndrome. Government of India is now implemented country wide measles–rubella vaccination program in order to eliminate measles and control of rubella and CRS. In December 2016, Indian Council of Medical Research and the Ministry of Health and Family Welfare, Govt of India initiated the laboratory-supported surveillance of congenital rubella syndrome (CRS) in five sentinel sites in five Indian states. Infants of 0–11 months who were fulfilling the criteria of suspected CRS were enrolled. The case definition for CRS was kept in the line of WHO's definition.

Suspected CRS: The criteria of suspected CRS were defined as an infant meeting any one of the following five criteria: (a) **structural heart defects** (excluding patent ductus arteriosus/patent foramen ovale) that is confirmed by echocardiography; (b) **hearing impairment** (confirmed by brainstem evoked response audiometry (BERA), or auditory steady-state response); (c) one or more of the following **eye signs** such as, cataract, microphthalmos, microcornea, congenital glaucoma, and

pigmentary retinopathy; (d) maternal history of suspected or confirmed rubella infection during pregnancy or (e) strong clinical suspicion.

Clinically confirmed CRS: The detection of two clinical signs from group A or one each from group A and group B in an infant is termed as clinically confirmed.

Group A: Cataract(s), congenital glaucoma, pigmentary retinopathy, congenital heart defect, or hearing loss.

Group B: Microcephaly, developmental delay, meningoencephalitis, splenomegaly, purpura, radiolucent bone disease, or jaundice with onset within 24 hours after birth.

Laboratory-confirmed CRS. The presence in an infant of one condition from Group A (as above) and detection of rubella IgM antibody and/or persistently detectable rubella IgG antibody on at least two occasions at age 6–12 months, in the absence of receipt of rubella vaccine.

During 2016–18 CRS surveillance, 21.2% were labelled as laboratory confirmed CRS, of which 78.8% had cardiac defects, 59.9% showed one or more signs as per criteria and hearing impairment was noted in 38.6% infants and 24.1% died over a period of two years.

Bibliography

1. https://www.who.int/ihr/elibrary/manual_diagn_lab_mea_rub_en.pdf

2. https://www.who.int/immunization/position_papers/PP_rubella_July_2011_summary.pdf

3. https://www.who.int/immunization/sage/meetings/2016/october/1_MTR_Report_Final_Color_Sept_20_v2.pdf

4. Lambert N, Strebel P, Orenstein W, Icenogle J, Poland GA. Rubella. Lancet. 2015 Jun 6;385 (9984):2297–307.

5. Leung AKC, Hon KL, Leong KF. Rubella (German measles) revisited. Hong Kong Med J 2019 Apr;25(2):134–41.

Coronaviruses

Coronaviruses are a group of viruses that infect wide range of animals including humans. The name "corona" has been given due to the crown-like appearance observed under electron microscope by the surface spikes present on the envelop. Some of the members of the Coronaviridae family were known as human pathogens, but as mild respiratory pathogens. The virus family came to limelight for the first time after the emergence of the novel corona virus "Severe acute respiratory syndrome virus" (SARS CoV) in the year 2002 and SARS-CoV-2 which emerged in December 2019 and caused the COVID-19 pandemic and turned out to be the most dreadful pandemic the mankind has ever faced in the modern age.

The family Coronaviridae is taxonomically located within the order Nidovirales, and has a subfamily Coronavirinae. The subfamily Coronavirinae has four genera; Alphacoronavirus, Betacoronavirus, Deltacoronavirus and Gammacoronavirus.

Alphacoronaviruses and betacoronaviruses are of bat origin, gammacoronaviruses and deltacoronaviruses are of bird or swine origin. Till date seven human coronaviruses have been identified. Four of which mostly causes mild upper respiratory tract infections; HKU1, HCoV229E, HCoV NL63, HKU OC43 and three are responsible for pneumonia and severe respiratory distress syndrome; SARS CoV, MERS CoV and SARS-CoV-2. HCoV229E and HCoV NL63 belong to genus Alpha-

coronavirus and remaining five HKU1, HKU OC43, SARS CoV, MERS and SARS CoV2 belong to the genus Beta-CoV. SARS related CoV and MERS related CoV belong to the lineage B and C of the Betacoronavirus genus respectively. However, MERS related CoV is most closely related to the bat coronaviruses of lineage C.

This chapter includes SARS Coronavirus and MERS CoV. SARS-CoV-2 and COVID-19 is discussed in Chapter 24.

A. SEVERE ACUTE RESPIRATORY SYNDROME CORONAVIRUS (SARS CoV)

Emergence of SARS: During November 2002, an outbreak of acute community acquired pneumonia occurred in the Guangdong province of China and by February 2003, more than 300 such cases were reported in that region.

The disease took a pandemic turn when a physician came from the outbreak affected area to Hong Kong Special Administrative Region (HKSAR) and stayed in a hotel. He transmitted the infection to several other residents of that hotel. These secondary cases when went back to their respective countries developed the clinical manifestations. This led to the occurrence of similar cases (as that of the Guangdong province of China) in the USA, UK, Canada, Singapore, Vietnam, Philippines and also in China.

The disease came to the limelight when a WHO physician, Carlo Urbani who was working in Vietnam reported the outbreak of severe pneumonia to the World Health authorities. In March 2003, WHO issued a global alert and the clinical syndrome was named "severe acute respiratory syndrome" (SARS). The unknown pathogen was then detected to be a novel Coronavirus and named severe acute respiratory syndrome (SARS) virus.

Since then, cases were reported from 37 countries worldwide affecting a total of 8098 cases and 774 deaths with a mortality rate of ≈10%. The last human chain transmission was declared to be over in July 5th 2003.

Origin of SARS Virus

SARS virus is zoonotic in origin. Bats are the reservoir of SARS virus. The virus is believed to be transmitted from bat to civet cat, raccoon, dog and possibly to several other animals.

These animals are thought to be the source of virus to humans. Once the virus has infected the human host, transmission of virus from infected person to other individual through respiratory route has acted as the primary mode of transmission amongst the human host (Fig. 15.1).

SARS CoV was isolated from Chinese horseshoe bat which was genetically similar to the SARS CoV of human and palm civet. Bats act as reservoir of several other viruses without manifesting the disease and have been associated with zoonotic transmission.

The virus has also been isolated from several animals like palm civet, raccoon dogs and Chinese ferret badgers. These animals are available in the Chinese wet market, providing an ecologically suitable environment for transmission of virus to humans.

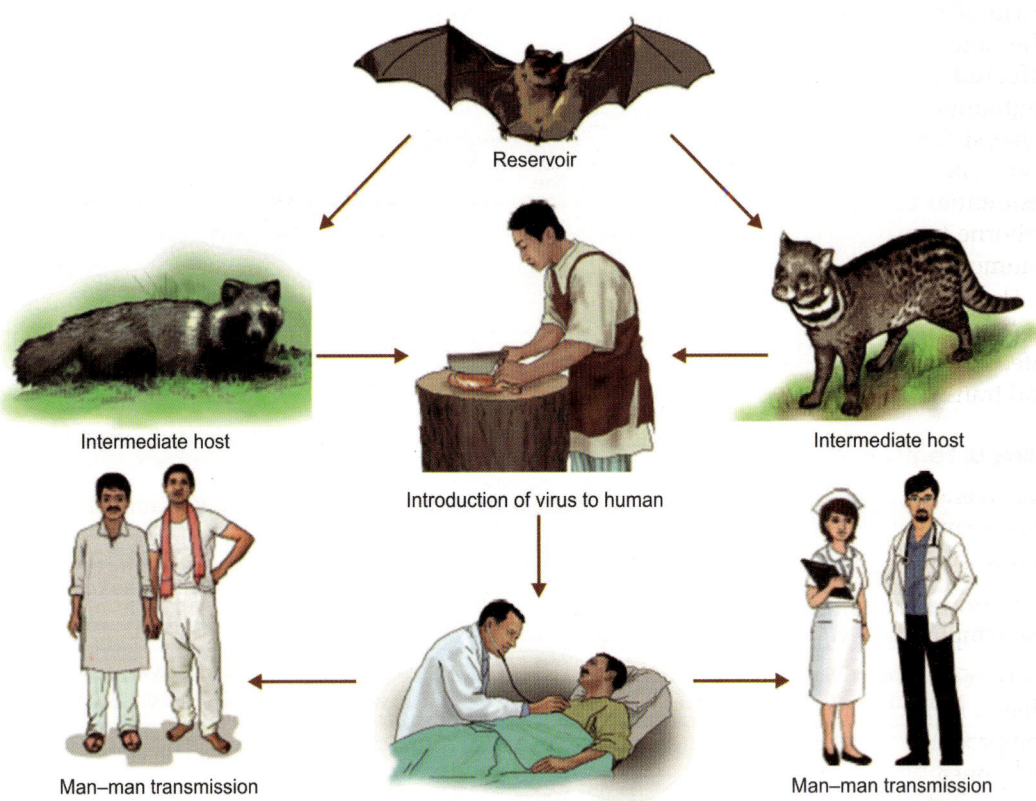

Reservoir

Intermediate host

Intermediate host

Introduction of virus to human

Man–man transmission

Man–man transmission

Fig. 15.1: Emergence and transmission of SARS virus

The initial SARS cases occurred in animal handlers. Infected human developed pneumonia-like illness and the disease then transmitted from person-to-person through respiratory route.

Modes of Transmission

Transmission through infected droplets is the main mode of person-to-person transmission. A large number of secondary cases are seen within the family members, health care workers and other close contacts. Direct or indirect contact with infected fomite is the major way of droplet infection.

In health care setting, handling the invasive procedure, nebulizer, suction, intubation or any procedure generating a large amount of respiratory droplets are mostly associated with disease transmission.

Airborne transmission has been documented in a localized residential setting in Hong Kong. The source of infection was traced to an infected patient's toilet from where the contaminated aerosols generated through exhaust fan were transmitted to the upper floors infecting hundreds of individuals in the residential complex. However, in general, airborne transmission is not considered as a common mode of transmission.

The presence of virus in the fecal sample and occurrence of diarrhea in around 40% infected patients raises the possibility of feco-oral transmission.

Clinical Features

The average incubation period is 2–10 days. Disease mainly manifests as prodromal phase and respiratory phase.

Prodrome phase: Disease starts with influenza like symptoms, such as fever, malaise, myalgia.

Early respiratory phase: After a few days of illness, respiratory symptoms start with dry non-productive cough, mild shortness of breath, tachypnea. Unlike other atypical pneumonia, features of upper respiratory tract such as rhinorrhea and sore throat are less common.

Chest physical signs are less marked as compared to chest radiograph like other atypical pneumonia. During the early respiratory phase, chest CT shows prominent ground glass consolidation in periphery and subpleural region of lower zone.

Late respiratory phase: In around two-thirds of cases, respiratory symptoms gradually progress to severe dyspnea and hypoxia, deterioration in chest sign with development of diarrhea during the second week of illness. In around 10–20% of cases, disease progresses and respiratory failure develops. Chest radiograph shows the progression of original consolidation to unilateral or bilateral consolidation. The case fatality rate is near 10%.

Pathogenesis

SARS CoV enters the target cells through binding to host cell receptor angiotensin-converting enzyme 2 (ACE-2) present on the surface of the target cell. The surface spike protein (S protein) of SARS CoV mediates the binding and fusion with the host cell. ACE-2, a metallopeptidase, which acts as the receptor for SARS CoV is present on the human lungs epithelial cells, enterocytes of small intestine and tubular epithelial cells of kidney indicating the tropism of SARS CoV for these organs. The receptor ACE-2 is also present in endothelial cells of vessels and arterial smooth muscle cells.

Besides ACE-2, lymph node and dendritic cell specific ICAM3 grabbing non-integrin (LSIGN and DC-SIGN) have also been found as the receptors for SARS CoV.

The disease pathogenesis in SARS CoV is multifactorial.

Role of ACE-2: ACE-2 by its normal function inhibits ACE and angiotensin II, both of which are responsible for inducing the lungs failure. SARS CoV when binds to ACE-2 downregulates it, leading to inhibition of the inhibitory action of ACE-2 on ACE and angiotensin II which in turn leads to lungs failure.

Direct viral effect: Replication of virus in the infected cells causes direct damage to the cells and induces local inflammation leading to injury of type II pneumocytes.

Immune cells: The role of the immune cells in disease pathogenesis is not very conclusive. Infection of the immune cells probably contributes to the disease pathogenesis by disseminating the infection to various organs. Destruction of infected immune cells may lead to immune suppression which exacerbates the severity of disease.

Increased level of cytokines and chemokines has been found in severe cases of SARS CoV patients. Overexpression of gene expression of various chemokines also has been demonstrated in SARS CoV infected cells. Based on these observations, it has been hypothesized that the infected pneumocytes and dendritic cells induce the release of proinflammatory cytokines and chemokines that are responsible for causing lung injury.

Antibody against the surface protein of SARS CoV has been found to cross-reacting with the pulmonary epithelial cells. Autoantibodies against the pulmonary epithelial cells have been found in SARS patients. These autoantibodies have been shown to induce the cytotoxic injury to the epithelial and endothelial cells.

LAB DIAGNOSIS

Nasal/nasopharyngeal or throat aspirate or swab is the primary sample for detection of SARS CoV. Specimens collected in swabs are put in to the viral transport medium (VTM). Stool, urine and tissue sample can also be collected depending on the clinical condition. All samples are to be transported to the laboratory in cold chain.

Serum is tested for demonstration of four-fold rise of antibody titer.

Isolation of virus, virus nucleic acid detection, antigen detection and demonstration of fourfold rise in antibody titer confirm the diagnosis. Virus culture and neutralizing

antibody detection require biosafety level-3 facility and also time consuming. Therefore, viral nucleic acid detection is the most common and preferred method of diagnosis.

Viral nucleic acid detection: This is done by conventional reverse transcriptase PCR or by real-time PCR. The latter (real-time PCR) shows higher sensitivity and specificity as compared to conventional PCR. Virus Orf1b gene or nucleoprotein gene is the common target used.

As the viral load is higher during the second week of illness, sensitivity of PCR increases during this period. Therefore, if a sample is negative during early part of illness, the test should be repeated after a few days. Positive reaction by conventional PCR should be repeated with a second sample or a second PCR using a different target to confirm the result.

Virus antigen detection: Enzyme immune assay has been developed to detect the virus N protein using the specific antibody from serum samples. The sensitivity of this test is maximum during the 1st week of illness which gradually decreases with the appearance of antibodies.

Antibody detection: Detection of antibody can be done by immunofluorescence (IF) method or neutralization assay. Detection of antibody is most helpful during the second week of illness. Demonstration of fourfold rise in titer is considered as confirmatory in retrospective confirmation. IgM antibody detection is usually not considered as a reliable diagnostic method in respiratory viral infection.

B. MIDDLE EAST RESPIRATORY SYNDROME CORONAVIRUS (MERS CoV)

Emergence of MERS CoV: After a decade of discovery of the SARS CoV, another novel coronavirus was isolated from respiratory specimen of a patient in Saudi Arabia in September 2012. This patient was a 60-year-old patient admitted with pneumonia and renal failure in June 2012. The virus was initially named human coronavirus EMC (Erasmus Medical Center) which was later changes to Middle East respiratory syndrome coronavirus (MERS CoV) as per the recommendation of coronavirus study group of International Committee for Taxonomy of Virus. Within a few days, MERS CoV was isolated from another pneumonia patient admitted in the UK who was a Qatar national and had a travel history to Saudi. Both the cases were epidemiologically unrelated. The nucleotide sequence of virus isolated from both the patients was found to have 99.5% similarity with each other.

After the initial discovery, two more cases were identified retrospectively which had occurred in a health care associated cluster in April 2012.

Since then the infection has been reported from various countries across the globe. Majority of the cases have been reported from Arabian Peninsula. Cases have also been reported from several countries in Europe, Asia, Africa, and the USA. However, all the cases reported from outside the Middle East countries either had history of recent visit to Saudi Arabia or contact with infected person who had acquired the infection from there. Since September 2012 till mid-July 2016, near 1800 cases have been reported with more than 600 deaths amounting to a mortality rate of 30–40%.

Several clusters of cases have been reported within the family and amongst the health care workers both in Middle East and outside. The largest cluster outside the Middle East has occurred in Korea. The index case had visited Middle East and developed symptoms after his return. A total of more than 180 cases were reported with more than 30 deaths.

Origin of MERS CoV

It is thought that MERS CoV has originated from bats. Coronavirus related to MERS virus (group 2C CoV) has been isolated from several species of bats in Africa. Coronavirus that was isolated from bats in Middle East showed 100% sequence similarity with MERS CoV.

However, so far no human cases have shown direct or indirect epidemiological link with bats. This excludes the direct role of bat in MERS CoV transmission in humans.

Camels have now been proven as the intermediate animal reservoir. Several studies have shown MERS CoV serological infection in dromedary camels of the Middle East as well as in Africa from where the Middle East camels have originated. MERS virus has also been isolated from nasal samples of sick camels and viral RNA has been detected from nasal as well as rectal and fecal samples.

The 100% sequence similarities found in the virus that is isolated from a dromedary camel with that of the virus isolated from a patient who had close contact with a camel having upper respiratory tract infection.

The exact route of infection from camel to human is not known. Respiratory droplets are thought to be the most possible route. Ingestion of unpasteurized camel milk also can transmit the infection. Until now sustained man-to-man transmission of MERS CoV is not seen. Usually the infection first comes from the infected camel to susceptible individual from whom other persons get the infection. Most of the man-to-man transmissions occur only up to secondary cases and very few tertiary cases (Fig. 15.2).

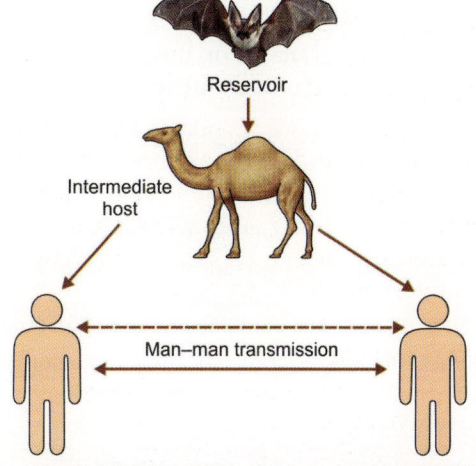

Fig. 15.2: Transmission of MERS virus

CLINICAL FEATURES

The average incubation period is around 5 days. The symptoms start with flu-like illness with fever, chill, rigor, cough, sore throat and other symptoms of upper respiratory tract infection. The disease progresses rapidly to pneumonia almost a week after the onset of symptoms. Radiograph of the chest starts with unilateral lesion and progresses to multifocal bilateral lesion of the lower lobes. The features are non-specific and cannot be differentiated from other viral pneumonia.

Extrapulmonary symptoms also can occur in MERS virus infection. Renal involvement is the commonest of them followed by involvement of gastrointestinal system. Other manifestations such as pericarditis, hepatic dysfunction may be there. Hematological abnormalities show either leucopenia or leukocytosis with thrombocytopenia.

In general, severity is mainly seen in primary cases with advanced age and underlying comorbid conditions. Diabetes mellitus and chronic diseases such as chronic renal/cardiac or pulmonary are the important comorbid conditions associated with severe MERS CoV infection.

The difference between SARS and MERS CoV are depicted in Table 15.1.

LAB DIAGNOSIS

Samples: Upper respiratory samples such as nasopharyngeal/nasal or throat aspirate or swab are collected. Samples from lower respiratory specimen (bronchoalveolar lavage, tracheal aspirate) are preferred in pneumonia patients. If the patient has trachea intubation, then tracheal aspirate can be collected. However, because collection of lower respiratory specimen usually involves invasive procedure, samples from upper respiratory tract are considered as standard system.

The approach to diagnosis of MERS CoV is in line with SARS CoV. Detection of MERS CoV can be done by RT-PCR, virus isolation, antigen detection and demonstration of four-fold rise in antibody titer.

Nucleic acid detection: The commonly used target genes are upstream E gene (upE), orf1a and orf1b. Up E assay is used for screening followed by detection of orf1a or orf1b for confirmation. In case of discrepant result, the sample can be tested for other genes such as N, S or RdRp.

According to WHO criteria, sample positive for two different genes by RT-PCR are considered as positive.

Antibody detection: Four-fold rise in antibody titer in acute and convalescent sample collected 3–4 weeks apart is considered diagnostic. Because of requirement of convalescent sample, this test mainly helps in retrospective diagnosis. Testing of antibody in single serum sample is not recommended.

Antigen detection: Test is used for detection of virus antigen in the affected tissue.

Table 15.1: Differences between SARS and MERS		
	SARS CoV	*MERS CoV*
Virus	Lineage B of the Betacoronavirus genus	Lineage C of the Betacoronavirus genus
Transmission	No repeated animal to human transmission Sustained man-to-man transmission	Repeated animal to man transmission Less efficient man-to-man transmission
Clinical features	No association of comorbid condition with disease severity	Comorbid condition is associated with disease severity
Case fatality rate	≈10%	≈40%
Present status	No more SARS case	MERS CoV cases still occurring

Immunochromatographic test (ICT) showing >90% sensitivity and specificity has been developed for use in nasal samples of camels.

Virus isolation: MERS CoV can grow in various monkey kidney cell lines such as Vero, LLCMK2. Cytopathic effect showing round refractile cells appears around 5–6 days. The final identification of infected cells is done by RT-PCR.

TREATMENT AND PREVENTION

There is no specific antiviral available for SARS or MERS coronaviruses. Ribavirin, an antiviral commonly given for RNA viruses, has been given but the role is doubtful and hence not recommended. Clinical management is mostly given with supportive therapy and broad spectrum antibiotics. No vaccine is available presently for SARS CoV.

Bibliography

1. Al-Hazmi A. Challenges presented by MERS coronavirus, and SARS coronavirus to global health. Saudi J Biol Sci 2016;23(4): 507–11.

2. Cameron MJ, Bermejo-Martin JF, Danesh A, Muller MP, Kelvin DJ. Human immunopathogenesis of severe acute respiratory syndrome (SARS). Virus Res 2008;133(1):13–19.

3. Chan JF, Lau SK, To KK, Cheng VC, Woo PC, Yuen KY. Middle East respiratory syndrome coronavirus: another zoonotic betacoronavirus causing SARS-like disease. Clin Microbiol Rev 2015;28(2):465–522.

4. Chen J, Subbarao K. The Immunobiology of SARS*. Annu Rev Immunol 2007;25:443–72.

5. Cheng VC, Lau SK, Woo PC, Yuen KY. Severe acute respiratory syndrome coronavirus as an agent of emerging and reemerging infection. Clin Microbiol Rev 2007;20(4): 660–94.

6. Cheng VC, Chan JF, To KK, Yuen KY. Clinical management and infection control of SARS: lessons learned. Antiviral Res 2013;100(2): 407–19.

7. Christian MD, Poutanen SM, Loutfy MR, Muller MP, Low DE. Severe acute respiratory syndrome. Clin Infect Dis 2004;38(10):1420–27.

8. Cunha CB, Opal SM. Middle East respiratory syndrome (MERS): a new zoonotic viral pneumonia. Virulence 2014;5(6):650–54.

9. Fehr AR, Perlman S. Coronaviruses: An overview of their replication and pathogenesis. Methods Mol Biol 2015;1282:1–23.

10. Gralinski LE, Baric RS. Molecular pathology of emerging coronavirus infections. J Pathol 2015 Jan;235(2):185–95.

11. Gu J, Korteweg C. Pathology and pathogenesis of severe acute respiratory syndrome. Am J Pathol 2007;170(4):1136–47.

12. Guo Y, Korteweg C, McNutt MA, Gu J. Pathogenetic mechanisms of severe acute respiratory syndrome. Virus Res 2008;133(1):4–12.

13. Hui DS, Memish ZA, Zumla A. Severe acute respiratory syndrome vs. the Middle East respiratory syndrome. Curr Opin Pulm Med 2014;20(3):233–41.

14. Lee H, Ki CS, Sung H, Kim S, Seong MW, Yong D, Kim JS, Lee MK, Kim MN, Choi JR, Kim JH; Korean Society for Laboratory Medicine MERS-CoV Task Force. Guidelines for the Laboratory Diagnosis of Middle East Respiratory Syndrome Coronavirus in Korea. Infect Chemother 2016;48(1):61–69.

15. Milne-Price S, Miazgowicz KL, Munster VJ. The emergence of the Middle East respiratory syndrome coronavirus. Pathog Dis 2014;71(2): 121–36.

16. Mishra B. Combating the spread of Middle East respiratory syndrome coronavirus: Indian perspective. Indian J Med Microbiol 2016;34: 135–36.

17. Mohd HA, Al-Tawfiq JA, Memish ZA. Middle East Respiratory Syndrome Coronavirus. (MERS-CoV) origin and animal reservoir. Virol J 2016;13(1):87.

18. Momattin H, Mohammed K, Zumla A, Memish ZA, Al-Tawfiq JA. Therapeutic options for Middle East respiratory syndrome coronavirus (MERS-CoV)—possible lessons from a systematic review of SARS-CoV therapy. Int J Infect Dis 2013;17(10):e792–98.

19. Peiris JS, Yuen KY, Osterhaus AD, Stöhr K. The severe acute respiratory syndrome. N Engl J Med 2003;349(25):2431–41.

20. van den Brand JM, Smits SL, Haagmans BL. Pathogenesis of Middle East respiratory syndrome coronavirus. J Pathol 2015;235(2):175–84.

Gastroenteritis Viruses

Gastroenteritis is defined as the inflammation of gastrointestinal tract mucosa and is characterized by diarrhea and vomiting. It affects all age groups but is more important in children because of high morbidity and mortality.

Worldwide it causes 3.5 to 5 billion cases and 1.5 to 2.5 million deaths in children every year.

Viruses are responsible for near 70% of acute gastroenteritis cases in children. Amongst the viral agents, rotavirus and noroviruses are the two most important viral agents of gastroenteritis worldwide. Table 16.1 shows the list of viral agents responsible for gastroenteritis.

A. ROTAVIRUS

Rotavirus was first discovered in animals. In humans, the virus was first detected by electron microscopy in duodenal biopsy of a child with acute gastroenteritis. The shape of the virus appeared like a wheel under electron microscopy which led to the naming of virus as "Rotavirus" (**"Rota" means wheel**).

VIRUS

Rotavirus belongs to the family Reoviridae.

Structure of virus (Fig. 16.1): Rotavirus (RV) is a non-enveloped virus of 100 nm particle having icosahedral symmetry. The fully infectious virus particle consists of three concentric layers of proteins also termed triple-layered particle (TLP).

Three layers from outside inwards are outer capsid, inner capsid and internal core which are made up of VP7, VP6 and VP2, respectively. These layers surround the inner genome

Table 16.1: Viral agents of gastroenteritis

Viruses	Family	Epidemiological feature
Common agents		
Rotavirus	Reoviridae	Children <5 years
Norovirus	Caliciviridae	All age group
Sapoviruses	Caliciviridae	Infant and young children
Astrovirus	Astroviridae	Children <12 months: Outbreak in day care center
Adenovirus group F (Adenovirus 40, 41)	Adenoviridae	Children
Rare agents		
SARS coronavirus	Coronaviridae	Diarrhea in >30%
Human coronaviruses (OC43, 229E)	Coronaviridae	
Toroviruses	Coronaviridae	Nosocomial diarrhea in children <2 years
Aichivirus	Picornaviridae	
Non-F adenovirus	Adenoviridae	

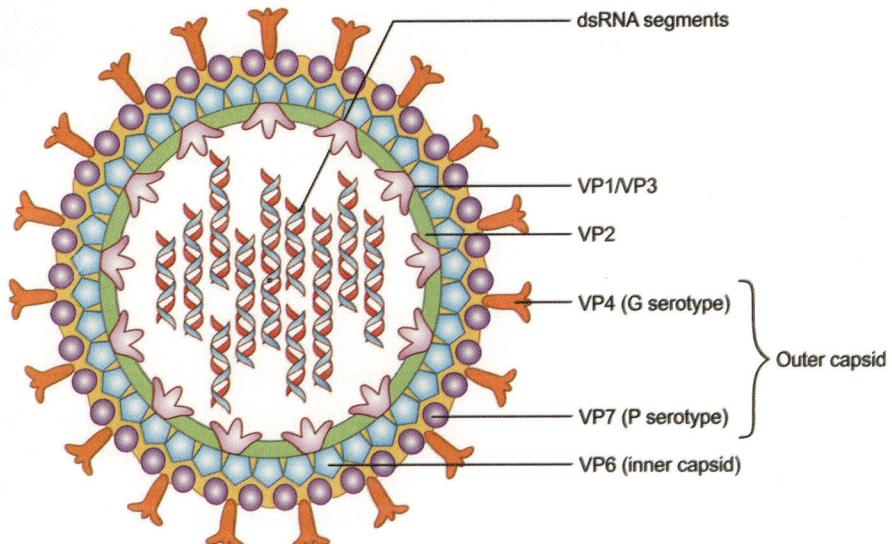

dsRNA segments

VP1/VP3

VP2

VP4 (G serotype)

} Outer capsid

VP7 (P serotype)

VP6 (inner capsid)

Fig. 16.1: Structure of rotavirus

which is a **double-stranded RNA having 11 segments**. Each gene segment encodes for single protein.

Outer capsid proteins: VP4 and VP7 are structural proteins, present in the outer capsid. VP7 forms the smooth layer of the outer shell. VP4 forms the spikes on it. They are the outermost components of the virus and act as the neutralization antigens and target of neutralizing antibody. That makes them important candidate for vaccine preparation.

- **VP7:** It is a glycosylated protein. Defines the **G serotypes**. There are currently 27 G serotypes, of which 12 are human strains.
- **VP4:** It is a protease cleaved protein. It defines the **P serotypes**. P types have been divided into serotypes and genotypes.

Inner capsid protein: VP6 contains the group-specific antigen. Based on VP6, RV has been divided into **8 groups A to H**. Group A rotaviruses are most common human pathogen responsible for diarrhea in children.

- Inner core: This layer is made up of VP2 protein.
- VP1 (RNA dependent RNA polymerase) and VP3 (a methyl transferase), together

makes a complex and situated on the inner aspect of VP2. VP1 and VP3 are required for viral replication.

- NSP4 is a non-structural protein. It is also called viral enterotoxin. It acts a virulence factor for the virus.
- Replication of the virus in cell culture requires prior treatment with proteases which cleaves the outer capsid spike protein VP4 and increases the infectivity of the virus.

PATHOGENESIS

Pathogenesis of rotavirus diarrhea is multifactorial.

Malabsorption Mediated Diarrhea

- **Destruction of epithelial cells:** Rotavirus (RV) infects the differentiated enterocytes at the top of the villi, leads to vacuolization and destruction of epithelial cells which in turn leads to disruption of function of enterocytes.
- **Action of rotavirus enterotoxin NSP4:** This viral enterotoxin is released from the infected cells and exerts the paracrine effect on its neighboring cells. It activates the cell

signaling pathway by which there occurs intracellular mobilization of calcium and chloride secretion.

- Villus ischemia and activation of enteric nervous system also plays role in causing malabsorption.

Pathogenesis of vomiting: Enterochromaffin cells of the gut when gets infected with RV, stimulates the secretion of serotonin. This in turn activates the vagal afferent nerves and stimulates the part of the brainstem that controls the vomiting.

Role of Viral Proteins in Pathogenesis

- VP4 and VP7: Mediate viral entry to host cells.
- VP3, VP6, NS2 and NS3: Efficient viral replication
- NSP1 and NSP3: Extraintestinal spread
- NSP4: Enterotoxin

In neonates and immunosuppressive individuals, RV can infect and replicate in various extraintestinal sites like liver, biliary tract and pancreas leading to biliary atresia and pancreatitis.

CLINICAL FEATURES

Rotavirus in children: Symptomatic infection of rotavirus usually occurs in children. Rotavirus causes moderate to severe degree of diarrhea associated with vomiting and fever. Diarrhea is usually preceded by vomiting. Electrolyte disturbance due to diarrhea and vomiting leads to dehydration. Fever generally is of mild grade. Severe dehydration and cardiovascular failure are mainly responsible for rotavirus diarrheal death in developing countries.

RV usually causes severe diarrhea in certain group of immunosuppressed individuals like bone marrow or stem cell transplant recipients. Rarely it has been associated with chronic symptoms in immunocompromised patients.

The role of rotavirus in causation of intussusceptions is still not clear. However, it is presently accepted that wild RV can cause intussusception but with extreme rarity.

Temporal association of rotavirus has been reported with necrotizing enterocolitis and hemorrhagic gastroenteritis in neonates.

Rotavirus in adults: In adults, repeated infection can occur due to rotavirus. However, most of the infections are asymptomatic in nature. Mild symptoms can occur rarely. In contrast to other parts of the world, RV is the second cause of acute gastroenteritis in China, next to norovirus.

EPIDEMIOLOGY

Rotavirus is one of the important causes of acute gastroenteritis amongst children below 5 years age leading to near 5% deaths worldwide. This accounts for approximately 453,000 deaths worldwide annually, of which more than one-fifth (22%) is contributed by India. It is estimated to account for near 2.5 millions hospitalizations globally and 10 lakhs in India. Majority of rotavirus deaths (80%) occurs in Southeast Asian countries and Africa. Annual rotavirus deaths per year in India, China and Pakistan are 1,22,000; 27,000; and 20,000, respectively.

Rotavirus is seen both in developed and developing countries including North America, Europe, Australia, Japan amongst the developed world, and Asia, Africa and South America amongst the tropical developing countries. In both developed and developing countries, the age and symptom pattern is same but due to early and effective management the death rate in developed world is much less as compared to that of developing countries.

Molecular Epidemiology

Group A rotavirus: Group A rotaviruses are genetically divergent. It consists of 27G types and 35 P types. Of these, 12 G (1 to 6, 8 to 12, and 20) and 15 P types (1 to 6, 8 to 11, 13, 14, 19, 25, and 28) have been isolated from human beings. G1 is most frequently detected serotype from each continent.

G1–G4 constitute 97% of all G types isolated from Asia, North America and Australia. G9 occasionally emerges as the predominant strain in several continents.

Amongst the combination of G and P types, five combinations are predominant throughout the world—**G2P[4], G1P[8], G3P[8], and G4P[8] and G9P[8]**.

Group B rotaviruses: Group B rotaviruses otherwise known as **adult diarrheal rotavirus** (ADRV) are responsible for causing several large outbreaks of severe gastroenteritis in adults in China and other south Asian countries like India and Bangladesh. Large outbreak affecting near 20,000 people has been reported in China. Cholera-like symptoms with severe watery diarrhea is the main symptom. Contaminated water has been implicated as the source of infection.

Rotavirus diarrhea also occurs quite often in children in day care centers and health care settings.

Infectiousness of RV: Transmission of rotavirus occurs through feco-oral route. Rotaviruses are highly infectious in nature as compared to other diarrheal pathogens.

The high infectivity occurs due to:
- Low infective dose: One tissue culture infective dose can cause infection in susceptible host.
- A large number of viruses shedding in diarrheic stool (up to 10^{11}/mL).
- Survival of rotavirus in adverse environmental conditions and resistance to physical inactivation helps in efficient transmission.
- Short incubation period: 1–2 days.
- Shedding of virus in stool starts before the onset of diarrhea and continues after cessation of diarrhea.

Transmission through respiratory route also has been proposed because:
- Occurrence of rotavirus outbreak without evidence of feco-oral transmission.
- High seroprevalence of rotavirus during first year of life even in developed countries.
- Occasional occurrence of respiratory symptoms in rotavirus infected patients. Isolation of RV from upper respiratory tract.

However, there is no definite proof for respiratory mode of transmission.

Role of animal rotavirus strains: Rotaviruses are host-specific but genetic recombinations can occur between RVs of same as well as different hosts. Various reassortant strains, human-bovine, human-pig and human-simian strains, have been isolated from humans particularly from high endemic countries like India and Brazil.

Seasonal pattern: In topical developing countries, rotavirus infection is seen throughout the year. In southern India, rotavirus infection is seen throughout the year, but in north India it is mostly observed during the cold dry months of the year.

In temperate climate countries, peak rotavirus infection occurs during winter months of January to April.

IMMUNE RESPONSE

Presence of maternal antibody in the neonate protects from severe rotavirus infection. In neonate, severe rotavirus diarrhea is thus seen between 3 and 24 months of age, 7 and 15 months being the peak. This possibly reflects the decreasing titer of maternal antibody.

Repeated infection in neonates also protects them from subsequent severe diarrhea.

Humoral immunity plays an important role in protection of severe rotavirus infection. Serum antibody is one of the major indicators of protection after natural and vaccine-induced infection.

- VP6 is the main immunodominant antigen. IgG or IgA antibody against this acts as the measure of protective immunity.
- Antibodies against the surface antigens VP4 and VP7 also play important role in protection against infection. Their role in vaccine-induced immunity is not clear. Though it is believed that antibody

against the surface protein is essential for vaccine-induced protection, vaccine producing low titer of serum antibody also has been found to be effective.

- Local gut immunity also considered as important for protection.

DIAGNOSIS

Detection of virus or its components like viral antigen or genome from diarrheic stool is the mainstay of diagnosis.

Detection of antibody against RV is done to check the immunity status of the individual after vaccination or infection.

Stool sample during early phase (1–4 days) is the best sample because of high number of virus.

Detection of Virus

- Direct detection of virus in the stool sample was previously done by electron microscopy (EM). Because of very typical wheel-like shape of rotavirus, it can be diagnosed by EM. However, as the method involves expensive equipment and tedious procedure, it is no more used for routine diagnosis purpose.
- Isolation of rotavirus can be done by cell culture method which further can be used for serotyping of the virus. However, this is time consuming and technically demanding, so not a preferred method for patient diagnosis. However, cultivation of virus is important to study the virus.

Antigen Detection

Detection of viral antigen by enzyme immune assay is routinely used for its high sensitivity, specificity as well as for its convenience. Commercial ELISA systems are available for group A and also for group B and C. It requires at least 10^4 to 10^7 virus particle/mL of specimen to be positive.

Viral RNA Detection

- Viral RNA can be detected by reverse transcriptase polymerase chain reaction (RT-PCR). This test is 1000 times more sensitive than ELISA and also can be used for genotype detection.
- Dot-blot hybridization was used as a good sensitive method. This was based on the principle of hybridization between the labeled rotavirus single strand RNA and heat denatured double strand rotavirus RNA immobilized to nitrocellulose membrane. The technique was more sensitive than ELISA but no more preferred after the wide use of RT-PCR.
- Older methods: Gel electrophoresis of RV RNA, countercurrent immunoelectrophoresis, latex agglutination were previously used and not used presently because of availability of better tests.

Detection of Antibody

- IgA, IgG and IgM against rotavirus can be detected to check the immune status and generally not required for diagnosis. Rotavirus antibody detection is done commonly by ELISA or neutralization assay.
- Complement fixation, immune adherence hemagglutination, hemagglutination inhibition, immunofluorescence, etc. were used previously.

TREATMENT

The primary goal of treatment is fluid replacement and electrolyte maintenance. Oral rehydration salt therapy is recommended by World Health Organization (WHO) and widely used for treatment of acute gastroenteritis. Intravenous administration is applicable whenever oral therapy is not possible.

PREVENTION

Isolation of infected patient, hand hygiene and environmental disinfection are the primary key to prevent the spread of virus.

- Surface decontamination with hypochlorite with free chlorine content of minimum 20,000 parts per million should be used instead of phenolic disinfectants.
- Ethyl alcohol hand sanitizer is able to destroy the virus.

VACCINE

Rotavirus vaccine has come a long way and presently several licensed vaccines are available. Many countries have already adapted the vaccination in their national program.

The need of vaccine to prevent rotavirus is self-explanatory from the high number of mortality and morbidity associated with it in children below 5 years age.

Important Factors in Rotavirus Vaccine

Protective Immunity by Natural Infection

- Natural infection provides immunity from severe infection.
- Decrease in severity with repeated infection.
- Severity of infection is almost nil after second infection.

Circulating rotavirus serotypes: Multiple G and P serotypes and genotypes circulate worldwide. This was thought to be important for vaccine preparation, so that vaccine can be made based on the locally prevailing serotypes. Presently it is evident that both monovalent and polyvalent vaccines are able to provide effective immunity.

Jennerian approach: Edward Jenner had used animal virus, cowpox virus, for preparation of vaccine against the smallpox that is caused by the variola virus which is antigenically similar another poxvirus. This approach is based on the principle that animal viruses that share antigenic relatedness and non-pathogenic for human host, when infect the human host do not produce disease but able to mount protective immunity against the pathogenic human strains because of cross-protection. The concept of using an antigenically related but non-pathogenic virus for production of protective antibody is called Jennerian approach. This approach has been tried in rotavirus vaccine preparation with animal rotaviruses as they share a major common antigen.

Presently Licensed Vaccine

Monovalent Vaccine (RV1)

- This is a live attenuated vaccine prepared from a virulent human rotavirus strain 89-12, of serotype G1 P[8].
- Attenuation has been done by several passages in tissue culture.
- It is given orally, 2 doses at 6 weeks and 10 weeks.
- It provides protection against G homotypic and heterotypic strains as well as P homotypic strains.
- Protective efficacy against severe disease in Asia and Africa is ≈60% and in developed countries ≈95%.
- Manufactured by GlaxoSmithKline, with brand name **Rotarix (RV1)**.

Polyvalent Bovine-human Reassortant Vaccine (RV5)

- This is a **pentavalent reassortant vaccine** which consists of
  - VP7 of G1, G2, G3 and G4 in 4 strains and VP4 of P1A[8] in 1 strain.
  - All five strains are reassortant with WC3 strain which provides the genetic backbone.
- It is given orally, in 3 doses at 2, 4, and 6 months age.
- Protective efficacy against severe disease in Asia and Africa is ≈50% and in developed countries ≈85%.
- Manufactured by Merck with the brand name **RotaTeq (RV5)**.

Both the vaccines are licensed in more than 100 countries worldwide.

Both the vaccines show rare temporal association with intussusceptions, <1:50,000.

Indian Nursery Strain 116E

- This is a live attenuated monovalent vaccine which is now licensed in India.
- The vaccine is prepared from the strain **116E G9 P8[11]**, a naturally occurring **human-bovine reassortant strain**. This strain was first found in nursery of All India

Institute of Medical Sciences, New Delhi. It contains 10 genes from human rotavirus and VP4 from bovine rotavirus.

- The strain was found to cause subclinical infection and protects from severe rotavirus for up to 3 years.
- The vaccine is given in 3 doses along with DPT with a vaccine efficacy against severe gastroenteritis of 56.4%.
- The vaccine has been launched in several states across India as first phase with a plan to expand the program to all over the country.
- It is manufactured by Bharat Biotech with the brand name of Rotavac.

B. NOROVIRUS

Noroviruses are one of the leading causes of acute gastroenteritis in all age groups, in both immunocompetent and immunocompromised individuals and have been associated with outbreaks in hospitals, schools, and cruise ship. This is the first viral agent identified as the cause of acute gastroenteritis. Norovirus was previously known as Norwalk virus as it was identified in stool samples collected during a school outbreak in Norwalk.

VIRUS

It belongs to the family Caliciviridae and genus Norovirus. It is a single stranded, positive sense RNA virus contained within a non-enveloped protein coat having cup-shaped depressions.

Norovirus genus is divided into six genogroups (GI to GVI) and >40 genotypes based on the amino acid sequence of the VP1 capsid protein. Genogroup I, II and IV cause infection in humans. Genogroup II causes maximum number of cases followed by GI and GIV.

EPIDEMIOLOGY

In developing countries, noroviruses cause ≈2,00,000 deaths every year in children below 5 years of age. In developed countries, it is responsible for ≈64,000 severe diarrheal cases each year and in USA it is considered as the single most important cause of acute gastroenteritis in adults requiring hospitalization.

Infection can occur throughout the year but more during cold dry months of winter. Because of more incidences during winter, previously it was known as **"winter vomiting disease"**.

Modes of Transmission

Humans are the only natural host of human noroviruses. Transmission of virus occurs through feco-oral route. Like rotavirus, it also has a high potency to spread which is mainly because of:

- Low infectious dose
- High secondary attack rate >30%
- Spread of virus mainly occurs through person to person contact, fomites, droplets and environmental contamination.
- Long duration of virus shedding: Starts before onset of symptoms and prolongs after cessation of diarrhea.
- Capacity to survive in adverse environmental conditions:
  - Wide range of temperature: Freezing up to 60°C
  - Survives in vegetables, oysters and can be transmitted through raw vegetables or fruits or improperly cooked food.
  - Survives in drinking water, environmental surfaces.
- Repeated infection can occur in the same individual due to lack of long-lasting protective antibody and high genetic diversity.

Outbreaks

Norovirus infection predominantly occurs as outbreaks. This can be due to food-borne, water-borne, and outbreak in health care setting.

Foodborne outbreak: Fruits and oysters get contaminated with noroviruses when treated with contaminated water. These food items when eaten raw or improperly cooked, can lead to infection. Food items like raspberry, lettuce, shellfish have been implicated in various norovirus outbreaks. Transmission through food can also occur through preparation of food by infected food handler.

Waterborne outbreaks: Several waterborne outbreaks due to norovirus have been reported. Discharge of waste water into river and various water bodies are thought to play important role in contaminating the water. Outbreaks have been linked to drinking water sources as well as recreational water.

Outbreaks in closed settings like day care centers, school and health care settings are common due to noroviruses. This is mainly due to ability of the virus to survive in environment for a long period, low infectious dose and also susceptibility of individuals. Spread of virus occurs through person to person contact, fomites and also through aerosol generated while vomiting.

Sporadic disease due to norovirus also occurs commonly, second only to rotavirus in children. This is commonly seen amongst the family members or in family clusters.

CLINICAL FEATURES

Norovirus affects all age groups. Average incubation period is 2 days (10–50 hours). Clinical symptoms in most of the cases start with sudden onset of vomiting and diarrhea, vomiting is more common in children and diarrhea is more common in adults. In general, symptoms are mild and self-limiting in nature and resolve within 1–3 days. In extremes of age, the disease may be severe with fatal outcome. Outbreak with severe disease has been more commonly associated with GII.4 strains than non-GII.4. Old age (>85 years) is more significantly associated with fatal outcome.

Norovirus can also cause asymptomatic infection. This is more commonly seen in children of developing countries. However, asymptomatic infection also has been reported in children of developed countries and in adults.

In immunocompromised individuals, more specifically in hematopoietic stem cell and solid organ transplant recipients, symptoms are more severe and lasts for a prolong period. Excretion of virus may occur for a long period.

DIAGNOSIS

Diarrheal stool sample collected within 48–72 hours of symptom onset is preferred. Vomitus also can be used as an alternative sample.

Antigen detection: Antigen detection of GI and GII genogroups in stool sample is commonly employed for diagnosis. Several ELISA systems are commercially available. Specificity of these tests is high (85 to 100%), however, the sensitivity is variable and has been reported 40 to 80% in various studies.

Rapid immunochromatographic test: Several commercially available rapid tests are available for GI and GII antigen detection in stool samples. These tests have good specificity of >80% but suffers from variable sensitivity ranging from <50 to 80%. Advantage of these tests is applicability in field setting.

Molecular test: Conventional reverse transcriptase PCR (RT-PCR) and real-time PCR have been developed for detection of viral RNA from clinical samples as well as from food and water. These tests also can be used for genotyping and further for sequencing, thus employed for outbreak investigation. Molecular tests are presently considered as the reference test for diagnosis of norovirus infection.

Multiplex PCR/RT-PCR: Presently commercial multiplex RT-PCR tests are available for detection of acute gastroenteritis agents which includes viral, bacterial and parasitic pathogens. The sensitivity of >95% for detection of GI and GII noroviruses has been reported by various studies with specificity of 100%.

TREATMENT AND PREVENTION

Management of diarrhea and vomiting follows the standard treatment protocol of oral/intravenous fluid and electrolyte maintenance.

Avoiding the contaminated food and handling by infected food handler is not a practical solution for prevention of norovirus outbreaks.

Maintenance of personal hygiene, hand hygiene, environmental disinfection and eating properly cooked food are the key measures to prevent the spread of norovirus.

VACCINE

Due to high prevalence of norovirus infection both in children and adults and its potential to cause outbreaks have given a thought for development of vaccine for this virus.

Vaccines containing GI and bivalent vaccine containing GI and GII.4 in the form of virus-like particles (VLP) have been tried in humans with reduction in the number of vomiting and diarrhea in the vaccine group.

High genetic diversity, lack of knowledge in viral immunity, and inability to cultivate the virus are the major problems associated with the development of vaccine against noroviruses.

Bibliography

1. Chow CM, Leung AK, Hon KL. Acute gastro-enteritis: From guidelines to real life. Clin Exp Gastroenterol 2010;3:97–112.

2. Cunliffe N, Zaman K, Rodrigo C, Debrus S, Benninghoff B, Pemmaraju Venkata S, Han HH. Early exposure of infants to natural rotavirus infection: a review of studies with human rotavirus vaccine RIX4414. BMC Pediatr 2014;14:295.

3. Desselberger U. Rotaviruses. Virus Res 2014; 190:75–96.

4. Esona MD, Gautam R. Rotavirus. Clin Lab Med 2015;35(2):363–91.

5. Evan J Anderson and Stephen G Weber. Rotavirus infection in adults. Lancet Infect Dis 2004; 4: 91–99.

6. Ghazanfar H, Naseem S, Ghazanfar A, Haq S. Rotavirus vaccine—a new hope. J Pak Med Assoc 2014;64(10):1211–16.

7. Jadhav S, Gautam M, Gairola S. Role of vaccine manufacturers in developing countries towards global healthcare by providing quality vaccines at affordable prices. Clin Microbiol Infect 2014; Suppl 5:37–44.

8. Kahn G, Fitzwater S, Tate J, Kang G, Ganguly N, Nair G, Steele D, Arora R, Chawla-Sarkar M, Parashar U, Santosham M. Epidemiology and prospects for prevention of rotavirus disease in India. Indian Pediatr 2012;49(6):467–74.

9. Karafillakis E, Hassounah S, Atchison C. Effectiveness and impact of rotavirus vaccines in Europe, 2006–2014. Vaccine 2015;33(18): 2097–2107.

10. Karin Bok, Kim Y. Green. Norovirus gastroenteritis in immunocompromised patients. N Engl J Med 2012;367: 2126–32.

11. Kollaritsch H, Kundi M, Giaquinto C, Paulke-Korinek M. Rotavirus vaccines: a story of success. Clin Microbiol Infect 2015;21(8):735–43.

12. Lee PI, Chen PY, Huang YC, et al. Recommendations for rotavirus vaccine. Pediatr Neonatol 2013;54(6):355–59.

13. Miles MG, Lewis KD, Kang G, Parashar UD, Steele AD. A systematic review of rotavirus strain diversity in India, Bangladesh, and Pakistan. Vaccine 2012;30 Suppl 1:A131–39.

14. Parashar UD, Nelson EA, Kang G. Diagnosis, management, and prevention of rotavirus gastroenteritis in children. BMJ 2013;347: f7204.

15. Pollard SL, Malpica-Llanos T, Friberg IK, Fischer-Walker C, Ashraf S, Walker N. Estimating the herd immunity effect of rotavirus vaccine. Vaccine 2015;33(32):3795–800.

16. Ramani S, Kang G. Burden of disease & molecular epidemiology of group A rotavirus infections in India. Indian J Med Res 2007; 125(5):619–32.

17. Ramani S, Atmar RL, Estes MK. Epidemiology of human noroviruses and updates on vaccine development. Curr Opin Gastroenterol 2014; 30(1):25–33.

18. Rao TS, Arora R, Khera A, Tate JE, Parashar U, Kang G; Indian Rotavirus Vaccine Working Group. Insights from global data for use of rotavirus vaccines in India. Vaccine 2014;32 Suppl 1:A171–78.

19. Robilotti E, Deresinski S, Pinsky BA. Norovirus. Clin Microbiol Rev 2015;28(1):134–64.

20. Roger I Glass, Umesh D Parashar, Mary K Estes. Norovirus gastroenteritis. N Engl J Med 2009; 361:1776–85.

21. Sasirekha Ramani, Robert L Atmara, Mary K Estesa. Epidemiology of human noroviruses and updates on vaccine development. Curr Opin Gastroenterol 2014; 30(1): 25–33.

22. Tate JE, Parashar UD. Rotavirus vaccines in routine use. Clin Infect Dis 2014;59(9):1291–301.

23. White PA. Evolution of norovirus. Clin Microbiol Infect 2014;20(8):741–5.

24. Yen C, Tate JE, Hyde TB, Cortese MM, Lopman BA, Jiang B, Glass RI, Parashar UD. Rotavirus vaccines: Current status and future considerations. Hum Vaccin Immunother 2014;10(6): 1436–48.

Polio: Current Perspective

Poliovirus is the causative agent of the disease known as poliomyelitis (polio: Gray; myelos: Spinal cord), a disease that predominantly involves the central nervous system.

VIRUS

Poliovirus is a member of the genus Enterovirus of the family Picornaviridae ("Pico" in Italian means small and "rna" for the RNA type of nucleic acid). Polioviruses are small viruses, 20–30 nm in diameter. It contains single stranded, positive sense RNA genome of 7500 nucleotide length inside the capsid of icosahedral symmetry and is non-enveloped.

Biological Properties

- Acid stable: Stable at acidic pH 3–5 of stomach.
- Thermo stable: Stable at room temperature for days, and at 4°C for weeks. Inactivated at 42°C.
- Resistant to most of the lipid solvents: Ether, chloroform.
- Resistant to common disinfectants: Ethanol, isopropanolol, quaternary ammonium compounds.

These properties help the virus to replicate in the intestine and also enables them to survive in the environment.

The virus is inactivated by 0.3% formalin, UV light, and 0.5 ppm free residual chlorine.

Polioviruses are of three antigenic types or serotypes 1, 2, and 3. These three serotypes can be differentiated by monoclonal antibodies or by genome sequencing.

The prototype strains of three different poliovirus serotypes are:
1. Poliovirus serotype 1: Brunhield
2. Poliovirus serotype 2: Lansing
3. Poliovirus serotype 3: Leon

All the three serotypes can cause paralytic poliomyelitis.

EPIDEMIOLOGY

Global scenario: In 2016, 80% of world's population live in polio free areas. In 1988, more than 125 countries were affected with near 350,000 cases of poliomyelitis. After the initiation of global polio eradication program, the number of cases has decreased to >99% with only 74 cases in two countries in 2015.

Polio free status: World Health Organization (WHO) is the authority to declare the polio status of a region. Declaration of polio free status is given to the entire region (not to any individual country), when all the countries of a particular WHO region shows the absence of wild poliovirus transmission for at least three consecutive years (≥3 years of last wild poliovirus isolation) in presence of sensitive certification standard surveillance.

Of the six WHO regions, four have been certified so far as polio free or polio eradication. The year of certification of various regions is as below:
- WHO region of Americas: 1994
- WHO Western Pacific region: 2000
- WHO European region: 2002
- WHO South East Asian region: 2014

Eradication Status of Poliovirus Types

Transmission of wild poliovirus (WPV) type 2 has been stopped since 1999. The last detection was in 24th October, 1999. WPV2 was declared as eradicated in September 20th, 2015 by Global Certification Commission.

The detection of WPV3 was last recorded in 10th November, 2012 in Nigeria. More than three years have passed since the last report of WPV3. WPV1 is the only wild poliovirus type which has been reported presently.

Polio status in two endemic countries: Presently two countries in the world are considered as endemic for poliovirus.

Pakistan: In 2014, a major outbreak occurred in Pakistan due to WPV1 with increased number of cases and wider geographic distribution of transmission. The number of WPV cases reported in 2015 and 2016 was 54 and 20, respectively.

Afghanistan: Endemic transmission of WPV1 still occurs in parts of Afghanistan. Being the neighboring country of Pakistan, the numbers of WPV1 cases have also showed an increase in 2014 either due to primary cases or secondary to cross-border transmission from Pakistan. Improvement in the surveillance system and immunization activity is required to break the chain of indigenous transmission of WPV. The number of WPV cases reported in 2015 and 2016 were 20 and 13, respectively.

Present Indian scenario: In February 2012, India was removed from the list of active polio endemic countries of world. In 2014, Southeast Asia region (which includes India) was declared by WHO as polio free region.

The last polio case in India was detected in January 2011 in Howrah district, West Bengal in an 18-month-old girl which was of Poliovirus type 1. Last case of poliovirus type 3 was reported in October 2010 from Jharkhand and poliovirus type 2 was reported in October 1999 from Uttar Pradesh.

History of polio scenario in last decade: In 2005 for the first time, the number of polio cases dropped to below 100. A total of 66 cases were reported from 35 districts. However, from 2006 again the number of cases increased and from 2006 to 2009 the number was more than 500 every year. In 2010, after intense vaccination coverage, the number again came down to 42. Subsequently in January 2011, the last case was detected from West Bengal.

PATHOGENESIS

Poliovirus enters through oral cavity and enters into the cells expressing the poliovirus receptor (PVR). The virus gets implanted in the tonsils, local lymph nodes and Peyer's patches of ileum. Replication occurs at the implanted sites leading to minor viremia. Through blood, virus goes to the reticulo-endothelial cells, causing major viremia. At this stage, host mounts the immune response and type specific neutralizing antibodies are appeared. In majority of the cases, virus replication gets halted without further progression of disease process. In a few cases, virus replication continues and virus travels to the neuronal cells. Replication occurs in the motor neuron of the anterior horn cells and brainstem causing paralytic poliomyelitis (Fig. 17.1).

CLINICAL FEATURES

Poliovirus infection remains asymptomatic in majority of cases (70–75%). The clinical manifestation when occurs can be of different grade of severity.

In around 20–25% cases, the disease is manifested as minor illness with mild fever and systemic manifestations like headache, nausea, vomiting.

Aseptic meningitis or non-paralytic poliomyelitis is the other form of manifestation where the patient develops fever, sore throat, vomiting and develops the signs of meningeal irritation after a gap of 1–2 days in the form of severe headache, vomiting, neck

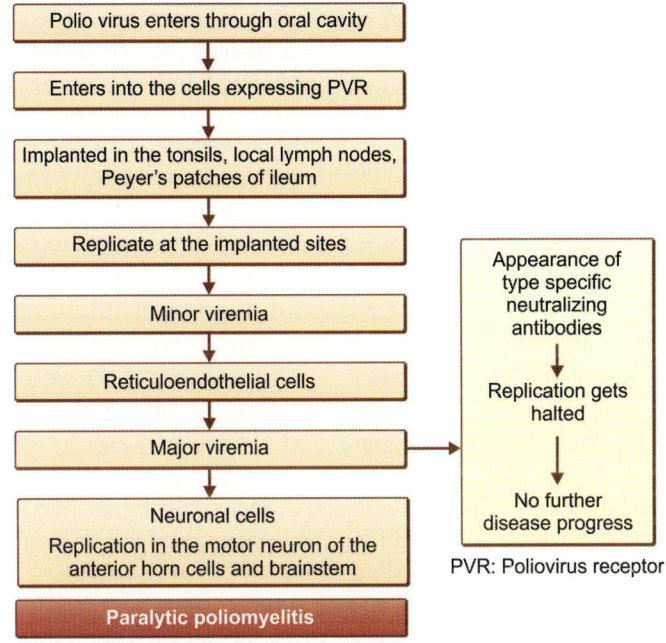

Fig. 17.1: Steps of pathogenesis of poliovirus

stiffness. Symptoms generally last for a week and the patient recovers completely within a few days.

Paralytic poliomyelitis, the most severe form of disease, starts with minor illness, then after a gap of few days patient suddenly develops flaccid paralysis of the extremities. Paralysis is asymmetric in nature which starts at the proximal part of the extremities, then descends downwards causing descending paralysis. Depending on the site of motor neuron damage, whether spinal or in the brainstem, the paralytic involvement can be spinal, bulbar or mixed type. Bulbar type is most fatal in nature because of respiratory muscle involvement. Paralytic poliomyelitis is seen in 1% of infection.

Post-polio syndrome, a late clinical condition due to poliovirus infection. The entity was realized only during 1980s. It is a late manifestation of acute paralytic poliomyelitis. It occurs in persons who have completely or partially recovered from the acute paralytic poliomyelitis. The symptoms manifest as a new onset of muscle weakness with pain and atrophy. Mostly the weakness develops in the same muscle that was affected during the acute polio. The average duration between the acute paralytic polio and post- polio syndrome is 25–30 years. The exact cause of post-polio syndrome is not known. It is believed that the weakness occurs due to attrition of the motor units that innervate the muscle.

LABORATORY DIAGNOSIS

Samples: Poliovirus can be detected from stool, throat swab and spinal fluid.

Laboratory methods followed for testing of poliovirus are:

- Virus isolation
- Intratypic differentiation
- Genome sequencing
- Serology

Virus isolation: At least two stool samples, throat swab or cerebrospinal fluid (CSF) are collected at an interval of 24 hours from suspected case of poliomyelitis within 14 days of onset of illness.

Sensitivity of virus isolation in different samples: Stool—high; throat swab—intermediate; CSF—low.

WHO recommended cell lines for poliovirus isolation are:

- L20B: Genetically engineered mouse cell line expressing the poliovirus receptor.
- RD or rhabdomyosarcoma cell line.

Samples are inoculated into both the cell lines.

The cell lines inoculated with the clinical samples are examined for the appearance of characteristic cytopathic effect (CPE) that is rounded, refractile cells with detachment from the surface.

Samples showing no CPE after seven days are given a second passage and examined for another seven days before being reported as negative.

Isolation of poliovirus from the stool sample of a suspected poliomyelitis case does not substantiate the causal association, whereas isolation of poliovirus from CSF is considered as diagnostic.

Isolated virus is then subjected for identification, intratypic differentiation and genome sequencing.

Identification of poliovirus: The isolated poliovirus strains are identified by microneutralization using type specific polyclonal antisera.

The test is put in a microtiter plate. The isolated strain of poliovirus is mixed with antisera of different poliovirus types and is incubated for 1 hour. The cell lines are then inoculated with the virus-antisera mixture. The type of antiserum which prevents the development of CPE identifies the type of the isolated poliovirus.

Intratypic differentiation: This is done to identify the isolated poliovirus type as wild virus or vaccine virus.

Methods used: ELISA, hybridization probe and polymerase chain reaction have been developed by the Center of Disease Control and Prevention (CDC).

ELISA: Poliovirus type specific antibody coated wells are incubated with the isolated and typed poliovirus. After incubation, it is treated with type specific cross adsorbed rabbit antisera. The bound rabbit antiserum is detected by adding peroxidase labeled anti-rabbit IgG.

Probe hybridization: RNA of the isolated and typed poliovirus is extracted and immobilized onto the nitrocellulose paper. This is then hybridized with digoxigenin labeled enterovirus group, sabin type specific and wild virus genotype specific probes. The bound probe is then detected by colorimetric reagent.

PCR: PCR using the enterovirus group specific, sabin type specific, and wild virus type specific (P1, P2 and P3) primers is employed for intratypic differentiation.

Genome sequencing: Nucleotide sequencing of the isolated strain differentiates between vaccine and wild type poliovirus and considered as the confirmatory method. Presently CDC uses this method for intratypic differentiation.

Serology: Fourfold rise of titer between acute and convalescent serum sample indicates the acute poliovirus infection. However, the limitations are:

- Antibody may not be detected in immune compromised individuals.
- It does not differentiate between the immune response due to vaccination or infection or whether the infection due to VDPV (vaccine derived poliovirus) or wild poliovirus.

VACCINE

Oral polio vaccine (OPV): OPV is made up of live attenuated sabin virus strain. The administration of OPV mimics the milder form of natural infection. The immune response is same as that of natural infection in terms of humoral and local immunity. The vaccine allows the replication of the vaccine virus in the intestine, but the progress is restricted due to development of neutralizing antibody. OPV thus permits the shedding of virus in the throat as well as in the intestine and in the stool. This was beneficial in the initial phase of vaccination to increase the herd immunity through feco-oral transmission of the vaccine virus in the community, particularly in developing country. However, due to virus shedding in the stool, OPV is associated with VDPV which needs to be stopped at this juncture of polio eradication.

OPV Formulations

Trivalent OPV (tOPV) is the commonest formulation available. It contains the attenuated strains of all three poliovirus serotypes. The proportion of three serotypes in vaccine is 10:1:6, respectively. The rate of seroconversion in children to the three poliovirus serotypes (1, 2, and 3) is 73, 90 and 70% respectively after the completion of the primary vaccination.

Bivalent OPV (bOPV): This is available in the combination of serotypes 1 and 3 as these two serotypes are circulating all over after the eradication of serotype 2 in 1999.

Monovalent OPV (mOPV): Monovalent form of OPV is available for serotype 1 (mOPV1) and serotype 3 (mOPV3).

In general, the rate of seroconversion against each serotype is higher with monovalent preparation followed by bivalent and trivalent preparation (mOPV > bOPV > tOPV).

Inactivated polio vaccine (IPV): IPV mainly induces humoral immunity. As the virus strains are inactivated, they do not replicate in the intestine and thus the induction of intestinal immunity is limited. However, IPV has been seen to reduce the duration and amount of virus shedding in stool.

The seroconversion rate is near 100% for each serotype. IPV is available as stand alone vaccine or in combination with other primary immunization vaccine.

It is not associated with VDPV.

Vaccine associated paralytic polio (VAPP): Paralytic polio occurs due to oral polio vaccine administration in two settings. First, when the vaccine strain undergoes mutation during replication in the intestine and there is revert back of the neurovirulence property. Second, when the vaccine (OPV) is administered to a B cell deficient child.

In the first scenario, it occurs due to the oral polio vaccine strain which genetically gets changed during replication in the intestine and revert back its neurovirulence. It occurs in (i) the child vaccinated with the first dose of OPV and (ii) susceptible close contacts of the recipient who is excreting the virus.

VAPP occurs in 1 in 2.7 million doses of OPV administration and does not spread to cause outbreaks.

Vaccine derived poliovirus (VDPV): Vaccine derived poliovirus is a genetically different strain that is originated from the oral polio vaccine strain.

VDPV can be of three types: Circulating VDPV (cVDPV), immunodeficiency related VDPV (iVDPV) and ambiguous VDPV (aVDPV).

Circulating VDPV (cVDPV): Circulating VDPV strain emerges in poorly immunized population. A fully immunized person is protected against both wild virus and VDPV. Due to poor coverage of immunization, population remains susceptible to virus, and the excreted VDPV strain continues to transmit from person-to-person. Longer is the circulation, higher is the chance of genetic changes. It takes a few months to circulate to emerge as cVDPV.

cVDPV can lead to outbreak and has the potential to become endemic when circulates for a long period. The strain can also get transmitted to other susceptible country or locality.

Outbreak due to cVDPV can be prevented by giving 2–3 rounds of supplementary immunization.

Prevalence of cVDPV: It has caused 24 outbreaks in last 10 years, with more than 750 cases of paralytic polio, affecting 21 countries.

To prevent the emergence and circulation of VDPV, OPV has to be stopped.

Table 17.1: Poliovirus laboratory network and environmental surveillance sites, Southeast Asian region

Type of lab	No. of lab	City, Country
Global specialized polio lab	1	Entero virus research center, India: Mumbai
Regional reference lab	2	Sri Lanka: Colombo; Thailand: Bangkok
Intratypic differentiation lab	15	Bangladesh: Dhaka; Democratic People's Republic of Korea: Pyongyang; India: Ahmedabad, Bengaluru, Chennai, Delhi, Kolkata, Lucknow, and Mumbai; Indonesia: Bandung, Jakarta, and Surabaya; Myanmar: Yangon; Sri Lanka: Colombo; Thailand: Bangkok
Primary virus culture lab	11	India: Ahmedabad, Bengaluru, Chennai, Delhi, Kolkata, Lucknow, Kasauli, and Mumbai; Indonesia: Bandung; Sri Lanka: Colombo; Thailand: Bangkok

Immunodeficiency related VDPV (iVDPV): Immune deficient individuals are not able to mount immune response which leads to continuous replication of vaccine virus in the intestine. These individuals excrete immunodeficiency related VDPV. The excretion of iVDPV is usually stopped within six months.

Thirty-three cases of iVDPV have occurred so far.

Ambiguous VDPV (aVDPV): These strains are VDPV that are isolated from individuals with no known immunodeficiency.

Polio end game plan: The strategy plan recommendation of polio eradication and end game strategic plan 2013–2018 of WHO is as follows:

• Vaccination with "only OPV" should be stopped.
• OPV containing P2 should not be used any more.
• At least single dose of IPV must be introduced before six months of switch over from trivalent OPV (tOPV) to bivalent OPV (bOPV) by the end of 2015.
• In 2016, switch over from use of tOPV to bOPV should be done in countries using only OPV.
• With introduction of IPV, complete withdrawal of all OPV use by 2019–2020.

Polio surveillance: Polio surveillance is done by two methods:
• Acute flaccid paralysis (AFP) surveillance
• Environmental surveillance

AFP surveillance is the gold standard for detection of poliomyelitis cases. The purpose of AFP surveillance is to isolate, identify and characterize the poliovirus isolated from the AFP cases.

This includes (i) finding and reporting of AFP case, (ii) stool sample collection and transport to designated lab for testing, (iii) isolation and identification of poliovirus in the lab, and (iv) to determine the origin of the isolated virus.

Environmental surveillance is an additional measure to detect the presence of poliovirus in the sewage and other environmental samples. Detection of poliovirus in the environmental samples indicates the transmission of virus in absence of paralytic cases.

Amongst the Southeast Asian countries, India is the only country to have environmental surveillance in addition to AFP surveillance. Presently environmental surveillance system in India includes a total of 21 sites in various cities of Maharashtra, Delhi, Kolkata and Punjab.

The list of poliovirus laboratory network of Southeast Asian region is shown in Table 17.1.

Bibliography

1. Bandyopadhyay AS, Garon J, Seib K, Orenstein WA. Polio vaccination: Past, present and future. Future Microbiol 2015;10(5):791–808.
2. Burns CC, Diop OM, Sutter RW, Kew OM. Vaccine-derived polioviruses. J Infect Dis 2014 Nov 1;210 Suppl 1:S283–93.

3. http://www.polioeradication.org/dataandmonitoring/Surveillance.aspx

4. http://www.polioeradication.org/polioand prevention/thevirus/vaccinederivedpolioviruses.aspx.

5. http://www.polioeradication.org/resourcelibrary/strategyandwork.aspx

6. http://www.searo.who.int/entity/immunization/topics/polio/eradication/sea-polio-free/en/

7. http://www.cdc.gov/vaccines/pubs/surv-manual/chpt12-polio.pdf

8. http://www.who.int/immunization/diseases/poliomyelitis/endgame_objective2/oral_polio_vaccine/VAPPandcVDPVFactSheet-Feb2015.pdf

9. O'Connor PM, Allison R, Thapa A, Bahl S, Chunsuittiwat S, Hasan M, Khan Z, Sedai T. Update on polio eradication in the World Health Organization Southeast Asia Region, 2013. J Infect Dis 2014 Nov 1;210 Suppl 1: S216–24.

Hand-Foot-Mouth Disease

Hand-foot-mouth disease (HFMD) is an infectious disease, commonly affects children below five years of age. The disease is so named because of its characteristic clinical presentations of vesicular rash in hand, foot and mouth.

AGENT

HFMD is caused by several viruses of Picornaviridae family and genus Enterovirus. The genus Enterovirus consists of 12 species; nine enteroviruses (Enterovirus A–F, J and H) and 3 Rhinovirus (Rhinovirus A–C).

Human enteroviruses belong to the Enterovirus species A–D. **Coxsackievirus A16 (CVA16), and Enterovirus 71 (EV 71)** are the two major genes of HFMD. Both EV71 and CVA16 belong to the species Enterovirus A. Other enteroviruses that are also associated with the causation of HFMD are CVA6, CVA9, CVA10, echoviruses and several coxsackie B viruses.

The members of enterovirus are small, non-enveloped, single stranded, positive sense RNA viruses. The genome is translated as a large polyprotein which is composed of four capsid proteins (VP1–VP4) and seven non-structural proteins (2A–C and 3A–D). VP1 is the major surface protein. It contains neutralizing epitopes, used as the major protein for genomic analysis and candidate antigen for vaccine preparation.

HFMD OUTBREAKS

During the last 10 years, the disease has emerged in Asia Pacific region with several large outbreaks in China, Singapore, Hong Kong, Korea, Vietnam, Japan, Malaysia, Taiwan and Thailand. In China, a major outbreak occurred in Anhui province in 2008, after which the disease has affected more than 12 million cases and 3000 deaths. HFMD is now a notifiable disease in many of these countries. Outbreaks have been reported due to EV71, CVA16 and recently due to CVA6. Outbreaks due to EV71 and CVA6 are associated with more severe cases with high rate of mortality.

Changing etiology pattern has been noted in some countries, where one agent remains predominant for one or more years and then gets replaced with another agent in subsequent years.

In India, the first outbreak of HFMD was reported in 2003 from Calicut, followed by in 2007 from West Bengal and thereafter from several other states. So far all the outbreaks have reported mild form of disease. CVA16 has been confirmed to be the causative agent in one of the outbreak.

HFMD disease can occur at any time of the year. However, the disease more commonly shows a seasonal peak during the late spring and early summer; April to August with a peak at May or June. In some years, two peaks are observed; major one during April to August and minor peak during September to October.

PATHOGENESIS

Virus is present in oropharyngeal secretion, vesicular fluid and excreted in feces of the infected person. Transmission of virus occurs primarily through person-to-person or through feco-oral route by (i) direct contact with infected oral or vesicular fluid, (ii) contact with infected droplets or fomites or (iii) ingestion of contaminated water.

Virus causing HFMD initially gets implanted in the oropharynx or ileum, from where it reaches the regional lymph nodes. Replication of virus at this site leads to viremia which causes seeding of virus to skin, mucous membrane, CNS, and other tissues. Replication at these sites leads to major viremia and clinical manifestations.

CLINICAL FEATURES

The incubation period varies from 2 to 7 days. The disease affects mostly children of below 5 years of age and adults are rarely affected. The symptoms start with general manifestations of fever, malaise, loss of appetite and sore throat.

After 1–2 days of fever, lesions initially appear in the oral cavity. Lesions in hand and feet can occur simultaneously or after the appearance of lesions in the oral cavity. Along with hand and feet, the other sites of lesions are knee, elbow and buttocks. Lesions are painful and pruritic. To begin with, these are maculopapular in nature which progresses to become vesicular with erythematous border. Symptoms in most of the cases subside within 7–8 days.

COMPLICATIONS

Complications are more commonly seen with EV71 and CVA6. In general, coxsackievirus A16 causes mild disease, whereas complications are more common due to EV71. Complications in EV71 cases occur in 20–30% of cases and mostly associated with neurological and cardiopulmonary manifestations such as meningitis, meningoencephalitis, encephalitis, acute flaccid paralysis, pulmonary hemorrhage, and myocarditis.

Coxsackie AV6 has also been associated with a relatively severe course of disease as compared to the classical cases of HFMD and considered to be a more virulent strain. In pediatric patients, CVA6 has been reported to cause severe form of skin lesions in the form of vesiculobulous, erosive and purpuric lesions. In adults with CVA6 purpuric lesions in palm and sole may resemble secondary syphilis.

LAB DIAGNOSIS

The diagnosis of HFMD is mainly based on the typical clinical presentation. Laboratory confirmation is required in atypical presentation and for genetic analysis of the strains. Confirmations of etiological agent though not important for individual patient but is helpful from a larger perspective in order to understand the molecular epidemiology of virus and to predict the severity of the outbreak.

Throat swab, vesicular fluid, stool samples are required for isolation of virus and viral genome detection by RT-PCR. Serum sample is required for detection of antibodies. Samples for virus isolation and RT-PCR should be transported at 4°C to the laboratory.

Molecular detection: Diagnosis of enterovirus infection by molecular method is presently the most commonly used practice for patient diagnosis and confirmation of outbreaks. Detection of viral RNA can be done by RT-PCR directly from clinical samples or from culture supernatant. Primers are available both pan enterovirus genome and for type specific viruses like CA16 and EV71.

Isolation: Isolation of enterovirus is done by cell culture system or suckling mice host. No single cell line support the growth of all types of enteroviruses. Various cell lines, such as human rhabdomyosarcoma (RD), HEp-2, human colorectal cancer cell line (Caco-2), and human pulmonary adenocarcinoma cell line (A549) have, therefore, been recommended for isolation of the HFMD agent. Appearance of cytopathic effect is confirmed by presence of viral antigen or RNA by immunofluorescence or PCR using type specific antisera and primer, respectively.

Serology: Demonstration of four-fold change in the antibody titer by complement fixation or neutralization can diagnose the acute disease.

Neutralization assay based on the inhibition of cytopathic effect by serial dilution of patient's sera is the conventional method used for measurement of antibody titer against enteroviruses. The method is not used for patient diagnosis and mostly used for seroprevalence assay or retrospective confirmation of outbreak.

Presently IgM ELISA is commonly used for diagnosis. The test is available for CVA16 and EV71. IgM antibody detection is positive in around 50% of cases during the first few days of illness which increases to near 100% towards 8th day onwards. The test is commonly used these days as this is convenient and rapid. However, high antigenic similarities between different enteroviruses can lead to false positives.

VACCINE

Three Enteroviruses, 71 vaccines have been developed in China, of which one has got the licensure in December 2015 and the other two have completed phase III clinical trial.

All the three vaccines are inactivated vaccine prepared from the whole virus (EV71 C4 genotype) grown on cell culture using the principles used for the preparation of polio vaccine.

The vaccine that has been licensed by China Food and Drug Administration (CFDA) is the first EV71 vaccine to get the licensure. It has been manufactured by Chinese Academy of Medical Sciences. Vaccine has shown 97% efficacy in preventing infection by EV71 in a large clinical trial which included more than 12,000 children. Vaccine is given intramuscularly in two doses at six months and second dose after 4 weeks.

Bibliography

1. Fu Y, Sun Q, Liu B, Xu H, Wang Y, Zhu W, Pan L, Zhu L. Epidemiological characteristics and pathogens attributable to hand, foot, and mouth disease in Shanghai, 2008–2013. J Infect Dev Ctries 2016;10(6):612–18.

2. Guo C, Yang J, Guo Y, et al. Short-term effects of meteorological factors on pediatric hand, foot, and mouth disease in Guangdong, China: a multi-city time-series analysis. BMC Infect Dis 2016; 16(1):524.

3. http://www.wpro.who.int/china/mediacentre/releases/2015/20151208/en/

4. Kim BI, Ki H, Park S, Cho E, Chun BC. Effect of Climatic Factors on Hand, Foot, and Mouth Disease in South Korea, 2010-2013. PLoS One 2016;11(6):e0157500.

5. Kim HJ, Hyeon JY, Hwang S, et al. Epidemiology and virologic investigation of human enterovirus 71 infection in the Republic of Korea from 2007 to 2012: a nationwide cross-sectional study. BMC Infect Dis 2016;16(1):425.

6. Kok CC. Therapeutic and prevention strategies against human enterovirus 71 infection. World J Virol 2015;4(2):78–95.

7. Li J, Sun Y, Du Y, et al. Characterization of Coxsackievirus A6- and Enterovirus 71-Associated Hand Foot and Mouth Disease in Beijing, China, from 2013 to 2015. Front Microbiol 2016;7:391.

8. Li JS, Dong XG, Qin M, et al. Outbreak of hand, foot, and mouth disease caused by coxsackievirus A6 in a Juku in Fengtai District, Beijing, China, 2015. Springerplus 2016;5(1):1650.

9. Mao Q, Wang Y, Yao X, Bian L, Wu X, Xu M, Liang Z. Coxsackievirus A16: Epidemiology, diagnosis, and vaccine. Hum Vaccin Immunother 2014;10(2):360–7.

10. Omaña-Cepeda C, Martínez-Valverde A, del Mar Sabater-Recolons M, et al. A literature review and case report of hand, foot and mouth disease in an immunocompetent adult. BMC Res Notes 2016;9:165.

11. Reed Z, Cardosa MJ. Status of research and development of vaccines for enterovirus 71. Vaccine 2016;34(26):2967–70.

12. Sarma N, Sarkar A, Mukherjee A, et al. Epidemic of hand, foot and mouth disease in West Bengal, India in August, 2007: a multicentric study. Indian J Dermatol 2009;54(1):26–30.

13. Sarma N. Hand, foot, and mouth disease: Current scenario and Indian perspective. Indian J Dermatol Venereol Leprol 2013;79(2):165–75.

14. Sasidharan CK, Sugathan P, Agarwal R, et al. Hand-foot-and-mouth disease in Calicut. Indian J Pediatr 2005;72(1):17–21.

15. Wang P, Goggins WB, Chan EY. Hand, Foot and Mouth Disease in Hong Kong: A Time-Series Analysis on Its Relationship with Weather. PLoS One 2016 Aug 17;11(8):e0161006.

16. Zhao J, Li X. Determinants of the Transmission Variation of Hand, Foot and Mouth Disease in China. PLoS One 2016 Oct 4;11(10):e0163789.

Rabies Virus

Rabies is an invariably fatal disease in human and animals due to acute encephalomyelitis. The disease is caused by the neurotropic viruses of Lyssavirus genus. It is a zoonosis of all mammalian species around the world. Human infections occur through the bite of the infected animals. The average estimated number of human deaths due to rabies is more than 60000 per year globally.

VIRUS

Rabies virus belongs to the order Mononega-virales, family Rhabdoviridae and genus Lyssavirus. According to the 2015 release of International Committee on Taxonomy of Viruses (ICTV), Lyssavirus genus includes 14 virus-species based on genetic distance, antigenic pattern, host range and geographic distribution (Table 19.1). Rabies virus is the type species and is responsible for almost all the cases of human rabies.

Rabies virus is a bullet-shaped virus, the average dimension of which is around 200 nm length and 75 nm diameter. The genome is a single stranded, non-segmented RNA of negative polarity of 12 kilobase pairs in length. It codes for five viral proteins; nucleoprotein (N), phosphoprotein (P), matrix protein (M), glycoprotein (G) and RNA dependednt RNA polymerase (L or large protein). The RNA genome along with N, P and L proteins form the ribonucleoprotein complex. The nucleo-capsid is present in a helical symmetry which is surrounded by the matrix protein (M protein).

Table 19.1: List of virus species of Lyssavirus genus

Virus species	Host
1. Aravan virus	Insectivorous bats
2. Australian bat Lyssavirus	Pteropodid bats and insectivorous bats
3. Bokeloh bat Lyssavirus	Insectivorous bats
4. Duvenhage virus	Insectivorous bats
5. European bat 1 Lyssavirus	Insectivorous bats
6. European bat 2 Lyssavirus	Insectivorous bats
7. Khujand virus	Insectivorous bats
8. Ikoma virus	Unknown
9. Irkut virus	Unknown
10. Lagos bat virus	Pteropodid bats
11. Mokola virus	Unknown
12. Rabies virus	Bats (new world) and carnivora
13. Shimoni bat Lyssavirus	Insectivorous bats
14. West Caucasian bat Lyssavirus	Insectivorous bats

The outermost layer is the envelop which contains surface projections of glycoproteins (G protein).

EPIDEMIOLOGY

Rabies is still a huge public health problem in canine infested developing countries which together contribute the near total deaths due to rabies in humans. Human rabies is present all over the world, in all continents including 150 countries and territories excepting Antarctica.

Canine rabies is still endemic in major parts of Asia, Africa, and Latin American countries. According to the estimation in 2010, the average numbers of annual death due to rabies is 61000. Of this, near one-fourth is contributed by India only (largest number by any single country) and near 60% by all Asian countries. Most of these deaths (near 85%) occur in rural areas and in children.

Several countries in the world have been declared free of canine rabies, such as the United States of America (USA), Canada, Western Europe, Japan, Malaysia and a few Latin American countries. Australia has been free from carnivore rabies. Whereas, Pacific islands are free from rabies and rabies-related viruses.

Rabies virus is transmitted by bite of rabid dog or other mammals. Dogs are the primary vector, contributing to more than 99% of all human rabies cases. Transmission through inhalation has been associated with laboratory accidents during vaccine preparation and exposure to insectivorous bats in the cave. Ingestion of raw milk from rabid animal possesses risk of infection, but there is so far no documented report available. Humans are the dead end host, but rare transmission of virus from human-to-human has been reported through transplantation of infected organs, human bite, kissing, transplacental route, and through human breast milk.

PATHOGENESIS

Rabies virus enters the human body through bite of infected animals and is inoculated in the skin and muscle. It cannot enter through intact skin and enters only through broken skin with intact mucosa. The source of virus is the salivary fluid of the rabid animal. The initial replication of the virus occurs in the muscle and then the virus binds to the nicotinic acetylcholine receptor present at the postsynaptic neuromuscular junction. Along the peripheral nerves, the virus spreads centripetally via fast retrograde transport along motor axons towards the spinal cord and brainstem. The rate of travel through peripheral nerve is approximately 50–100 mm per day. Once it reaches the CNS, extensive virus replication occurs in the neurons and clinical disease develops. Then the virus disseminates through the axonal transport to all over the CNS. From CNS, the virus then transverses centrifugally along sensory and autonomic nerves to other tissues of the body such as salivary gland, adrenal gland, heart, cornea and skin (Fig. 19.1). By the time patient develops the symptoms, virus is widely distributed throughout the CNS and possibly to extraneural organs.

The average incubation period is 2–3 months, which ranges from 5 days to more than 1 year depending on the amount of virus inoculums, density of the motor end plate at the site of animal bite and the proximity of site of virus inoculation to the CNS.

PATHOLOGY

Mild inflammatory changes occur with mononuclear cell infiltration in the leptomeninges, perivascular regions and brain parenchyma. **Negri body is the most pathognomonic** finding of rabies. These are intracytoplasmic eosinophilic inclusion bodies present in the neuronal cells and composed of viral RNA and proteins. The common sites of negri bodies are pyramidal neurons of hippocampus and Purkinje cells of the cerebellum. However, negri bodies are seen only in small number of rabies cases. Unlike other viral encephalitis, neuronophagia and neuronal apoptosis are not common in rabies encephalitis.

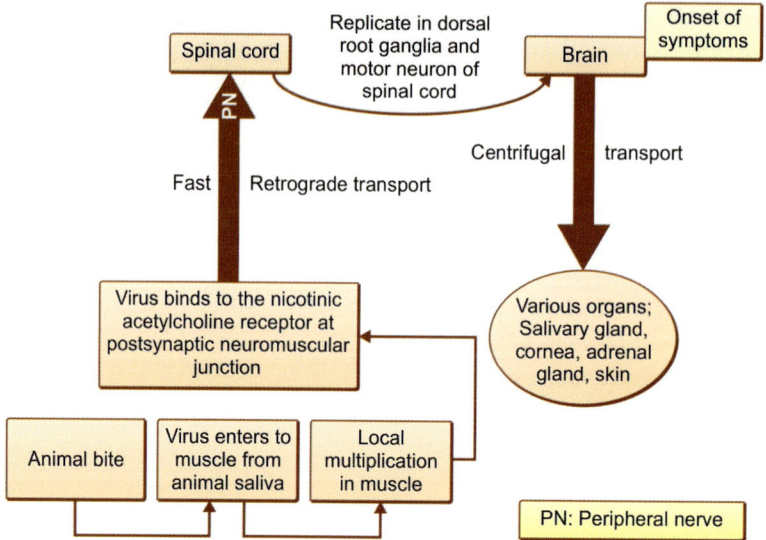

Fig. 19.1: Pathogenesis of rabies virus

CLINICAL FEATURES

Clinical symptoms of rabies start with the prodromal symptoms which are usually manifest with febrile illness with change in mood. The common prodromal symptoms are fever, headache, nausea and vomiting. The mood change usually manifests in the form of anxiety, irritability, restlessness, and agitation. The first specific symptom is pain, pruritus, itching or tingling sensation at the site of bite, which occurs in around half of the patients.

The acute neurologic form of human rabies can be manifested in one of the two forms—furious or encephalitic form and dumb or paralytic form. The former type is more common and seen in around 80% of patients, whereas the later is seen in only 20% of cases. This is probably because of the predominant involvement of brainstem, cranial nerves and limbic system in encephalitic form and medulla, spinal cord and spinal nerves in the paralytic form.

The **encephalitic form** is rapidly progressive in nature. The general symptoms of encephalitis such as fever, hallucination, seizure are same as that of other encephalitis. The rabies specific symptoms are autonomic dysfunction which manifest as hypersalivation, hydrophobia, aerophobia, foaming at the mouth, and period of hyperexcitability alternating with lucidity. The duration of lucidity gradually decreases. The term hydrophobia means "dread of water" which is painful, involuntary contractions of the diaphragm and other accessory respiratory muscles that is provoked by attempt to drink water or liquids. Similar reflex caused by stimulation of air. Finally patient develops complete paralysis and multiorgan failure and death ensues within days of development of classic rabies symptoms.

The **dumb or paralytic form** of rabies has a slow and protracted course. The neurological manifestations start as flaccid muscle weakness in the bitten limb which ascends gradually and leads to quadriparesis and ultimately affecting the respiratory and deglutition muscles. Involvement of sphincter is common leading to urinary retention and constipation. The classic features of encephalitic form are usually not seen in paralytic form and it can be confused with other neurological disorders like Guillain-Barré syndrome.

Some of the other members of Lyssavirus genus, most of which infect bat, can produce disease in humans indistinguishable from human rabies. These are: Australian bat Lyssavirus, European bat Lyssavirus 1 and 2, Irkut virus and Duvenhage virus and Mokola virus (which has been so far isolated from shrews). These viruses are called "rabies related viruses".

LAB DIAGNOSIS

Samples for confirmation of rabies are collected either during life or postmortem. Antemortem diagnosis is difficult as collection of samples is difficult in practical situation from a furious case of rabies and also because the negative result of the diagnostic test does not rule out the possibility of infection.

For intra-vitam diagnosis, saliva, cerebrospinal fluid (CSF), skin biopsy and serum are commonly collected. Collection of multiple saliva samples (3 no.) at the interval of 3–6 hours increases the sensitivity. Brain tissue though considered as the best sample, is usually not recommended for intra-vitam diagnosis. Samples are stored at –20°C.

For confirmation of infection postmortem, brain tissue is the most preferred sample. The tissue can be collected through cribriform plate at the bedside or in the field situation. Skin biopsy from nape of the neck also can be collected. Tissue samples can be preserved in glycerol and stored at –20°C. When tissue samples have been preserved in formalin, they should be taken out from formalin and kept in absolute ethanol for subsequent molecular testing and antigen detection.

Rabies viral infection is commonly diagnosed by detection of virus or its components (i.e. antigen detection, viral RNA detection and virus isolation), or by detection of rabies virus antibody in the serum. Demonstration of negri body (intracytoplasmic rabies virus inclusion body in neuronal cells) in neurons by Seller's stain is not considered these days because of its low sensitivity and availability of better alternative diagnostic tests.

Detection of rabies antigen: This can be done by various methods.

Direct fluorescent antibody testing (DFAT): The test is considered as the gold standard for rabies diagnosis. It is used for detection of rabies virus antigen in the brain tissue of the patient using polyclonal or monoclonal antibody. The fixed tissue or impression smear made on slide is incubated with the antirabies polyclonal antibody or broadly reactive monoclonal antibody tagged to fluorescein isothiocyanate. The test has 99% sensitivity and specificity, but the performance depends largely on the quality of the antirabies conjugate and expertise of the observer and not reliable in degraded sample.

Other antigen detection tests based on enzyme immune assay, rapid immunodiagnostic test and immunohistochemistry have been shown to be comparable with DFAT. These tests are rapid and does not require fluorescence microscope. But most of these tests are not available commercially and still under evaluation.

Detection of viral RNA: Presently the method is recommended for intra-vitam diagnosis. The test is applicable in all types of samples including tissues and fluids. Reverse transcriptase polymerase chain reaction (RT-PCR) and real-time RT-PCR are the commonly used methods for viral RNA detection. Other molecular amplification techniques have also been developed. The methods are highly sensitive and specific for viral RNA detection in a stringent quality control lab.

Virus isolation: Rabies virus can be isolated in cell culture or animal inoculation methods. Mice inoculation test (MIT) is done by intracerebral inoculation into 3–4 weeks old mice which are observed for 28 days for development of symptoms. Suckling mice of <3 days old can also be used when rapid results are required. Mice brain is harvested for detection of viral antigen or viral RNA detection. The sensitivity of MIT is comparable to that of cell culture. However, due to animal protection issues, it is recommended by World Organization for animal health to replace the method of animal inoculation by cell culture wherever possible.

The cell culture method uses mice neuroblastoma cell line. Unlike MIT which takes around 28 days, cell culture is a rapid and inexpensive method where result is available in 24–48 hours. The method is as sensitive as MIT.

Detection of antibody: Presence of rabies antibody in serum or CSF in a non-vaccinated individual gives an indirect evidence of infection. Antibody appears in serum or CSF only after 8–10 days of symptoms. The test is, therefore, not helpful for patient's diagnosis as majority of the patients dies within 6–7 days of onset of symptoms.

PREVENTION

Rabies is a 100% preventable disease when vaccination and immunoglobulin are administered as per the recommendations of world health authorities such as WHO or CDC.

Types of vaccine: There are two types of vaccines available for rabies, nerve tissue vaccine and cell culture or embryonated egg-derived vaccine.

Cell culture or embryonated egg-based vaccine: Rabies virus is propagated in different cell substrates, such as human diploid cell line (HDCV), Vero cells, primary chick embryo cell or embryonated duck eggs. After propagation, virus is harvested and then undergoes further steps of processing, such as concentration, purification, inactivation and lyophilization. All preparations are highly safe and effective, and chick embryo cell and Vero cell preparations are less expensive than HDCV.

Nerve tissue vaccine: These vaccines are prepared by propagating the virus in the nerve tissues of different animals (sheep, suckling mice) followed by different methods of inactivation. These vaccines are less immunogenic, and associated with severe adverse effects, require multiple doses. Because

of its association with serious adverse reaction, since 1984 WHO has advised to stop the preparation and use of neural vaccine and to replace them with cell culture or embryonated egg-based vaccines.

Types of vaccination: Rabies vaccination is indicated in two types of situations—pre-exposure prophylaxis and post-exposure prophylaxis.

Pre-exposure prophylaxis: Pre-exposure vaccination is indicated in laboratory personnel working with rabies virus, animal handlers or persons coming in contact with animals, individuals travelling to or living in the area at risk.

Post-exposure prophylaxis: Post-exposure vaccination is indicated in persons who are bitten by any suspected rabid animal. Vaccination is advised when there is breach in the skin at the site of animal bite (category II and III of animal exposure), whereas in category I there is no breach in the skin and vaccination is not recommended. Category II which includes minor abrasion is advised for vaccination and category III which includes single or multiple bites is advised for vaccination along with immunoglobulin.

Vaccination schedule: The types of vaccine and route of administration are same for both pre-exposure and post-exposure prophylaxis. The numbers of doses are more with post-exposure than pre-exposure prophylaxis (4/5 as compared to 3).

Vaccine can be given either intramuscularly or intradermal route following the same schedule. The dose of vaccine by IM route is 0.5 mL or 1 mL, and 0.1 mL by ID (Table 19.2).

Immunoglobulin: Rabies immunoglobulin (RIG) is recommended in category III exposure and also in severely immunocompromised patients. RIG is available in 3 preparations—human RIG (HRIG), equine RIG (ERIG) and highly purified F(ab') 2 fragments of equine RIG.

RIG is administered just before or shortly after the first dose of vaccination and given at

Table 19.2: Dose schedule of rabies prophylaxis

Type of prophylaxis	Immunoglobulin/vaccine	Route	Schedule
Pre-exposure	CC/EEV	Intramuscular or intradermal	0, 7, 21/28 days
Post-exposure	CC/EEV	Intramuscular	0, 3, 7, 14 days
		Intradermal	0, 3, 7, 28 days (two doses each)

CC: Cell culture; EEV: Embryonated egg vaccine.

the site of animal bite and can be given within 7 days after the initiation of the primary series of vaccination.

The dose is 20 IU/kg body weight for HRIG and 40 IU/kg body weight for ERIG or F(ab') 2 fragments products. Maximum possible amount of RIG should be administered into or around the wound site and the remaining amount is given intramuscularly at the site distant from the vaccination site.

Novel approaches like genetically engineered live attenuated vaccines and cocktails of monoclonal antibodies are under different phases of trial.

Bibliography

1. Duong V, Tarantola A, Ong S, Mey C, Choeung R, Ly S, Bourhy H, Dussart P, Buchy P. Laboratory diagnostics in dog-mediated rabies: An overview of performance and a proposed strategy for various settings. Int J Infect Dis 2016;46:107–14.

2. Fooks AR, Banyard AC, Horton DL, Johnson N, McElhinney LM, Jackson AC. Current status of rabies and prospects for elimination. Lancet 2014;384(9951):1389–99.

3. http://www.ictvonline.org/virustaxonomy.asp accessed on 30th December 2016.

4. http://www.who.int/rabies/en/ accessed on 31st December 2016.

5. http://www.who.int/rabies/human/postexp/en/ accessed on 31st December 2016.

6. http://www.who.int/rabies/epidemiology/en/ accessed on 31st December 2016.

7. WHO Expert Consultation on Rabies: second report. WHO technical report series; no. 982. 2013.

8. Zhu S, Guo C. Rabies Control and Treatment: From Prophylaxis to Strategies with Curative Potential. Viruses 2016 Oct 28;8(11). pii: E279.

Chapter
20

Emerging Viruses

A. EBOLA VIRUS

Ebola virus belongs to the order Mononegavirales and family Filoviridae. According to the recent classification by International Committee for Taxonomy of Viruses (ICTV), the family Filoviridae has three genera: Ebolavirus, Marburgvirus and Cuevavirus. The genus Ebolavirus has five species: Bundibugyo ebolavirus (BEBOV), Reston ebolavirus (REBOV), Sudan ebolavirus (SEBOV), Tai Forest ebolavirus (TEBOV) (previously known as Ivory Coast ebolavirus) and Zaire ebolavirus (ZEBOV) (Fig. 20.1).

The name of the virus family Filoviridae has been derived from the Latin word *Filum* means thread-like which comes from the shape of the virus as long, narrow and twisted appearing like thread.

During the outbreak of hemorrhagic fever in 1976 in central part of Africa (Sudan and Zaire), the virus was identified for the first time and was named after the river Ebola located in Zaire (presently known as Democratic Republic of Congo or DRC). The outbreak had occurred in two neighboring countries, southern Sudan and northern Zaire affecting more than 300 people with 88% case fatality rate. The virus identified from two places was later found to be two different Ebolavirus species and were named Sudan ebolavirus (SEBOV) and Zaire ebolavirus (ZEBOV).

TEBOV has been identified in Ivory Coast from one individual who had conducted an autopsy in chimpanzee from Tai National Park Reserve and later recovered from the disease.

The latest human pathogenic Ebolavirus Bundibugyo ebolavirus (BEBOV) was detected in 2007 from the town Bundibugyo in Uganda.

Reston ebolavirus (REBOV) was first detected in 1989 in Reston, Virginia, USA in monkeys that were imported from Philippines

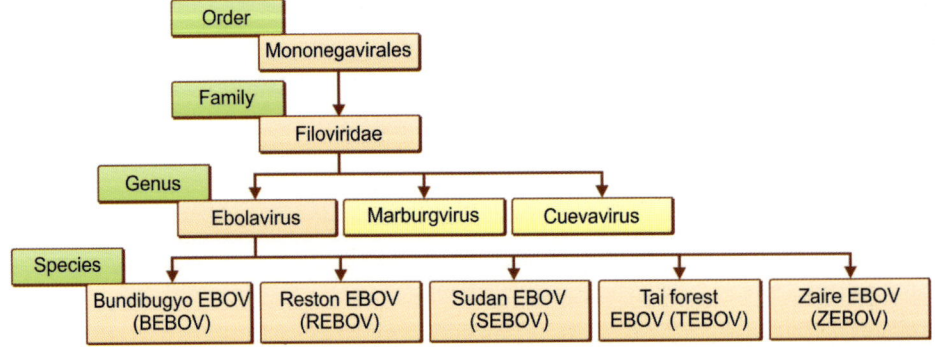

Fig. 20.1: Taxonomy of ebolavirus

and kept in quarantine. High mortality was observed amongst the monkeys but none of the animal handler developed the disease.

All the five species of Ebolavirus genus except Reston ebolavirus (REBOV) are pathogenic to humans. ZEBOV species is associated with most severe disease with near 90% mortality as compared to 53% of SEBOV and 25% of BEBOV.

VIRUS

As the name **"filo" suggests long and thread-like**, the length of the virus is 1000–1200 nm and may go up to 14000 nm with a uniform diameter of 80 nm giving an appearance of thread or rope. The genome consists of single stranded, negative sense, non-segmented RNA. The helical nucleocapsid is surrounded by envelop which possesses glycoprotein (GP) spikes of 10 nm length on its surface. The RNA genome encodes for seven genes which are nucleoprotein (NP), virion protein (VP) 35, VP40, glycoprotein (GP), VP30, VP24, RNA dependent RNA polymerase (L) and a small non-structural soluble form of glycoprotein (sGP). The function of each protein is as follows:

- NP, VP35, VP30 and RNA dependent RNA polymerase: Viral replication and transactivation.
- VP 24, VP 35: Type I interferon antagonist, important virulence factor.
- VP40 (matrix protein): Budding and release of virus particle.
- VP24 (minor matrix protein): Nucleocapsid formation.

- Glycoprotein (GP): Role in virus entry and immunogenicity, vaccine candidate.

Small soluble glycoprotein (sGP): Product of GP gene gets secreted from the infected cell in a large quantity. Possibly prevents the immune system from mounting an effective immune response.

Ebola hemorrhagic fever outbreaks: The first recognized outbreak of Ebola virus occurred during the year 1976 in Zaire, the place presently known as Democratic Republic of Congo (DRC). Outbreak started with an index case who had malaria-like symptoms. The spread of the disease was mainly due to the use of unsterilized needle affecting 318 cases with 280 deaths. The viral agent detected from the patient's blood sample was named Ebola. Similar outbreak was going on in the neighboring country, southern Sudan, which was also found to be ebola virus outbreak.

Since then several small and large outbreaks have occurred in the central part of Africa. Table 20.1 shows the major ebola virus outbreaks.

The latest ebola virus outbreak began in December 2013 in the rain forest of Guinea. The outbreak then spread to the neighboring countries, Liberia, Sierra Leone, Mali, Nigeria and Senegal and became the largest ever ebola virus outbreak of the world amounting to more than 28,000 cases with over 11,000 deaths. Human cases of ebola were reported for the first time outside African continent. Table 20.2 shows the countries affected during 2014–2015 outbreak.

Table 20.1: Major ebola virus outbreaks					
Year	Country	Ebola species	Cases	Deaths	%Mortality
1976	Sudan	Sudan	284	151	53
1976	Zaire (DRC)	Zaire	318	280	88
1996	DRC	Zaire	315	250	81
2000	Uganda	Sudan	425	224	53
2007	DRC	Zaire	264	187	71
2014	West African countries	Zaire	28616	11310	40

DRC: Democratic Republic of Congo (previous name: Zaire)

Table 20.2: Number of cases and deaths in 2014 ebolavirus outbreak in Western Africa

Country	Cases	Deaths
Guinea	3811	2543
Liberia	10675	4809
Sierra Leone	14124	3956
Nigeria	20	8
Senegal	1	0
Mali	8	6
Italy	1	0
Spain	1	0
UK	1	0
USA	4	1

EBOLA VIRUS TRANSMISSION

Ebola virus disease (EVD) is a zoonotic disease. Fruit bats are the reservoir of ebola virus.

Virus is transmitted from the reservoir bat to the non-human primates (monkeys or apes) or antelopes in the African forest.

The spillover of virus from the bat or infected animals to human occurs mainly through the practice of haunting, butchering and eating bush meat.

Once human being is infected, transmission occurs mainly through person-to-person contact. As the virus is secreted in blood and body fluids of patient, transmission of virus to the close contacts of the patient occurs either in the home amongst the family members or in health care setting in absence of proper barrier precaution (Fig. 20.2).

Use of unsterilized needle has been implicated as the source of transmission in the previous outbreak in Zaire, Africa.

Practice of touching the dead body has been responsible for a large number of secondary cases during the 2014 outbreak in West African countries.

Sexual mode of transmission is also possible as the virus has been detected in semen even after the recovery of symptoms.

Transmission through aerosol has not been substantiated for man-to-man transmission under natural condition.

Viral factors that facilitate the transmission are:

- The low infective dose ($\leq$10 virus particle).
- High viral load in the severely ill patient.
- Shedding of virus in patient's blood, body fluids, vomitus, stool and secretions.

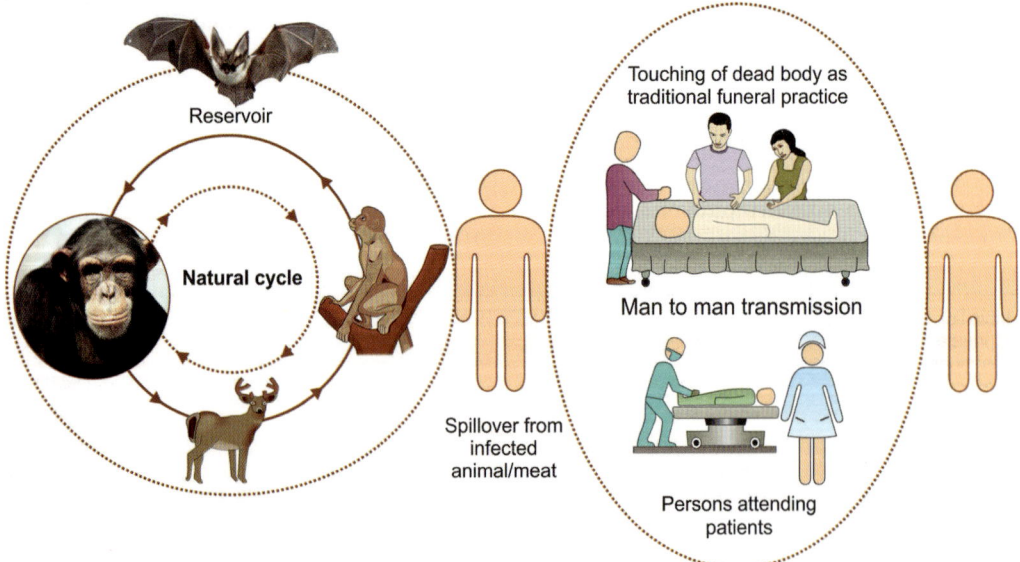

Fig. 20.2: Modes of transmission in ebola virus

PATHOGENESIS

Ebola virus enters the body through skin, mucosa or through parenteral route. Dendritic cells, monocytes and macrophage present in the skin are the primary target cells. After initial replication in these cells, virus disseminates to the regional lymph nodes, spleen, liver, adrenal glands. The migratory nature of these cells helps the virus spreading inside the body.

The infected dendritic and macrophage gets activated and releases proinflammatory cytokines and vasoactive substances. This in turn causes inflammation, coagulation and increases the endothelial cell permeability leading to vascular leakage. Various viral proteins VP40, GP1 and GP2 also have been implicated with increased endothelial cell permeability by activating the intracellular and vascular adhesion molecules.

Ebola virus utilizes various methods to evade the immune system. VP35 and VP24 are the main viral proteins responsible for this by inhibiting interferon type I. Small, soluble glycoprotein (sGP) confounds the immune system and prevents it from mounting immune response. It also inhibits the migration of leukocytes.

Apoptosis of lymphocytes leading to depletion of lymphocytes has been observed in fatal cases.

Liver cell necrosis leads to decrease in clotting factor and deranged coagulation. In addition to this, activation of coagulation cascade, platelet aggregation and consumption together with endothelial damage contributes to bleeding or blood clotting and multiorgan failure (Flowchart 20.1).

CLINICAL FEATURES

The average incubation period is 4–10 days; however, it can range from 2 to 21 days depending on the viral load.

Symptoms generally manifest in three phases:

Phase I (nonspecific illness): It starts with sudden onset of fever with malaise, myalgia, headache, anorexia, nausea, vomiting, diarrhea, abdominal pain.

Phase II: Hemorrhagic manifestations: Bleeding from various sites starts occurring in the form of mucosal bleeding, subcutaneous hemorrhage or internal bleeding in various organs. This is overtly manifested as

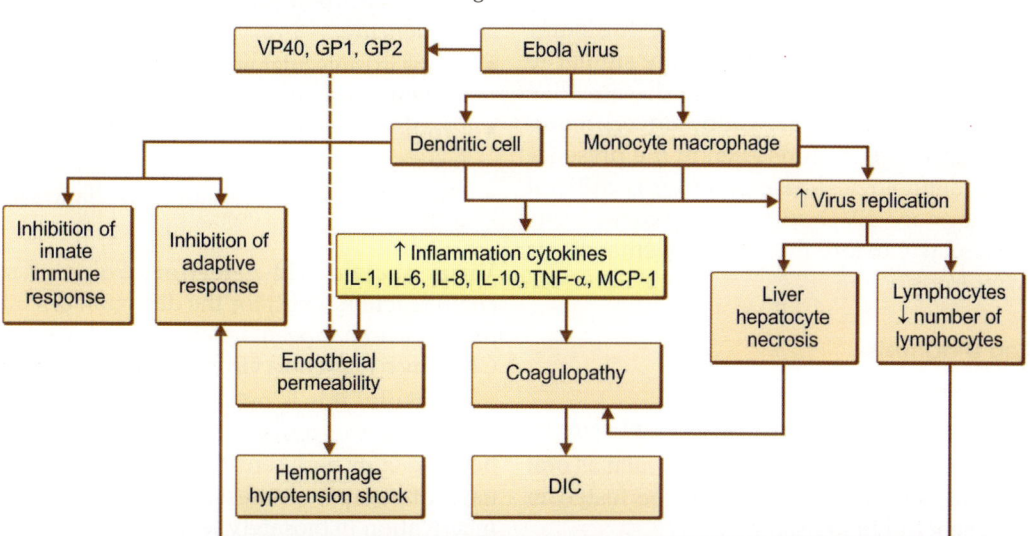

Flowchart 20.1: Pathogenesis of ebola virus infection

erythematous rash, petechiae and ecchymoses and mucosal bleeding from various sites.

[Absence of hemorrhagic manifestations during the initial phase of the illness are observed in majority of the cases during the last ebola outbreak in 2014 in Western Africa. In order to rectify the age old impression of association of hemorrhage with ebola virus infection, the disease is presently termed ebola virus disease (EVD) by replacing the previously used terminology "ebola hemorrhagic fever". However, it is the hemorrhagic manifestation that is responsible for disease transmission, severity, and high mortality.]

Phase III: This is the end stage of the disease process. Bleeding from multiple organs ultimately leads to organ failure, diffuse coagulopathy, metabolic disturbance leading to shock and death.

DIAGNOSIS

Epidemiological association is the key to suspect ebola virus infection.

Endemic countries: Possible contact with bat/animal or visiting any sick patients or funeral.

Non-endemic countries: Travel history to country where active ebola virus transmission is going on.

Differential diagnosis: During the early phase of illness, the symptoms are similar to several other common tropical diseases like malaria, enteric fever, yellow fever, leptospirosis, etc. Therefore, even in the presence of epidemiological association, it is difficult to diagnose EVD solely on clinical basis particularly before the onset of hemorrhagic symptoms.

Lab Diagnosis

Samples

- *Blood:* Blood is the most important sample.
- *Body fluids:* Other body fluids, such as oral fluid, semen, urine can also be tested in absence of blood sample.

- *Tissue:* Tissue samples from various affected organs can be tested in absence of blood sample or for postmortem diagnosis.

All the clinical samples are tested after inactivation with either gamma irradiation or heat inactivation. Chemical treatment with guanidium isothiocyanate (GITC) is done before PCR testing to degrade the viral proteins.

Methods for EBOV Detection

Virus isolation: Isolation of virus by cell culture method is the conventional gold standard for confirmation of presence of virus.

African green monkey kidney cell line Vero E6 is commonly used. Identification of virus is done either by visualization of virus by electron microscopy or virus antigen detection by immunofluorescence method.

Disadvantages

- The process is time consuming and requires the facility of biosafety level 4.
- The test is less sensitive as compared to PCR.
- Not recommended as a diagnostic test in a clinical setting.

Advantages

- Helps in providing the virus isolates for sequencing.
- Helps in molecular epidemiological studies and virus research.
- Provides positive control for diagnostic tests.
- Used as a tool to evaluate new diagnostic tests.

Virus nucleic acid detection: Polymerase chain reaction (PCR) for the detection of viral nucleic acid has been developed and used by CDC since 1995 ebola virus outbreak in Kikwit.

Currently this is one of the standard and preferred methods for EBOV diagnosis because of its high sensitivity, specificity and ease of sample processing with chemical inactivation in biosafety level 2 facilities.

PCR test is done from blood and body fluids, such as oral fluid, semen, ocular fluid, breast milk, etc.

Both conventional RT-PCR and real-time PCR (for viral load quantitation) have been developed by CDC.

Various viral genes have been used as the target gene for detection. These are, L, GP, VP40 and NP genes. Target using L gene can detect all fioviruses, whereas other gene targets are species specific.

L and NP genes have been reported to detect the viral RNA in convalescent samples for several weeks after the cessation of symptoms.

RT-PCR may be negative during the first three days of illness due to low level of virus RNA in the blood. Therefore, a negative PCR result at this phase of illness should be repeated after a few days and does not rule out the infection.

Virus RNA can be detected 1–3 days before the antigen test becomes positive and persists 1–3 days after the disappearance of antigen.

Significance of viral load: Quantitative value of viral load determination can be done by real-time PCR.

- Viral load determination in convalescent patient helps in taking decision in discharging the ebola positive patients.
- A CT value >35 in convalescent patient is considered non-infectious.
- CT value <40 is considered as positive.
- The exact clinical significance of low viral load in convalescent patients is not known as regarding their transmission potential.
- Viral load >10^7 RNA copies/mL has been associated with high mortality.

Detection of antibody: This can be done by indirect fluorescence antibody test (IFAT) or ELISA. The later has almost replaced the IFAT which requires biosafety containment level 4.

IFAT: This test was in use since 1977. It detects the antibody by using the viral antigen prepared from the cell culture. The processing requires the biosafety level 4, poorly sensitive and specific. The test has largely been replaced by ELISA.

ELISA: Ebola virus specific IgM and IgG antibody can be detected by ELISA system. The high sensitivity, high specificity and high throughput applicability have made this test feasible for patient diagnosis. The test can be done in samples inactivated by gamma irradiation. Therefore, ELISA can be done in a biosafety level 2 facility.

Capture ELISA format is preferred for detection of IgM antibody. For detection of IgG, indirect ELISA has been developed by CDC.

In general, IgM antibody appears during 2–11 days after the onset of illness and persists beyond 30 days. IgG appears mostly during the second week of illness and persists for years.

Detection of ebola virus antigen: ELISA for detection of ebola virus antigen in the serum sample has been used by CDC. Antigen is detected by using pool of mouse monoclonal antibodies.

Antigen can be detected from day 1 of illness, and most of the symptomatic patients are positive by day 3 of illness. Positive antigen test indicates the acute illness.

The ELISA available for antigen detection has not been extensively used in clinical setting because of less availability and widely used RT-PCR.

Rapid immunochromatographic test: During the last ebola virus outbreak in 2014, a rapid immunochromatographic test has been developed for the detection of ebola virus antigen in field setting as point of care test.

- Test can be performed from whole blood (both from finger prick and venipuncture blood), plasma and serum samples.
- It detects the VP40 antigen of ebola virus with 91.8% sensitivity and 84.6% specificity as compared to PCR.
- Test result is given in 20–30 minutes.

- Positive rapid test gives a presumptive diagnosis.
- Three different rapid tests have been approved by WHO.

ANTI-EBOLA VIRUS DRUGS

TKM-Ebola or Ebola SNLP: This drug is made up of small interfering RNA particle against ebola virus proteins packaged in lipid nanoparticle. It targets three Zaire ebola virus proteins—L polymerase, VP35 (polymerase complex protein) and VP24 (membrane associated protein).

The TKM-Ebola Phase I clinical trial was conducted to study the safety, tolerability in healthy adult volunteers. It was on partial hold by FDA in 2014 which was later removed and trial was permitted with lower dose. However, during phase II clinical trial in 2015 in Sierra Leone, because of non-encouraging result, company decided to stop further trial.

ZMapp: It is a combination of three mono-clonal antibodies grown in tobacco plant. It targets to inhibit the spread of infection in the body. The clinical trial (PREVAIL II) was launched in March 2015. Seventy-two adult and children ebola virus infected patients from Guinea, Liberia and Sierra Leone were enrolled. Mortality in patient group receiving only ZMapp was less in comparison to the group receiving only standard patient care, but difference was not significant.

Favipiravir: A pyrazinecarboxamide derivative used against several RNA viruses. The drug has been approved in Japan for use against influenza. A multicentric nonrandomized trial was conducted during December 2014 to April 2015 to assess the safety and effectiveness of the drug to reduce mortality. The trial did not show any beneficial role of the drug in reducing mortality in ebola virus disease.

VACCINE

No vaccine against ebola virus has been approved so far. However, several vaccine candidates are presently under phase III clinical trial and appear to be promising. The recombinant vesicular stomatitis-ebola virus vaccine has completed its phase III trial.

Vescicular stomatitis virus (rVSV-ZEBOV): The candidate vaccine is a recombinant replication competent vesicular stomatitis virus genetically engineered to express surface glycoprotein of ZEBOV. In a phase III vaccine efficacy trial conducted in Guinea that included 7651 people of 90 clusters using ring vaccination strategy (after identification of a new confirmed ebola case, clusters of all his contacts and contacts of contacts). The vaccine showed 100% efficacy.

ChAd3-EBO-Z: This is a recombinant replication defective chimpanzee adenovirus type 3 encoding Zaire ebola virus glycoprotein (GP).

Vaccine has been prepared by Glaxo-SmithKline (GSK) and tested by National Institute of Allergy and Infectious Disease (NIAID).

Preclinical trial in cynomolgus monkeys showed 100% efficacy. Currently phase II and phase III are going on in several African countries.

MVA-BN Filo: This is a multivalent vaccine containing glycoproteins from Zaire EBOV, Sudan EBOV and Marburgvirus. Currently under phase III clinical trial to study safety, immunogenicity, tolerability with a heterologous prime boost regimen using three different batches of Ad26 and single batch of MVA-BN Filo in healthy adults.

Bibliography

1. Beeching NJ, Fenech M, Houlihan CF. Ebola virus disease. BMJ 2014;349:g7348.

2. Bociaga-Jasik M, Piatek A, Garlicki A. Ebola virus disease—pathogenesis, clinical presentation and management. Folia Med Cracov 2014;54(3): 49–55.

3. Broadhurst MJ, Brooks TJ, Pollock NR. Diagnosis of Ebola Virus Disease: Past, Present, and Future. Clin Microbiol Rev 2016;29(4):773–93.

4. Feldmann H, Geisbert TW. Ebola haemorrhagic fever. Lancet 2011;377:84962.

5. Geisbert TW. First Ebola virus vaccine to protect human beings? Lancet 2016;6736(16)32618–6.

6. Henao-Restrepo AM, Camacho A, Longini IM, et al. Efficacy and effectiveness of an rVSV-vectored vaccine in preventing Ebola virus disease: final results from the Guinea ring vaccination, open-label, cluster-randomised trial (Ebola Ça Suffit!). Lancet 2016;6736(16) 32621–6.

7. http://www.cdc.gov/mmwr/volumes/65/su/su6503a14.htm

8. Judson S, Prescott J, Munster V. Understanding ebola virus transmission. Viruses 2015;7(2): 511–21.

9. Marcinkiewicz J, Bryniarski K, Nazimek K. Ebola haemorrhagic fever virus: Pathogenesis, immune responses, potential prevention. Folia Med Cracov 2014;54(3):39–48.

10. Mishra B. Threat of Ebola: An update. Indian Journal of Medical Microbiology 2014; 32(4): 364–70.

11. Na W, Park N, Yeom M, Song D. Ebola outbreak in Western Africa 2014: What is going on with Ebola virus? Clin Exp Vaccine Res 2015;4(1): 17–22.

12. Olszanecki R, Gawlik G. Pharmacotherapy of Ebola hemorrhagic fever: A brief review of current status and future perspectives. Folia Med Cracov 2014;54(3):67–77.

13. Osterholm MT, Moore KA, Kelley NS, et al. Transmission of Ebola viruses: what we know and what we do not know. MBio 2015;6(2): e00137.

14. Smith DW, Rawlinson WD, Kok J, Dwyer DE, Catton M. Virological diagnosis of Ebolavirus infection. Pathology 2015;47(5):410–3.

15. Trad MA, Naughton W, Yeung A, et al. Ebola virus disease: An update on current prevention and management strategies. J Clin Virol 2017; 86:5–13.

16. Zawilińska B, Kosz-Vnenchak M. General introduction into the Ebola virus biology and disease. Folia Med Cracov 2014;54(3):57–65.

B. ZIKA VIRUS

Zika virus, a member of Flavivirus, was known as a causative agent of mild fever in a limited geographical area. Recently the virus has gained importance because of its increase in geographical range as well as expanding clinical spectrum including neurological involvement and severe congenital malformation.

VIRUS

Zika virus is a mosquito borne Flavivirus belongs to the spondwoni serocomplex of Flavivirus genus within the family Flaviviridae. It is an enveloped virus with icosahedral symmetry containing single stranded, positive sense RNA genome of 10,794 base pair in length. The genome has a single open reading frame (ORF) flanked by 5′and 3′ non-coding region (5′NCR and 3′NCR). The open reading frame codes for a single poly protein which gets cleaved into three structural proteins and seven non-structural proteins. Structural proteins are capsid, premembrane/membrane and envelop and non-structural proteins are NS1, NS2A, NS2B, NS3, NS4A, NS4B, and NS5.

Phylogenetic studies have shown that the Zika virus has two major lineages: African and Asian. African strains have two groups: Prototype MR766 Ugandan cluster and Nigerian cluster. Asian lineage consisted of Malaysian strains, 2007 Micronesian strains. Based on the phylogenetic results, it has been suggested that the virus had originated from East Africa, then moved to West Africa through different introductions. From West Africa, it came to Asia in around 1940 and then circulated in different region. The current American isolates are similar to Asian genotype and most closely related to Yap, Cambodia, Thailand and French Polynesia. The circulation of Zika virus of different genotypes might have impact on diagnostic assay and designing of vaccine.

DISCOVERY AND EMERGENCE OF ZIKA VIRUS

In 1947, Zika virus was detected for the first time from the blood samples of a caged rhesus monkey kept for surveillance of yellow fever in Zika forest, Uganda. Subsequently in 1948 the virus was detected in the pool of mosquito collected from Zika forest. The virus was first isolated in human in 1954 from a 10-year-old girl in Nigeria.

During 1950–1960, serological studies suggest the endemicity of Zika virus in several parts of Africa and Asia. However, the positive results for Zika virus antibody by hemagglutination could be due to cross-reactivity with other flaviviruses.

Emergence of Zika Virus

In 2007, an outbreak of dengue-like illness in Yap state, Micronesia was confirmed to be due to Zika virus. This was the first Zika virus outbreak outside Africa prior to which only 14 human cases were reported. Zika virus IgM antibody and RNA was detected from serum of the patients. More than 70% of population of Yap state was found to be infected with the virus and near 18% had clinical illness attributable to Zika virus.

Subsequent to this in 2013, Zika virus outbreak occurred in French Polynesia with a bigger magnitude affecting near 11% of its population with around 30,000 cases. Neurological manifestations were noted for the first time during this outbreak.

In 2014, Zika virus outbreaks were reported from several South Pacific islands.

In 2015, Zika virus outbreak occurred in several states of Brazil, with more than 10 lakhs cases were reported by late December, 2015. During the later part of 2015, Zika virus spread to several other countries and territories of America and Caribbean. During the Brazil outbreak, Zika virus was found to be associated with fetal microcephaly in babies born to infected mother.

During 2015, Zika virus outbreak was also reported from Africa where previously only sporadic cases used to occur.

In February 2016, considering the link of foetal microcephaly and congenital Zika syndrome, WHO declared Zika virus infection as a Public Health Emergency of International Concern (PHEIC).

Zika virus in India: The earliest evidence of Zika virus infection in India dates back to 1954. Blood samples from healthy adults from south, east and central parts of country were screened for Zika virus antibody by neutralization assay of which 33 were positive. Maximum number of positives were from Gujarat state.

No further report of Zika virus infection was reported possible because it was not looked for. However, after the outbreaks in French Polynesia and Brazil, report of imported cases from USA and France and subsequent declaration by WHO as a Public Health Emergency of International Concern (PHEIC) in 2016 and advised to all countries to set up surveillance for Zika virus and microcephaly or congenital Zika syndrome, India started testing for Zika virus infection.

The **surveillance** activity on Zika virus testing reported the first case in November 2016 in Gujarat in a postpartum female, followed by few more cases from the same region. **Sporadic cases** have been reported from Gujarat, Tamil Nadu. **Seropositivity in donors** were detected in Gujarat, Maharashtra, Madhya Pradesh, Hyderabad and Chennai. Two **Zika virus outbreaks** have also been reported, one from Rajasthan and one from Madhya Pradesh during September-October 2018. In both the outbreaks, more than 100 cases were reported from each site including about 50% pregnant women. The third outbreak is going on in Thiruvananthapuram, Kerala in July 2021 where near 50 cases have been detected.

Zika virus surveillance in India: As per the instruction of Ministry of Health and Family Welfare (MOHFW), Govt. of India, Zika virus surveillance system was established by Indian Council of Medical Research (ICMR). Surveillance network comprised of National Institute of Virology, Pune as the apex institute and 56 trained Virus Research Disease Laboratories (VRDL) from west, east, south and central states of the country.

Surveillance was conducted mainly by three major strategies:

- Human case surveillance
- Vector surveillance
- Fetal microcephaly

Human case surveillance: The strategy for human case surveillance was followed by testing minimum number of 50–100 dengue/chikungunya negative samples every month. All samples to be tested by singleplex prescribed RT-PCR protocol or CDC trioplex RT-PCR.

Vector surveillance: ICMR establishes a vector surveillance for flaviviruses in Southern and Eastern coastal states and Port Blair. Several species of Aedes mosquitoes were tested for the presence of Zika virus by RT-PCR. Positivity in vector was found from outbreak locality but vector surveillance system could not alert any outbreak pre-emptively.

Fetal microcephaly surveillance: MOHFW and ICMR have established a network between the clinical departments of various medical colleges with VRDLs for sample testing of suspected microcephaly cases.

TRANSMISSION

Vector-borne Transmission

Zika virus was maintained in Africa in sylvatic cycle between non-human primate and mosquito *Aedes africanus* with a cyclic epizootic in monkeys. In places where there is no non-human primates are present, the virus is maintained through man–mosquito–man cycle. In Asia, sylvatic cycle does not exist and virus maintenance occurs trough transmission between man–mosquito–man (Fig. 20.1).

The virus was first detected in *Aedes africanus* mosquito in Africa. Subsequently it has been isolated from various other species, such as *Aedes polynesiensis, Aedes luteocephalus, Aedes vittatus, Aedes apicoargenteus, Aedes furcifer, Aedes albopictus* and *Aedes aegypti*. Vector competency has been experimentally proved in *A. aegypti*. In Yap state, *A. henselii* was the primary vector.

Non-Vector-borne Transmission

Mother to Fetus Transmission

Transmission of Zika virus from infected mother to fetus can be transplacental, perinatal or through breast milk.

Perinatal transmission has been reported in other flaviviruses such as dengue virus and West Nile virus. In two cases, perinatal transmission of Zika virus has been reported where Zika virus RNA was detected in serum of mother and infants and mother's milk.

Transplacental transmission was first noted during the outbreak in Brazil when babies of infected pregnant women were born with severe malformation. This was confirmed by the detection of Zika virus RNA in amniotic fluid, blood and brain tissues of the neonates born with severe microcephaly and died soon after birth.

Zika virus particle has been demonstrated in the breast milk but transmission through breast milk has not yet been documented.

Sexual Transmission

Zika virus RNA has been detected in semen of patients during acute and convalescent phase of illness. Several cases of sexual transmission have been reported in various outbreaks.

Blood Transfusion

During the French Polynesia outbreak, 3% of the donor's blood was found to be positive for Zika virus RNA. Transmission through blood transfusion is not yet documented. However, considering the epidemiology of other flaviviruses and asymptomatic infection in a large number of individuals, the possibility of transmission through blood transfusion remains for Zika virus. Figure 20.3 depicts the various modes of transmission of Zika virus.

CLINICAL FEATURES

Febrile Illness

The incubation period in Zika virus infection is 6–10 days. Majority of the patients present with mild febrile illness with rash, arthralgia, myalgia, non-purulent conjunctivitis, headache and malaise. Disease can have a biphasic presentation.

Neurological Complications

Adults

Temporal association of Zika virus infection with Guillain-Barré syndrome (GBS) was

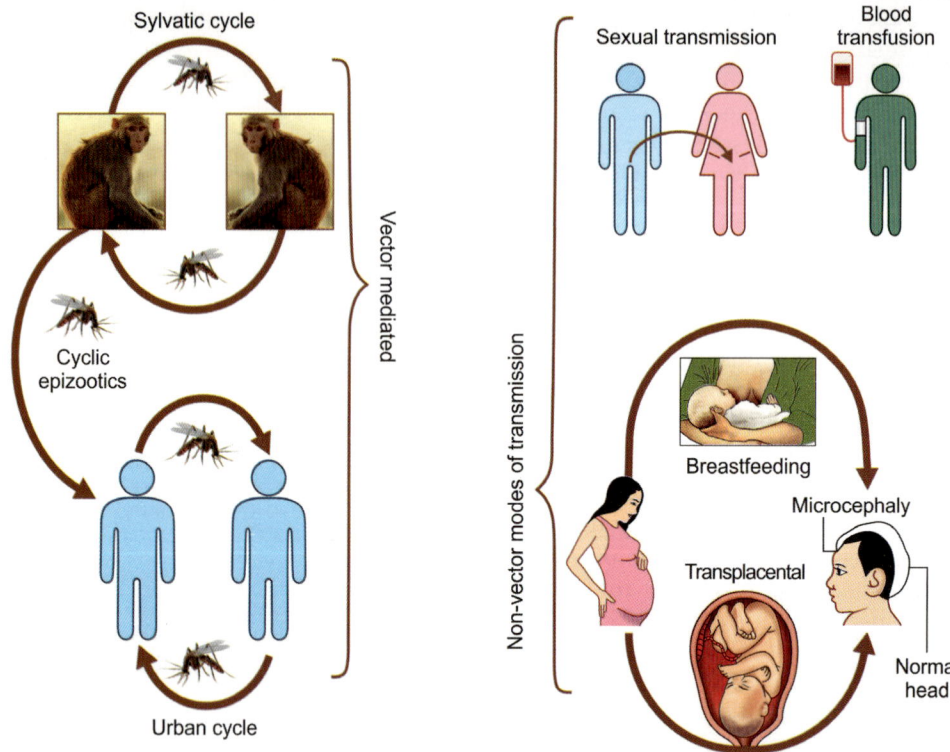

Fig. 20.3: Transmission of Zika virus

(This figure is published with permission from Journal of Postgraduate Medicine (Mishra B, Behera B. The mysterious Zika virus: Adding to the tropical flavivirus mayhem. J Postgrad Med. 2016 Oct–Dec; 62(4). 249–254)

observed during the Zika virus outbreak in the Pacifics and the Americas. The symptoms of GBS occurred around 7 days after the Zika-like illness and were more common in adult males.

Meningoencephalitis and myelitis were the other neurological symptoms associated with Zika virus infection.

Neonates

Severe congenital malformation with micro-cephaly has been reported in neonates born to pregnant women who had Zika virus infec-tion during pregnancy. Placental insufficiency, fetal growth restriction, intracranial calcifi-cation, abnormal amniotic fluid volume, cerebral and umbilical artery blood flow have also been reported. Miscarriage and neonatal death shortly after birth have also been reported.

In general, fetal brain anomalies have been more commonly seen in mother who had infection in their early part of pregnancy, whereas placental insufficiency, growth restriction and fetal death have been related to infection during the later part of pregnancy.

The association of Zika virus with the microcephaly was proved by the following:

- Presence of Zika virus RNA in the amniotic fluid and brain and other tissues of the affected neonates.
- Significantly high number of microcephaly in newborns of Zika virus infected mothers during pregnancy.

LAB DIAGNOSIS

Detection of Zika virus RNA or IgM antibody in serum samples are the mainstay of diag-nosis.

Blood, serum, urine, semen, amniotic fluid and tissue specimens are the main samples for Zika virus diagnosis.

Detection of Zika virus RNA: Zika virus RNA can be detected within the first week of illness by molecular tests such as reverse transcriptase PCR or real-time PCR. Virus RNA becomes negative towards the end of first week as the antibody starts appearing. Commercially available Zika virus PCR system is available.

Detection of IgM antibody: Diagnosis from single serum sample can be done by detection of specific IgM antibody during the second week of illness by using capture ELISA (MAC ELISA). Zika virus IgM MAC ELISA is commercially available.

Antigen detection: Virus antigen detection by immunohistochemistry using monoclonal antibody can be done in tissues.

Isolation of virus: Zika virus isolation can be done by mice inoculation or by cell culture using monkey kidney cell lines (Vero, LLCMK2) or mosquito cell lines (C6/36). However, the method is not often used for patient diagnosis.

Approach to diagnosis: Zika virus testing should be done in the following clinical settings:

- Occurrence of cluster of cases with dengue-like illness.
- Individuals returning from countries where Zika virus transmission is going on.

Zika virus RNA detection by reverse transcriptase polymerase chain reaction (RT-PCR) is preferred during the first week of illness and IgM detection by MAC ELISA during the second week of illness or when RT-PCR is negative. In case of IgM positivity, the sample should ideally be subjected for plaque reduction neutralization test (PRNT) in order to rule out the cross-reactivity with other flaviviruses.

TREATMENT

Patients are managed with symptomatic treatment with acitaminophen for fever and pain and antihistaminic for pruritic rash. Like other flavivirus infection, non-steroidal anti-inflammatory drugs are avoided.

PREVENTION

No vaccine has been developed for Zika virus infection. Vaccines using different vaccine platforms such as inactivated vaccine, DNA vaccine and mRNA vaccine are currently under trial. In addition to the mosquito bite avoidance and vector control strategy as applicable for other arboviruses, guidelines have been specifically addressed the prevention of sexual transmission for Zika virus infection. Individuals who reside or have returned from the Zika virus transmission area, have been advised to refrain from sexual activities or use of condom with their partners in order to reduce the risk of sexual transmission and thereby the risk of microcephaly or malformations in the neonate.

Bibliography

1. Bell BP, Boyle CA, Petersen LR. Preventing Zika Virus Infections in Pregnant Women: An Urgent Public Health Priority. Am J Public Health 2016;106(4):589–90.

2. Cao-Lormeau VM, Blake A, Mons S, et al. Guillain-Barré syndrome outbreak associated with Zika virus infection in French Polynesia: A case-control study. Lancet 2016;387(10027): 1531–39.

3. Mishra B, Behera B. The mysterious Zika virus: Adding to the tropical flavivirus mayhem. J Postgrad Med 2016; 62(4):249–54.

4. Musso D, Gubler DJ. Zika Virus. Clin Microbiol Rev 2016; 29(3):487–524.

5. Oster AM, Brooks JT, Stryker JE, Kachur RE, Mead P, Pesik NT, Petersen LR. Interim Guidelines for Prevention of Sexual Transmission of Zika Virus—United States, 2016. MMWR Morb Mortal Wkly Rep 2016 Feb 12;65(5):120–21.

6. Panchaud A, Stojanov M, Ammerdorffer A, Vouga M, Baud D. Emerging role of Zika virus in adverse fetal and neonatal outcomes. Clin Microbiol Rev 2016;29(3):659–94.

7. Petersen LR, Jamieson DJ, Powers AM, Honein MA. Zika Virus. N Engl J Med 2016; 374(16): 1552–63.

C. NIPAH VIRUS

Nipah virus (NiV) was first discovered during an outbreak of encephalitis in the pig farming villages in Malaysia and amongst the abattoir workers in Singapore during 1998–1999. Initially the outbreak was thought to be due to Japanese encephalitis virus (JEV) due to pig involvement. But some of the epidemiological features like adult cases instead of children, symptoms in pigs and clustering of cases were different from Japanese encephalitis which raised the possibility of a causative agent other than JEV. Finally, a new virus was detected from various samples and was named Nipah virus after the name of the village Sungei Nipah, from where the virus was first detected.

Outbreaks: The first NiV outbreak in Malaysia occurred amongst the pig farmers who were in close contact with the pigs. The outbreak affected a total of 265 cases of which 105 died with near 68% mortality. Encephalitis was the predominant symptoms in humans, whereas pigs had respiratory ailments and neurological manifestations. It was thought that the disease has been transmitted from the infected pigs to its close contacts. Pigs were probably acquired the infection from bats who were later found to be the reservoir of the NiV.

The virus then spread from Malaysia to Singapore through trade of infected pigs and caused outbreak in 1999 among the abattoir workers.

After 1999, no NiV outbreak has been reported from these countries.

NiV outbreaks in India: In 2001, a major outbreak of encephalitis occurred in Siliguri district of West Bengal. The outbreak was confirmed to be due to NiV. Pigs were not involved in this outbreak. Person-to-person mode of transmission was thought to be responsible for the outbreak.

In 2007, NiV outbreak was reported in Nadia, West Bengal infecting five people, all of them died.

During May 2018, Nipah virus outbreak occurred in Kerala. Twenty-three cases were detected as NiV positive with a case fatality ratio of 91%. The outbreak started in Kozhikode district. Index case was a 27-year-old and was admitted with fever, myalgia and altered sensorium later succumbed to death. Nosocomial spread was observed affecting 22 cases who contracted the infection from the index case. Of the 23 NiV positive cases, 21 patients died. All cases were confirmed by detection of NiV RNA by RT-PCR and IgM antibody. The lineage found in this outbreak was similar to Bangladesh lineage. The source of infection in index case was assumed to be close contact with bats.

NiV outbreaks in Bangladesh: In 2001, Bangladesh reported its first NiV outbreak in Meherpur, where the transmission was thought to be directly or indirectly from the reservoir bat. After that, Bangladesh has reported NiV outbreaks almost every year from different parts of the country. The number of infected cases reported in various outbreaks varies from 5 to 45 with a mortality rate of 43 to 100%. Most of these outbreaks have been associated with the drinking of raw date palm sap contaminated with bat saliva or excreta.

VIRUS

Nipah virus was found to be similar to another member of Paramyxovirus; hendra virus, that causes infection mainly in horses. Both Nipah virus and hendra virus are members of the genus Henipavirus and family Paramyxoviridae. The genome of Henipavirus is the largest among the paramyxoviruses and has near 18,250 nucleotides. Unlike other paramyxoviruses, NiV does not have hemagglutinin or neuraminidase surface proteins.

EPIDEMIOLOGY

Reservoir: Fruit bats or flying foxes of genus Pteropus are the natural reservoir of NiV. Virus has been isolated in bat saliva and urine. Serum of several fruit bat species has been found positive for antibody to NiV. Several species of fruit bats are present all over Southeast Asia. The migratory nature of these fruit bats possesses risk of emergence of NiV infection in new geographic area.

Amplifying host: Pig plays the role of amplifying host when infected. In Malaysian outbreak, pigs were possibly acquired the infection from bats through eating fruits contaminated with bat saliva or excreta. Unlike Japanese encephalitis, pigs also suffered from disease with respiratory and neurological symptoms. Man acquired the infection from infected pigs.

Serological evidences of NiV infection have been shown in several animals in the outbreak affected areas. NiV has been found from symptomatic dogs in natural condition. In experimentally infected cats, virus has been found to be excreted from several sites including the nasopharyngeal secretion and urine during the viremic phase. However, no animal other than pig has so far been associated with the transmission to human host.

NiV transmission cycle: NiV is transmitted mainly through bat excreta or bat saliva. Fruit or sap when contaminated with bat excreta gets infected with NiV. When these contaminated sap/fruits are taken by the human or pig they acquire the infection.

Once pigs or humans are infected, they develop neurological and respiratory symptoms. Possibly through respiratory secretion the virus gets transmitted from person-to-person or from pig to human (Fig. 20.4). Person-to-person transmission has occurred in intra-familial setting as well as in hospital setting amongst the health care workers.

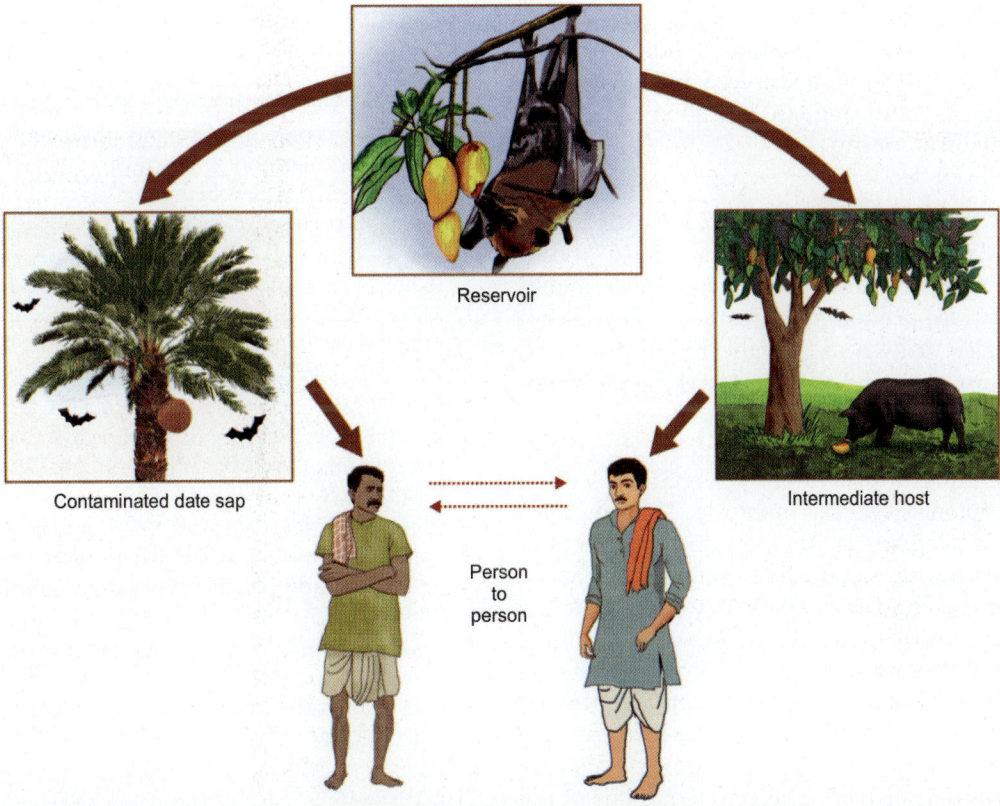

Reservoir

Contaminated date sap

Intermediate host

Person to person

Fig. 20.4: Nipah virus transmission

Molecular Epidemiology

Phylogenetic analysis of the NiV strains has shown that the virus entered to Southeast Asia in 1947. The strains have two lineages. The Indian and Bangladesh strains belong to one lineage: NiV-B and Malaysian lineage as NiV-M.

The Indian strains have been shown to be >99% nucleotide similarity with Bangladesh strains. Both Bangladesh and Indian strains are clustered and have been diverged from Malaysian strains. Indian and Bangladesh lineage has shown higher case fatality rate as compared to Malaysian lineage.

The strains of Malaysia and Cambodia belong to the same lineage.

Several studies have shown that multiple virus strains co-circulate in one geographic locality and co-evolve with the reservoir.

Modes of infection: In different NiV outbreaks, the virus has been found to be transmitted either through (i) respiratory route (from pig to human or from diseased person to its close contacts) or through (ii) ingestion of contaminated fruit or sap.

CLINICAL FEATURES

The incubation period of NiV is 4–15 days.

In pigs, respiratory and neurological symptoms are predominant features. The disease is also known as porcine respiratory and neurologic syndrome and barking pig syndrome.

In humans, fever, headache, dizziness, vomiting and myalgia are common initial symptoms in various outbreaks. The symptoms then progress to develop decrease level of consciousness and altered sensorium leading to coma and death. There may be acute or late onset of encephalitis. The duration of illness usually lasts for about 7–10 days. Indian outbreaks including the recent Kerala outbreak, respiratory symptoms including severe pneumonia and acute respiratory distress syndrome also have been noted as one of the predominant symptoms.

LAB DIAGNOSIS

Blood, CSF and urine are the main clinical samples required for diagnosis. Tracheal secretions can be collected in patients with respiratory symptoms. In postmortem cases, tissue samples can be taken.

NiV is a biosafety level-4 pathogen. As per the WHO guideline, the samples can be processed in biosafety level-2 after inactivation.

Serology, RT-PCR and virus isolation are the mainstay of diagnosis. Diagnosis of acute infection is done by detection of IgM antibody by capture ELISA or viral RNA by RT-PCR. IgG antibody detection in absence of IgM, however, indicates past infection.

Viral antigen can be demonstrated in the infected tissue by immunohistochemistry. Isolation of virus can be done in Vero cell line and the growth of NiV in the cell line is indicated by syncytium production.

Bibliography

1. Angeletti S, Lo Presti A, Cella E, Ciccozzi M. Molecular epidemiology and phylogeny of Nipah virus infection: A mini review. Asian Pac J Trop Med 2016;9(7):630–34.
2. Arankalle VA, Bandyopadhyay BT, Ramdasi AY, Jadi R, Patil DR, Rahman M, Majumdar M, Banerjee PS, Hati AK, Goswami RP, Neogi DK, Mishra AC. Genomic characterization of Nipah virus, West Bengal, India. Emerg Infect Dis 2011;17(5):907–9.
3. Chadha MS, Comer JA, Lowe L, Rota PA, Rollin PE, Bellini WJ, et al. Nipah virus-associated encephalitis outbreak, Siliguri, India. Emerg Infect Dis 2006;12:235–40.
4. Chua KB, Bellini WJ, Rota PA, Harcourt BH, Tamin A, Lam SK, et al. Nipah virus: A recently emergent deadly paramyxovirus. Science 2000; 288:1432–35.
5. Chua KB, Goh KJ, Wong KT, Kamarulzaman A, Tan PSK, Ksiazek TG, et al. Fatal encephalitis due to Nipah virus among pig farmers in Malaysia. Lancet 1999;354:1257–59.
6. Clayton BA, Middleton D, Arkinstall R, Frazer L, Wang LF, Marsh GA. The Nature of Exposure Drives Transmission of Nipah Viruses from

Malaysia and Bangladesh in Ferrets. PLoS Negl Trop Dis 2016 Jun 24;10(6):e0004775.

7. Gurley ES, Montgomery JM, Hossain MJ, Bell M, Azad AK, Islam MR, et al. Person-to-person transmission of Nipah virus in a Bangladeshi community. Emerg Infect Dis 2007;13:1031–37.

8. Harit AK, Ichhpujani RL, Gupta S, Gill KS, Lal S, Ganguly NK, Agarwal SP. Nipah/Hendra virus outbreak in Siliguri, West Bengal, India in 2001. Indian J Med Res 2006;123:553–60.

9. http://www.searo.who.int/entity/emerging_diseases/links/CDS_Nipah_Virus.pdf

10. Islam MS, Sazzad HM, Satter SM, Sultana S, Hossain MJ, Hasan M, Rahman M, Campbell S, Cannon DL, Ströher U, Daszak P, Luby SP, Gurley ES. Nipah Virus Transmission from bats to humans associated with drinking traditional liquor made from date palm sap, Bangladesh, 2011–2014. Emerg Infect Dis 2016;22(4):664–70.

11. Kulkarni DD, Tosh C, Venkatesh G, Senthil Kumar D. Nipah virus infection: Current scenario. Indian J Virol 2013;24(3):398–408.

12. Lo Presti A, Cella E, Giovanetti M, Lai A, Angeletti S, Zehender G, Ciccozzi M. Origin and evolution of Nipah virus. J Med Virol 2016;88(3):380–88.

13. Luby SP, Rahman M, Hossain MJ, Blum LS, Husain MM, Gurley E, et al. Foodborne transmission of Nipah virus, Bangladesh. Emerg Infect Dis 2006;12:1888–94.

14. Luby SP, Hossain MJ, Gurley ES, Ahmed B-N, Banu S, Khan SU, Homaira N, Rota PA, Rollin PE, Comer JA, Kenah E, Ksiazek TG, Rahman M. Recurrent zoonotic transmission of Nipah virus into humans Bangladesh 2001–2007. Emerg Infect Dis 2009;15:1229–35.

15. Nahar N, Paul RC, Sultana R, Gurley ES, Garcia F, Abedin J, Sumon SA, Banik KC, Asaduzzaman M, Rimi NA, Rahman M, Luby SP. Raw Sap Consumption Habits and Its Association with Knowledge of Nipah Virus in Two Endemic Districts in Bangladesh. PLoS One 2015 Nov 9;10(11):e0142292.

16. Nipah virus infection. http://www.searo.who.int/entity/emerging_diseases/links/CDS_Nipah_Virus.pdf

17. Soman Pillai V, Krishna G, Valiya Veettil M. Nipah Virus: Past Outbreaks and Future Containment. Viruses 2020 Apr 20;12(4):465.

D. KYASANUR FOREST DISEASE VIRUS

Kyasanur Forest Disease (KFD) is a **tick-borne** viral disease first identified in 1957 when an outbreak occurred in monkeys in Kyasanur forest of Shimoga district in Karnataka, India which was followed by outbreak of hemorrhagic fever in humans of that locality. The causative agent was found to be a virus that was named Kyasanur forest disease virus (KFDV). Virus was first isolated during 1957 outbreak. The disease was mostly restricted to five districts of Karnataka for a long time, and has started showing evidence of spread to other areas affecting more number of cases.

VIRUS

The agent KFD virus is a member of Flaviviridae family, genus Flavivirus and tick born virus group which was previously known as tick-borne encephalitis (TBE) serocomplex and Russian spring summer encephalitis complex. The members of TBE group are associated with neurological manifestations, whereas KFDV is mostly associated with hemorrhagic fever more than neurological symptoms.

TRANSMISSION CYCLE

Kyasanur forest disease virus is maintained in **tick-mammal and bird cycle**. The virus has been isolated from several species of ticks of which *Haemaphysalis spinigera* is considered to be the main vector. KFDV is maintained in ticks by **transstadial transmission** (virus passes through different developmental stages of the tick), and **transovarial transmission**.

In enzootic state, KFD virus is transmitted between small mammals and ticks. Rodents, shrews and ground birds act as the reservoir of the virus. These animals when infected develop low but sufficient level of viremia to infect the vector, without becoming ill due to infection. Rodents are ideal reservoir due to their short generation time.

Monkeys acquire the infection by bite of infected ticks. Infected monkeys develop high viremia level and thus facilitate the spread of infection. Black-faced langur and red-faced bonnet monkeys are the common species suffered from KFD. Infection in these monkeys can lead to serious illness and deaths commonly known as **monkey fever** or **monkey fall**, the two indicators of KFD activity.

The disease in monkeys occurs during dry months of **December to May** that coincides with increase in nymph population which are highly anthropophilic. Humans, who go to forest for collecting woods or grass or for any other reasons, are at high-risk of acquiring the infection by bite of nymphs. Humans are the dead end host (Fig. 20.5).

Geographic distribution: KFD is mostly restricted to five districts (Shimoga, Chikmagalore, Uttar Kannada, Dakshin Kannada and Udupi) of Karnataka state in India. Serological evidences have been found in Gujarat, West Bengal and Andaman Island. Since 2011, human cases have been reported from Chamrajnagara and

Tirthahalli districts in Karnataka, Nilgiri district in Tamil Nadu, Wayanad and Malappuram districts in Kerala and Palli in Goa. Variants of KFDV have been found from Saudi Arabia and China which were found to have the common ancestor with the Indian isolates.

The number of KFD cases each year varies from 40 to 1000 with an average of 400–500 cases. The mortality rate ranges from 4 to 15% in humans. The disease pattern has showed rising pattern in number of cases during 2003–2004 and again during 2012–2013 with a decline in between.

CLINICAL FEATURES

Incubation period generally ranges from 2 to 8 days. Symptoms start with sudden onset of fever, headache, myalgia and hypotension. The prodromal stage is followed by the stage of complication that is characterized by hemorrhagic manifestations such as petechial hemorrhage on the mucous membrane,

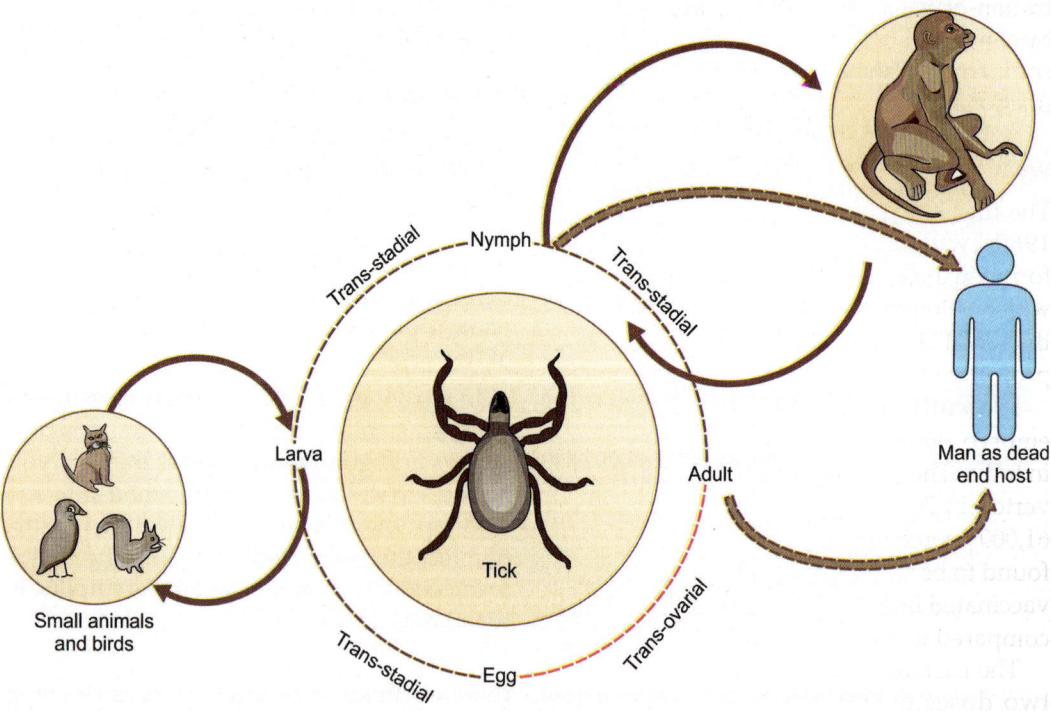

Fig. 20.5: Transmission cycle of Kyasanur forest disease

bleeding from nose and gum, conjunctival suffusion, hematemesis. In majority of the cases, the symptoms last for 7–14 days, however, in some of the cases a second phase of illness is observed with neurological manifestations like headache, tremor, photophobia and weakness.

LAB DIAGNOSIS

In an outbreak situation, detection of KFD virus can be done from human sample, dead monkey's tissue and from the tick pools. To confirm the disease in humans, blood samples are collected. Detection of viral RNA by **RT-PCR** and specific IgM antibody by **MAC ELISA** are commonly used for confirmation.

Isolation of KFD virus can be done by intracerebral inoculation in suckling mice or in Vero E6 cell line. IgM ELISA and RT-PCR have been developed by National Institute of Virology, Pune, India for diagnosis of KFD. In the past, the outbreaks were confirmed by virus isolation in mice system and demonstration of fourfold rise or fall antibody titer by conventional viral serological tests such as complement fixation test (CFT), hemagglutination inhibition (HAI) test or neutralization.

VACCINE

The first KFD vaccine was prepared in early 1960s which was a mice brain derived formalin inactivated vaccine using RSSE virus which belongs to the same antigenic group as that of KFD. However, it was found not to be protective against KFD, hence not in use.

Presently a formalin inactivated chick embryo fibroblast vaccine has been licensed in India. The vaccine was found to be seroconverted in 70% individuals. In a field trial in 61,000 people who received two doses, it was found to be safe and protective. Amongst the vaccinated individuals, 24 contracted KFD as compared to 325 in unvaccinated population.

The current vaccination strategy is to give two doses of vaccine, subcutaneously, at 4 weeks interval and a booster after 6–9 months of the primary vaccination. Considering the occurrence of cases during the months of December to March, it is recommended to give the first two doses and one booster dose by November month of the year.

Bibliography

1. Adhikari Prabha MR, Prabhu MG, Raghuveer CV, Bai M, Mala MA. Clinical study of 100 cases of Kyasanur Forest disease with clinicopathological correlation. Indian J Med Sci 1993 May;47(5):124–30.
2. Holbrook MR. Kyasanur forest disease. Antiviral Res 2012 Dec;96(3):353–62.
3. Kasabi GS, Murhekar MV, Sandhya VK, Raghunandan R, Kiran SK, Channabasappa GH, Mehendale SM. Coverage and effectiveness of Kyasanur forest disease (KFD) vaccine in Karnataka, South India, 2005–10. PLoS Negl Trop Dis 2013;7(1):e2025. doi:10.1371/journal.pntd.0002025.
4. Kyasanur forest disease. http://www.icmr.nic.in/pinstitute/niv/KYASANUR%20FOREST%20DISEASE.pdf
5. Mehla R, Kumar SR, Yadav P, et al. Recent ancestry of Kyasanur Forest disease virus. Emerg Infect Dis 2009;15(9):1431–37.
6. Mourya DT, Yadav PD, Mehla R, et al. Diagnosis of Kyasanur forest disease by nested RT-PCR, real-time RT-PCR and IgM capture ELISA. J Virol Methods 2012;186(1–2):49–54.
7. Murhekar MV, Kasabi GS, Mehendale SM, Mourya DT, Yadav PD, Tandale BV. On the transmission pattern of Kyasanur Forest disease (KFD) in India. Infect Dis Poverty 2015; 4:37.
8. Pattnaik P. Kyasanur forest disease: an epidemiological view in India. Rev Med Virol 2006 May-Jun;16(3):151–65. Review. Erratum in: Rev Med Virol 2008 May-Jun; 18(3):211.
9. Sadanandane C, Elango A, Marja N, Sasidharan PV, Raju KH, Jambulingam P. An outbreak of Kyasanur forest disease in the Wayanad and Malappuram districts of Kerala, India. Ticks Tick Borne Dis 2016 Sep 21. pii: S1877-959X(16)30148-0.
10. Tandale BV, Balakrishnan A, Yadav PD, Marja N, Mourya DT. New focus of Kyasanur Forest disease virus activity in a tribal area in Kerala, India, 2014. Infect Dis Poverty 2015; 4:12.

Chapter

21

Human Immunodeficiency Virus

A. VIRUS AND PATHOGENESIS*

The human immunodeficiency virus (HIV), the causative agent of acquired immune deficiency syndrome (AIDS), was isolated from a patient with lymphadenopathy in 1983 and was initially termed lymphadenopathy associated virus. They were subsequently named the human immunodeficiency virus (HIV) in 1986 by the International Committee on the Taxonomy of Viruses.

Taxonomy: The virus belongs to the genus Lentivirus of the family Retroviridae and has two types: HIV-1 and HIV-2.

Family: Retroviridae

Genus: Lentivirus

Types: HIV1 and HIV2

Based on the phylogenetic relatedness of their nucleotide sequences, HIVs are further classified into groups, subtypes, sub subtypes (clades), and recombinant forms which are discussed in detail in Chapter "Molecular epidemiology of HIV".

STRUCTURE AND GENOMIC ORGANIZATION

HIVs are enveloped single-stranded, positive sense RNA viruses, with icosahedral symmetry. Mature viral particles measure approximately 100 to 150 nm in diameter. The core contains two copies of single-stranded RNA, approximately 10 kb in length, which are surrounded by structural proteins that form the nucleocapsid and the matrix and by-products of the *pol* gene. The outer envelop contains several glycoprotein complexes. Each of this glycoprotein complex is a trimmer of two components; outer gp120 and a trnasmembrane component gp41. These surface glycoproteins mediate the binding of virus to CD4 and other co-receptors. Besides the viral glycoprotein, envelop may also contain certain host cell proteins, such as HLA class-I, HLA class-II and intercellular adhesion molecule-1 (ICAM-1). These host cell proteins get incorporated during the process of budding and may play a role in binding to other target cells. Below the envelop, the p17 matrix protein is present, which covers the capsid that consists of p24 proteins. The two copies of viral RNA genome are located at the innermost part of the virus particle which lies inside the capsid. Several enzymes that are required for viral replication are entangled to viral RNA in a complex form, of which reverse transcriptase and integrase are most important (Fig. 21.1).

The viral genome primarily comprises three genes (*gag, pol, env*) that encodes for the structural proteins and six other genes (*tat, rev, nef, vif, vpr,* and *vpu*) that code for various regulatory proteins which play a significant role in virus pathogenesis.

The *gag* **gene** (group antigen) encodes for nucleocapsid, capsid and matrix proteins including p24; *pol or* **polymerase gene** encodes for reverse transcriptase as well as protease,

*This chapter has been combinedly contributed by Dr Bijayini Behera and Dr Baijayantimala Mishra.

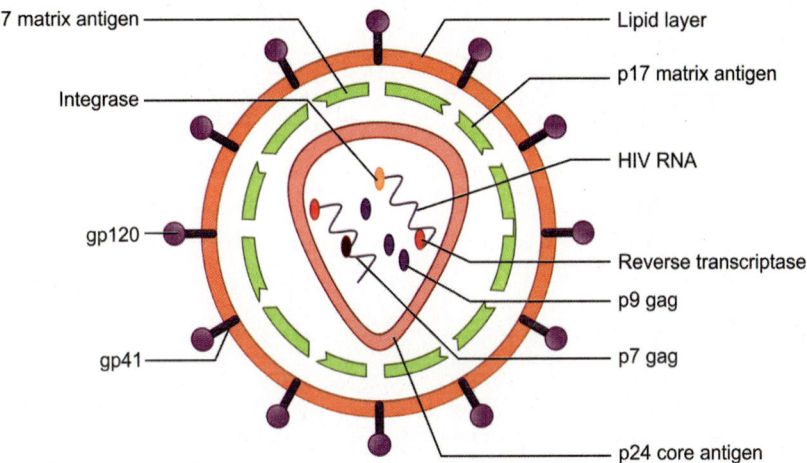

Fig. 21.1: Schematic diagram of HIV-1 structure

RNAse and integrase and *env* or **envelop gene** encodes for the envelop glycoproteins. These three genes in the sequence of *gag-pol-env* are flanked on both sides by **long terminal repeats (LTR)** which are present on both ends of the viral gnome.

HIV-2 differs structurally from HIV-1 in having a ***vpx*** gene in place of *vpu* gene and also the outer envelop and transmembrane glycoproteins are gp125 and gp36, and the core proteins are p16 and p26 instead of p17 and p24 in HIV-1.

REPLICATION

Receptors and Co-receptors

HIV replication begins with binding of envelop glycoprotein **gp120 to CD4** (expressed mainly on T lymphocytes, also on monocyte, macrophage, dendritic cells and Langerhans cells). Subsequent to this binding, there occurs a conformational change in gp120 which facilitates binding to other **co-receptors**, e.g. either **C-C chemokine receptor R5 (CCR5)** or co-receptor **X4 (CXCR4)**. CXCR4 binding occurs with **T cell tropic and syncytium** inducing viruses, whereas **macrophage tropic viruses bind through CCR5** and are **non-syncytium inducing**. Depending on the co-receptor used by the virus, it may be either a **R5—CCR5 tropic virus** or **X4—CXCR4 tropic virus**.

Subsequent to binding with suitable receptor and co-receptors, newly exposed transmembrane **gp41 promotes fusion** with the host cell membrane. After fusion with the host cell, the **pre-integration complex** (comprising of viral RNA, enzymes, accessory proteins, capsid and matrix proteins) traverses through the host cell cytoplasm to reach the nucleus. The viral reverse transcriptase enzyme catalyzes the reverse transcription of genomic RNA into DNA. The HIV hallmark **reverse transcriptase enzyme** possesses three distinct functions: **RNA-dependent DNA polymerase**, which facilitates synthesis of complementary DNA (cDNA); **ribonuclease H (RNase H)**, which degrades viral RNA from the cDNA–RNA complex; and a **DNA-dependent DNA polymerase**, which synthesizes a DNA strand complementary to the cDNA. During the pre-integration and reverse transcription steps, the pre-integration complex is vulnerable to several host proteins like TRIM5α (cytoplasmic tripartite motif containing protein 5α) and apolipoprotein B mRNA editing enzyme (catalytic polypeptide-like 3 [APOBEC3]). After the synthesis of genomic double-stranded DNA, it is integrated with the host cell chromosome via the viral integrase. The integrated HIV DNA genome is called the provirus, which can remain in a

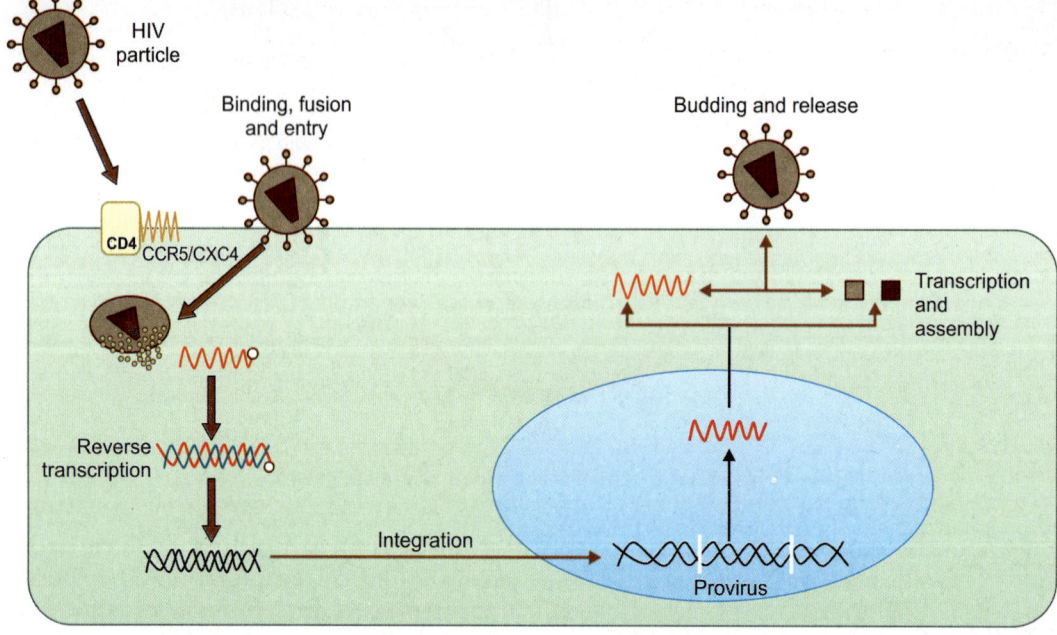

Fig. 21.2: H1V1 replication cycle

latent phase (transcriptionally inactive) for long periods. Subsequent transcription of provirus is dependent upon both cellular activation as well as viral regulatory proteins (*tat, rev, nef* and *vpu*). At the end of the replication cycle, assembly of HIV proteins, enzymes, and genomic RNA results in the formation of the virion which buds through the lipid bilayer of the host cell membrane and during budding acquires its external envelop. Host restriction factor tetherin inhibits this budding process. Figure 21.2 schematically depicts the various steps of HIV1 replication cycle. The functions of the various proteins during replication are summarized in Table 21.1.

PATHOGENESIS

Modes of Transmission

HIV is transmitted primarily via sexual transmission (vaginal, anal, or oral). Alternate modes include transmission via parenteral routes (unsafe blood and blood products transfusion, organ transplant, and intravenous drug use) and mother to child transmission is 20–45% (transplacental, childbirth and breastfeeding). It can also be transmitted by body fluids.

Role of Receptors and Co-receptors in HIV Pathogenesis

CD4, a 58 KDa monomeric glycoprotein, expressed on the surface of **T lymphocytes, monocytes, macrophages, dendritic cells, eosinophils**. It serves as the principal receptor for HIV. Binding of HIV gp 120 occurs within its V2 region. Apart from CD4, HIV utilizes either one of the **co-receptors CCR-5 or CXCR-4. CCR-5 and CXCR-4 (fusin)** belong to certain endogenous chemokine receptor families. **RANTES** (regulated on activation, normal T cell expressed and secreted) or chemokine (C-C motif) ligand 5 **(CCCL5)**, macrophage inflammatory proteins **(MIP) 1α and β** are the natural ligands for CCR-5. Stromal cell-derived factor 1 **(SDF1)**, also known as C-X-C motif chemokine 12 **(CXCL12)**, is the natural ligand for CXCR-4. HIV isolates are classified as either **T-tropic (X4 viruses)** if they utilize

Table 21.1: Summary of important functions of HIV viral proteins

Viral protein	Function
Tat: Transcriptional activator (p14)	• Increases transcription from the long terminal repeat region • Binds to transactivation response elements • RNA elongation
rev: Regulator of viral gene expression (p19)	• Regulatory protein • Binds to *rev* response elements (RRE) • Transportation of incompletely spliced viral RNAs from nucleus to cytoplasm
nef: Negative effector (p27)	• Downregulates CD4 and MHC class 1 expression from HIV infected cell surface • Essential for increased virus infectivity, high virus production and disease progression
vpu: Viral protein U	Facilitates the release of virus from host cell (budding)
vpr: Viral protein R (p15)	Facilitates viral replication in macrophages.
vif: Viral infectivity factor (p23)	Inactivates the host restriction factor APOBEC3 which degrades the viral DNA

CXCR 4 as the major co-receptor or **M-tropic (R5 viruses)** if they utilize CCR5 as the major co-receptor. The co-receptor tropism is partly explained by the affinity of the third variable region (v3 loop) of gp120 to either CCR-5 or CXCR -4. R5 viruses predominate in the early phase of the diseases, and there is transition to X4 viruses as the disease progresses.

The pathogenesis of HIV is a complex interplay of viral and host immunologic events and varies during different stages of the disease process. The typical course of an untreated HIV-infected individual can be broadly subdivided into primary infection, clinical latency and finally advanced disease. The pathogenic events during various stages of the disease are described below.

Immunopathogenesis of HIV Infection

Progressive deterioration of CD4+ T cells associated with a persistent state of immune activation is the hallmark of HIV infection.

The effects on CD4 cells, CD8 cells and B cells are described below.

CD4+ T cells: The progressive immunodeficiency during the course of HIV infection can be attributed to various degrees of immune dysfunction and immune activation. Multiple mechanisms of CD4+ T lymphocyte destruction (both structural as well as functional) have been observed in various stages of the disease.

During the early (primary) phase of infection, intense HIV replication occurs in gut associated lymphoid tissue (GALT) and the CD4+ Th17 cells which are the prime target of HIV and also abundant in the GALT. Depletion of CD4+ Th17 cells, which are important for defense against extracellular bacteria, leads to increased gut microbial translocation, and a state of persistent immune activation. Another subset of CD4+ T helper cells, called follicular helper CD4+ T cells, present in the germinal centers of lymphoid follicles are also believed to be early targets of HIV during initial infection.

As the disease progresses, there occurs a range of functional abnormalities of CD4+ T helper cells which include reduced impaired expression of IL7 and IL2 receptors, and decreased production of CD28 expressing cells. Cells without CD28 expression respond suboptimally to activation signals. Antibody-dependent cell-mediated cytotoxicity also contributes to overall decrease in number of

CD4+ T helper cells by causing destruction of uninfected CD4+ T helper cells also known as innocent bystander lysis.

The progressive decrease in the number of CD4+ T helper cells occurs due to a combination of plasma membrane disruption due to viral budding, Fas dependent and Fas independent apoptosis, and syncytia formation.

HIV gene products like _env, tat_ and _vpr_ enhance apoptosis. Recently, an inflammatory mode of cell death pyroptosis, involving caspase 1 and IL1β has been postulated in HIV mediated CD4 cell destruction.

CXCR-4 tropic HIV isolates induce fusion between adjacent T cell membranes to form a giant multinucleate cell called syncytium. Cell death of both HIV infected as well as uninfected T helper is accelerated during syncytia formation. The rate of destruction of CD4+ T helper cells exceeds the rate of reconstitution of CD4+ T helper cells pool. This is due to a number of factors like destruction of the fibroblastic reticular cell network, collagen deposition, and reduced utilization of the T cell survival factor interleukin 7 in the lymphoid tissue which ultimately leads to a state when CD4+ T helper cells count drops below critical level.

CD8+ T cells: HIV specific CD8+ CTL (cytotoxic T lymphocyte) response is able to contain the replication of virus only for a brief period. As the disease progresses, number of escape mutants keeps increasing which enables HIV to evade CTL response.

There are a lot of functional abnormalities of CD8+ T cells occur during the course of HIV infection like expression of activation markers HLA DR, upregulation of inhibitory receptors PD-1, reduced expression of IL-2 receptor (CD25), IL-7 receptor (CD127), and CD28. These functional abnormalities result in less secretion of CC chemokines such as RANTES (regulated on activation, normal T cell expressed and secreted) or chemokine (C-C motif) ligand 5 (CCCL5), macrophage inflammatory proteins (MIP) 1α and β. These CC chemokines inhibit viral replication. As

functional abnormalities of CD8+ T cells ensue, there is less secretion of chemokines and subsequently viral replication goes unchecked. Also the quantitative as well as qualitative decline in CD4+ T helper cells adversely affects CD8+ T cells.

B cells: In HIV infection, there is aberrant B cell activation which leads to hypergamma-globulinemia. Over the period of the disease, there is production of aberrant memory B cells with reduced expression of CD21, and upregulation of inhibitory receptors. All these eventually lead to functional exhaustion of B cells. Neutralizing antibodies directed against envelop glycoproteins gp120 and gp41 are able to neutralize HIV during early course of infection. Subsequently, due to lack of proof-reading capacity of the viral reverse transcriptase (RT) enzyme, there is generation of high degree of viral diversity, which accounts for escape mutants which escape the neutralizing antibodies. Although some individuals mount broadly neutralizing antibody response, capable of recognizing highly diverse stains, these antibodies develop very late in the course of the disease and their clinical significance is not very clear.

Protective antibodies raised against HIV are also implicated in disease progression. It has been shown that antibodies against gp41 facilitate infection of new cells through Fc receptors (anti-body enhancement). Anti- gp120 antibodies also participate in ADCC (antibody-dependent cell-mediated cytotoxicity) and destroy uninfected CD 4 cells by bystander killing.

The pathogenesis of HIV is a complex interplay of viral and host immunologic events and varies during different stages of the disease process. The typical course of an untreated HIV-infected individual can be broadly subdivided into primary infection, clinical latency and finally advanced disease. The pathogenic events during various stages of the disease are described below.

Primary Infection

HIV crosses mucosal barrier by binding to Langerhans cells, just beneath the surface or through microscopic abrasions in the mucosa. Resting CD4 cells as well as activated CD4 cells present in the lamina propria act as early amplifiers of HIV infection. A few dendritic cells (DCs) express a variety of C-type lectin receptors on their surface **(DC-SIGN)** that binds to the HIV gp120 envelop protein, allowing DCs to further spread the virus to CD4+ T cells. During this early time period which ranges from a few days to weeks, HIV cannot be detected in plasma by the currently available assays. This period is known as **"eclipse period"**. After establishment of early infection, viral dissemination occurs within days to weeks to the lymphoid organs. About 50% individuals experience **acute HIV syndrome** (acute mononucleosis-like symptoms) associated with high levels of viremia. The level of viremia during primary infection is not correlated with the subsequent rate of disease progression. However, the individual achieves a steady state of plasma viremia after almost a year, the level of which correlate with the disease progression, if remain untreated.

Establishment of Chronic and Persistent Infection

Despite the development of strong cellular as well as humoral immune response (as described above in immunopathogenesis subheading) following primary HIV infection, HIV escapes immune mediated clearance and is able to establish a chronic persistent type of infection which is the hallmark of HIV disease. HIV accomplishes this immune evasion primarily by generating viral diversity via mutation and recombination. The resulting quasi-species escapes recognition from cytotoxic T lymphocytes. Furthermore, viral proteins *nef*, *tat*, and *vpu* downregulate HLA class I molecules on the surface of HIV-infected cells, which in turn makes the CD8+ CTL unable to recognize and kill the infected target cell.

Several mechanisms are proposed for dysfunction of CD4+ T cells including direct infection and subsequent destruction by HIV,

as well as indirect effects, such as immune clearance of infected cells, and immune exhaustion due to aberrant cellular activation.

The following strategies are employed by HIV to evade neutralizing responses directed at the envelop glycoproteins: (i) Hypervariability in the primary sequence of the envelop, (ii) extensive glycosylation of the envelop, and (iii) conformational masking of neutralizing epitopes.

Thus, the evasion of cellular as well humoral arm of immune response constitutes a pool of latently infected cells (also described as post-integration latency) which serve as a reservoir of chronic persistent infection. The viral load gets stabilized when the immune system develops specific cytotoxic T cells against HIV. This point is reached after the acute phase of infection which takes approximately one year from the acquisition of HIV infection. The stabilized viral load achieved at this stage is called **"viral set point"** which is known to influence the subsequent progression to AIDS in untreated individuals.

In a small subgroup of HIV infected individuals (approx. 5–15%), even after 10 years of primary HIV infection, CD4 T cell counts remain in the normal range and they remain clinically stable without combination ART (cART or combined antiretroviral therapy). They were referred to as **"long-term nonprogressors."** However, on long-term follow-up, it has been observed that majority of the long-term non-progressors ultimately develops symptoms and require antiretroviral therapy.

A very small percentage of this group of long-term non-progressors, however, continue to maintain very low plasma viremia (<50 copies/mL plasma) with normal CD4 T cell counts. This group accounts for ≤1% of HIV infected individuals and are referred to as **'elite controllers'**. The host genetic factors, such as HLA type, heterozygosity for 32-bp deletion in chemokine receptor CCR5, mannose-binding lectin alleles, tumor necrosis factor c2 microsatellite alleles, Gc vitamin D-binding factor alleles have been proposed to be the primary factors responsible for elite controller status. Effective immunologic control of virus replication, and/or infection with an attenuated strain of HIV also has been considered as the attributing factors.

Advanced HIV Disease

Before the advent of combination antiretroviral therapy (cART), the median duration from primary infection to advanced HIV disease or AIDS was approximately 10 years. In untreated patients or in patients in whom virus replication is not adequately controlled by therapy, the CD4+ T cell count falls below a critical level (<200/µL) and the patient becomes highly susceptible to opportunistic infections and AIDS-associated neoplasms.

Bibliography

1. Colomer-Lluch M, Gollahon LS, Serra-Moreno R. Anti-HIV factors: Targeting each step of HIV's replication cycle. Curr HIV Res 2016; 14:175–82.
2. Levy JA. Pathogenesis of human immunodeficiency virus infection. Microbiol Rev 1993; 57(1):183–289.
3. Longo DL, Fauci AS. Infections due to Human immunodeficiency virus and other human retroviruses. In: Kasper DL, Fauci AS, Hauser SL, Longo DL, Jameson JL, Loscalzo J. Harrison's Principles of Internal Medicine. 19th ed. New York, NY: McGraw-Hill; 2015.
4. Maartens G, Celum C, Lewin SR. HIV infection: epidemiology, pathogenesis, treatment, and prevention. Lancet 2014; 384(9939):258–71.
5. Pantaleo G, Graziosi C, Fauci AS. New concepts in the immunopathogenesis of human immunodeficiency virus infection. N Engl J Med 1993; 328(5):327–35.
6. Sharp PM, Hahn BH. Origins of HIV and the AIDS pandemic. Cold Spring Harb Perspect Med 2011; 1: 006841.

B. LABORATORY DIAGNOSIS AND POST-EXPOSURE PROPHYLAXIS OF HIV

Early and accurate diagnosis of human immunodeficiency virus (HIV) infection is important both for patient management as well as to prevent the transmission of infection. The laboratory diagnostic techniques have evolved over the time with implementation of various testing strategies to achieve this objective.

Kinetics of HIV viral and serological markers: After the natural infection with HIV, various viral and host markers appear in the blood. Kinetics and time of appearance of these markers help in deciding the test as per the clinical situation (Fig. 21.3).

HIV RNA is the first marker to appear around 10–12 days post infection, followed by **p24 antigen** which appears mostly around 11–13 days. However, due to the less sensitivity of the p24 antigen detection test, in most of the cases, it is detected around 7 days later to RNA detection. The concentration of p24 antigen remains high for about a month and half during which period the marker is detectable. HIV p24 antigen is usually detectable till the appearance of antibodies after which it forms immune complex with the antibody and special technique needs to be employed to dissociate the complex. HIV specific IgM antibody appears after 3 weeks of infection and the titer reaches its peak after another 1–2 weeks. However, the IgM antibody response is not consistent and variable in different individuals. IgG antibody against HIV usually appears after 3–4 weeks after infection. In general, antibodies to HIV appear from 4th weeks onwards and the sensitivity of detection increases from near 60–65% after 4 weeks to 80% after 6 weeks, 90% after 8 weeks and 95% after 12 weeks.

LABORATORY DIAGNOSIS

HIV is mostly diagnosed either by direct method such as detection of virus or viral

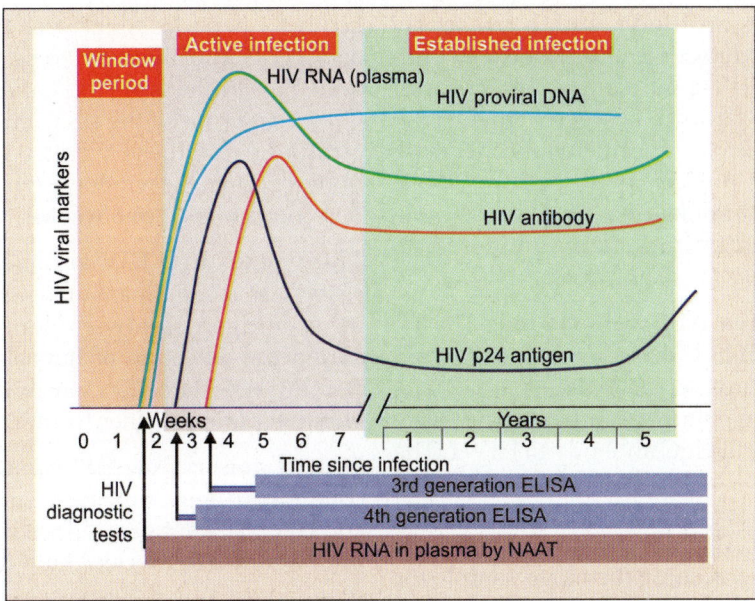

Fig. 21.3: Schematic representation of HIV viral markers

genome by molecular test or by detection of p24 antigen or by indirect host response of rising antibodies by various serological tests.

Serology: Serologic tests are the most commonly employed tests for diagnosis of HIV infection. The detection of antibodies is available in various formats. Rapid test format, enzyme linked immunosorbent assay (ELISA) in conventional micro-ELISA format and western blot are the commonly used test formats whereas it is also available in chemiluminescence platform, line immune assay and enzyme-linked fluorescence assay for detection of antibody as well as antigen.

Nucleic acid amplification test (NAAT): This is highly sensitive test and is used for detection of proviral DNA or viral RNA. PCR is the most commonly employed test. Various structural genes such as *gag, pol* or *env* are the common target genes used in PCR assay. These tests are particularly useful to diagnose during the window period (before appearance of antibody) or in newborn or infants of <18 months in whom the presence of antibody cannot be differentiated from maternal IgG.

Sample: All HIV testings are done in blood sample. **For serological assays,** either serum, plasma or whole blood is used. Saliva and urine also can be used for antibody detection. **For NAAT,** either whole blood in EDTA vial or dried blood spot (DBS) can be used. For viral load assay, plasma is separated from the whole blood collected in EDTA. Pretest counselling and written informed consent must be obtained in all cases.

Serological tests for HIV diagnosis: Serological tests for detection of HIV antibody is the main stay of HIV diagnosis.

Indication of HIV testing:
- Screening before blood transfusion/transplantation.
- Diagnosis in suspected asymptomatic/symptomatic individuals.
- Prevention of transmission from mother to child
- Post-exposure prophylaxis
- Epidemiological surveillance

Several generations of serological tests have been evolved over the time for detection of

HIV specific antibody which presently includes the detection of antibody to both HIV-1 and HIV-2 as well as detection of antigen. The purpose of development is to increase sensitivity, specificity and decrease the window period of detection. These different generations of tests are applicable mostly to ELISA and chemiluminescence assay platforms (Table 21.2).

First generation HIV antibody test: The 1st generation ELISA was developed in 1985 using the crude virus lysate from virus infected tissue culture as antigen. The assay used to detect the IgG antibody to HIV-1. The test was sensitive but had a large antibody negative window period of **up to 12 weeks** post infection. The high sensitivity of the test was useful in screening the blood sample for transfusion purpose but was associated with false positives which led to the development of second generation assay. False positives were reported in autoimmune diseases, non-HIV infections, pregnancy and several non-specified conditions.

Second generation HIV antibody test: These assays added the recombinant proteins or synthetic peptides to increase the sensitivity and specificity of the assay. Some manufacturers added the proteins of HIV-2 and HIV-1 group O protein to make the assay enable to detect antibodies to these viruses as well. This brought down the window period of antibody detection **from 12 to 6 weeks**.

Third generation HIV antibody test: These assays use recombinant or synthetic peptides in an antigen sandwich configuration. It added the detection of IgM along with the existing IgG detection which decreases the window period further **from 6 to 3 weeks**.

Fourth generation HIV test: This assay combinedly detects the HIV-1 and 2 antibody and HIV-1 p24 antigen decreasing the window period **from 3 to 2 weeks**. However, the assay result does not differentiate whether the positivity is due to presence of antibody to HIV-1 or 2 or antigen. The first 4th generation assay was developed in late 1990. First FDA approved 4th generation test came in 2010 and currently several FDA approved tests are available from different manufacturer. The sensitivity and specificity of all FDA approved tests are >99–100%.

Table 21.2: Summary of 1st to 5th generation HIV serological tests

	1st	2nd	3rd	4th	5th
Year of development	1985	1987	1991	1997	2015
Antigen type	HIV-1 crude whole virus lysate	HIV-1 crude lysate and recombinant antigens/peptide	HIV-1 and 2 Recombinant proteins/peptide	HIV-1 and 2 Recombinant proteins/peptides and monoclonal antibody to p24	HIV-1and 2 Recombinant proteins/peptides and monoclonal antibody to p24
Detects Ag/Ab	IgG Anti HIV-1	IgG anti HIV-1 and IgG anti-HIV-2 and Group O	IgG and IgM anti-HIV-1, anti-HIV-2 and Group O	IgG and IgM anti-HIV-1, anti-HIV-2 and Group O and HIV-1p24 antigen	IgG and IgM Anti-HIV-1, anti-HIV-2 and Group O and HIV-1p24 antigen
Sensitivity	>95%	>99%	>99.5%	>99.5%	>99.5%
Specificity	>95%	>99%	>99%	>99.5%	>99.5%
Result pattern	Single	Single	Single	Single: No separate result for HIV-1/2 antibody or p24 antigen	Separate result for anti-HIV-1 and 2 antibody and HIV-1 antigen

Fifth generation HIV test: This test like the 4th generation, detects the HIV-1 and 2 antibody and HIV-1 p24 antigen. The improvement here is it gives separate result for each analyte. Hence, the exact status of sample positivity regarding HIV-1 or 2 antibody or HIV p24 antigen is clear.

Rapid tests: Several rapid tests are available for HIV diagnosis. These tests can be done from serum, plasma as well as from the finger prick sample.

Assay types of rapid tests are (Figs 21.4 to 21.6):
- Immunochromatographic test (lateral flow assay)
- Immunoconcentration/dot blot (vertical flow assay)
- Immunocomb test
- Particle agglutination

First three types are more commonly used. Advantages of these tests are:
- No equipment is required. Test results are visibly recorded.
- Can be performed with very minimum training. So, can be done in remote field area by field worker and even self-testing can be done.
- Result in these tests is readable in 20–30 minutes. These criteria make these tests suitable for point of care test.

Presently both third and fourth generation tests are available in rapid test format for detection of HIV-1 and 2 antibody and detection of free HIV p24 antigen along with HIV-1 and 2 antibody, respectively. For detection of HIV-1 specific antibody, the device is coated with HIV-1gp120 and gp41 synthetic peptides and

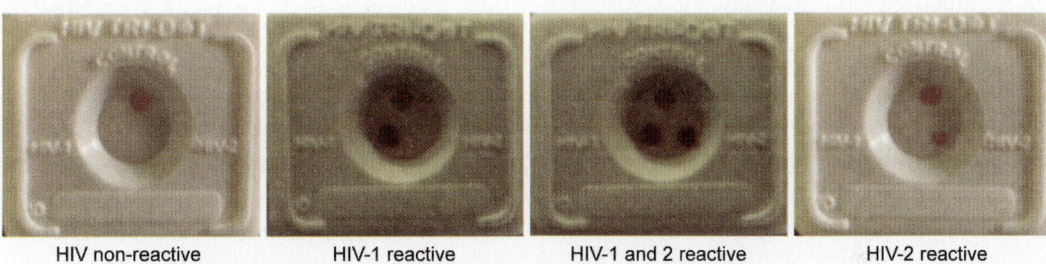

| HIV non-reactive | HIV-1 reactive | HIV-1 and 2 reactive | HIV-2 reactive |

Fig. 21.4: Interpretation of dot blot test for HIV diagnosis

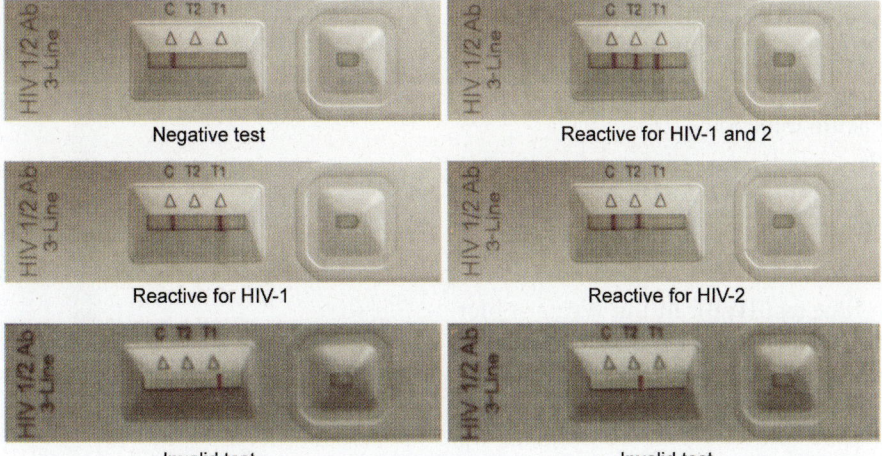

Fig. 21.5: Interpretation of immunochromatography test for HIV diagnosis

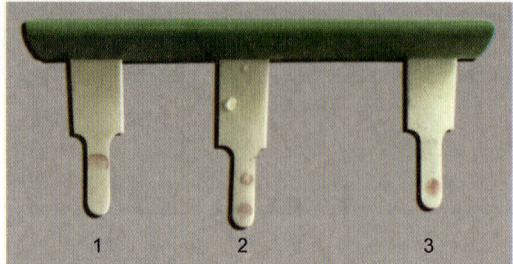

1: Negative test with control spot only
2: Positive test with both control spot and spot for HIV
3: Invalid test without control spot

Fig. 21.6: Interpretation of immunocomb test for HIV diagnosis

for detection of HIV-2 specific antibody it is coated with gp136 synthetic peptide.

In general, the sensitivity and specificity of rapid tests are comparable with that of ELISA. HIV tests in rapid test formats are available up to fourth generation assay. However, in a comparative assay between various 4th generation rapid tests, product of only one manufacturer was found to give additional 28% positivity than 3rd generation test result but products of all other manufacturer were not found to be suitable.

Reasons of false positive HIV antibody test:
- Pre-analytical error
- Hypergammaglobulinemia
- Influenza vaccination
- Recipient of HIV vaccine as a trial participant

Reasons of false negative HIV antibody test:
- Prior to seroconversion
- Early acute infection
- Infants
- Immunosuppressive conditions
- Hypogammaglobulinemia
- Advance AIDS

Western blot test (WB): In the Western blot assay, the various HIV proteins are blotted to nitrocellulose paper by electrophoresis. Separate Western blot strips for HIV-1 and 2 are available. The HIV-1 viral antigens are separated according to their molecular weight (from above downwards): gp160, gp120, p66, p55, p51, gp41, p31, p24, p17, and p15. On incubation with patient's serum, the HIV specific antibody present in the serum will bind to the corresponding proteins. On further addition of enzyme labelled secondary antibody and substrate, a colorimetric band will be produced at the site of antigen antibody reaction (Fig. 21.7).

Interpretation: Reactivity of HIV antibodies present in the patient sample with the different antigenic components on the WB strip should be interpreted as per manufacturer instructions.

According to the CDC guidelines: Positive when reactive to at least 2 of the following antigens: p24, gp41, gp120/160.

WHO recommendations: Positive when reactive to at least two envelop proteins (gp41, gp120/160). Presence of very weak p17 band is regarded as negative.

Reactivity to antigens not fulfilling WHO/CDC criteria is classified as indeterminate result and requires repetition of the test.

The HIV-1 Western blot was previously recommended by CDC and NACO (National AIDS Control Organization, Govt of India) to make a confirmatory laboratory diagnosis of HIV-1 infection. However, due to several reports on the false detection of HIV-2 as HIV-1 by

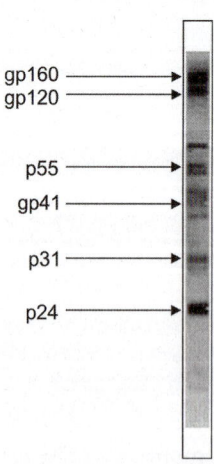

gp160
gp120

p55

gp41

p31

p24

Fig. 21.7: Western blot assay for HIV diagnosis

western blot test, the current CDC guidelines have removed HIV-1 Western blot as a confirmatory test from the recommended HIV testing algorithm. Considering the rapid, simple to perform reliable RDTs and enzyme immune assays WHO in its "2019 consolidate guideline on HIV testing services for changing epidemic" recommends countries *to move away from Western blotting*.

In general, antibody detection tests are not useful for diagnosis during the early part of infection because of window period of 3–4 weeks. The test is also not helpful to diagnose the infection in infants born to HIV positive mothers (till <18 months) as they may be positive due to transplacental transfer of maternal antibody.

Ultrasensitive p24 antigen detection test: p24 antigen can be detected in the whole blood, plasma or serum either in the free form or in the bound form with the bound antibody as antigen–antibody complex. After the appearance of antibody, p24 antigen is mostly present in the form of antigen–antibody complex and hence may not be detectable. Ultrasensitive antigen detection tests have been developed which includes dissociation of antigen from the antigen–antibody complex along with signal amplification to detect smaller amount of antigen. The performance of these tests is almost comparable with that of HIV PCR both in terms of sensitivity and time of detection. The advantages of antigen detection tests are, these tests do not require sophisticated lab facilities like molecular technique and can be done in a set up where enzyme immune assay can be performed.

Virus isolation: HIV is cultivated by co-cultivating peripheral blood mononuclear cells (PBMC) from an infected individual and a mitogen stimulated PBMC from a non-infected individual in presence of interleukin-2 (IL-2) at 37p C with 5% CO_2. Replication of HIV is detected by PCR or p24 antigen detection. This method is not used for diagnosis.

HIV TESTING STRATEGIES

The primary objectives of HIV testing are to make transfusion/transplant safety, to diagnose both symptomatic and asymptomatic individuals, to prevent parent to child transmission, post-exposure prophylaxis and epidemiological surveillance.

The strategies for HIV testing should be followed as per the country's guideline (e.g. National AIDS Control Organization in India). Testing guideline is also available by Center for Disease Control and Prevention (CDC, USA).

National AIDS Control Organization, Govt of India (NACO) has formulated different testing strategies according to the purpose of testing such as donor screening, surveillance and diagnosis in high-risk individuals or clinically suspected AIDS cases. The use of ELISA or rapid kits with a sensitivity of 99.5% and the specificity of 98% is recommended by NACO.

Common principles followed in testing strategies: In principle, these strategies are based on multiple sequential assays (ELISA/rapid tests) using different antigenic composition or different testing principle (e.g. indirect, sandwich) in subsequent assays. The testing algorithm usually constitutes of 2–3 serial assays which begin with the screening test and the subsequent tests are based on the result of the first assay.

The assay with highest sensitivity is used as the screening test (first test) and the subsequent tests which are used for confirmation are the tests with highest specificity.

The second and third assays should be able to differentiate between HIV-1 and HIV-2.

All tests should be conducted with counselling, confidentiality and informed consent.

In case final result is "Positive" (as per the strategy), post-test counselling should be done and patient should be referred to ART center. In case the final result as per the strategy is "Indeterminate", then follow-up testing is to be done after 2–4 weeks. If the result of follow-up test continue to be "Indeterminate", then the sample should be subjected to Western blot or PCR in the designated lab or it should be referred to national referral laboratory (NRL).

Any sample positive for HIV-2 only or both HIV-1 and HIV-2 is referred to state referral laboratory (SRL) for confirmation.

NACO testing strategy-1: This is employed in samples collected from donor with the aim of ensuring safety donation. One test with high sensitivity is used. Any sample tested to be reactive is discarded. However, if the donor has given consent to notify his result, then strategy 2 or 3 is used as applicable (Fig. 21.8a).

NACO testing strategy-2A: This strategy is used for sentinel surveillance where the testing is anonymous and unlinked. It uses a second test in case the first test comes reactive. The second test can be either ELISA or rapid (E/R) and should be based on a different test principle and/or different antigen from the first test (Fig. 21.8b).

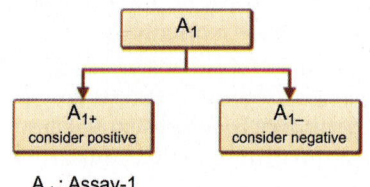

A_1 : Assay-1

Fig. 21.8a: NACO strategy-1 for HIV testing

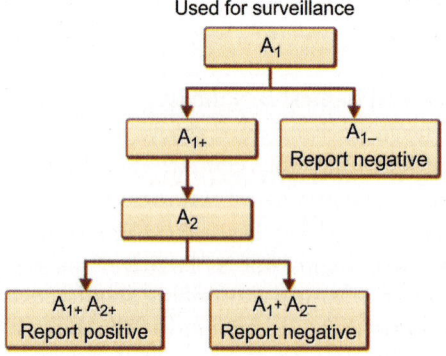

A_1: Assay-1; A_2: Assay-2

Fig. 21.8b: NACO strategy-2A for HIV testing

NACO testing strategy-2B: This strategy is employed in patients with symptoms of AIDS defining illness. Three tests are used. As per this strategy, two consecutive positive test is taken as "Positive" (Fig. 21.8c).

NACO testing strategy-3: This strategy is employed for diagnosis of HIV in asymptomatic individuals at ICTC (integrated counselling and testing center) and PPTCT (prevention of parent to child transmission) (Fig. 21.8d).

Testing strategy recommended by CDC: Center for Disease Control and Prevention (CDC) USA recommends the initial screening by the fourth generation antigen-antibody combination test. This has been recommended considering the fact that 4th generation assay has the potential to detect more number of acute HIV-1 infections than a 3rd generation antibody assay along with comparable number of established HIV-1 and 2 infections. The second assay is recommended only if the initial test is positive and aims to differentiate HIV-1 and HIV-2 antibodies. Third test by nucleic acid test (NAT) is recommended in samples positive for initial screening test (4th generation combo test) which are negative for HIV-1 and 2 antibodies detection in the second assay (Fig. 21.9).

Diagnosis in children: Early diagnosis of HIV in children is important because, treatment can be started early, and decision on breast-feeding can be taken.

A negative antibody detection test in an infant indicate absence of infection. Whereas, a positive antibody test can occur due to the presence of maternal antibody which may exist till 18 months of age. Hence, a positive antibody test does not confirm the infection.

HIV in infant or children should always be confirmed with a virological test. Detection of HIV DNA PCR is considered as the first choice of diagnosis. Detection of HIV RNA by real-time PCR (RTPCR) is equally reliable and more widely available. Ultrasensitive p24 antigen detection method can also be used for

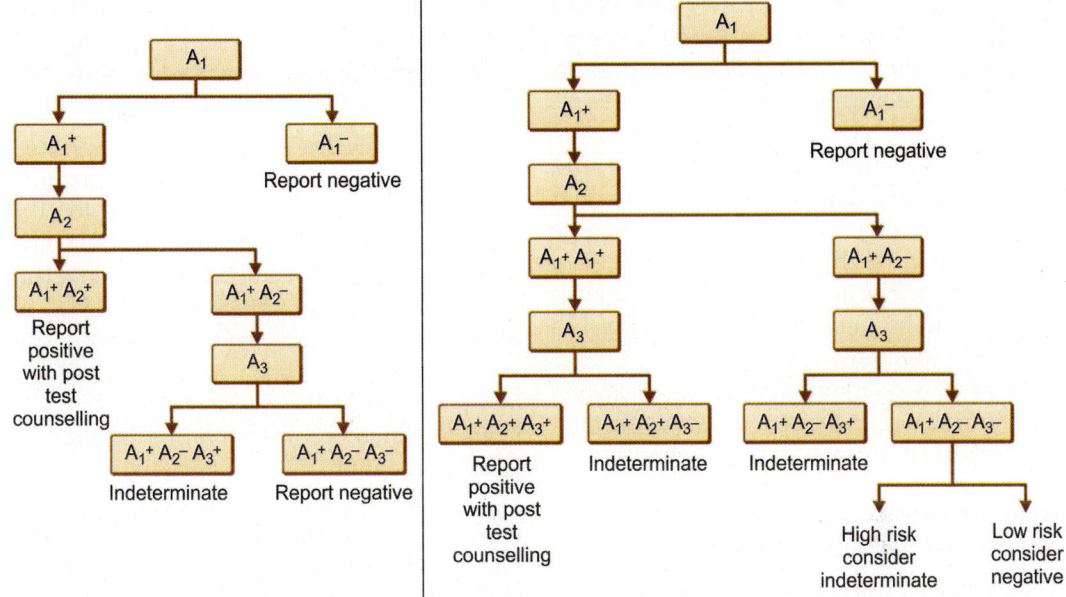

A₁: Assay-1; A₂: Assay-2; A₃: Assay-3

Fig. 21.8c: NACO strategy-2B: Diagnosis in individuals with symptoms of AIDS indicator disease

Fig. 21.8d: NACO strategy-3: Diagnosis in asymptomatic individual

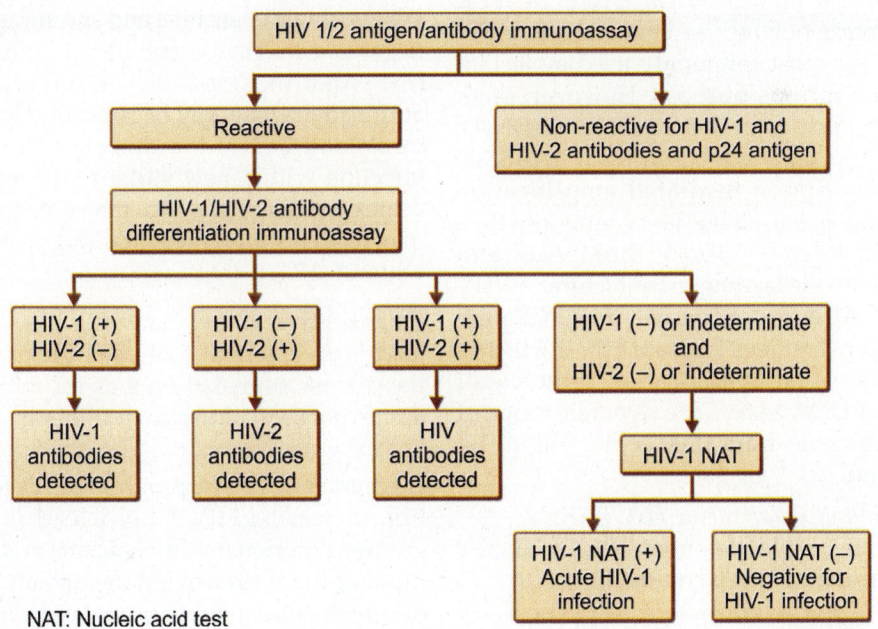

NAT: Nucleic acid test

Fig. 21.9: CDC recommended HIV testing algorithm (*Source*: Centers for Disease Control and Prevention. Laboratory testing for the diagnosis of HIV infection: updated recommendations. https://stacks.cdc.gov/view/cdc/23447)

confirmation of infection. However, considering the little advantage of cost of antigen over the more sensitive test PCR, its technical complexity, PCR test is the most preferred test.

In infants of less than 9 months age, virological tests are recommended at or after 6 weeks of age in a program setting though it can be done before 6 weeks as testing at 6 weeks yields high sensitivity of 98%. As a standard practice, any single positive test in a child should be confirmed by a repeat test in a second specimen.

In infant of >9 months age, serological test is done first. If comes reactive, then confirmed by virological testing (PCR or p24 antigen).

Molecular Assay (Qualitative and Quantitative)

The molecular assays used for the diagnosis of HIV can be broadly divided into two types—qualitative assay and quantitative assay.

Qualitative assays: These assays are used either for detection of HIV RNA or for detection of HIV proviral DNA. Reverse transcriptase polymerase chain reaction (RT-PCR) is the most commonly used molecular technique for this purpose. However, other methods such as branched DNA (b-DNA), nucleic acid sequence based assay (NASBA) and transcription mediated amplification (TMA) are also used but less commonly than RT-PCR. Roche, Abbott, BioMeriux and Siemens are the leading manufacturer of HIV RT-PCR. All are based on real-time PCR platform except Siemens' Versant HIV-1 which is based on signal amplification principle of branched DNA assay. The dynamic range of detection varies from <100 copies/mL to 10^8 copies/mL.

Indication of Qualitative HIV PCR:
- Screening of blood donors (NAAT: Nucleic acid amplification technique),
- During the window period particularly when p24 antigen is negative,
- Diagnosis in newborns born to HIV positive mothers, in children up to the age of 18 months

(due to the possibility of maternal HIV antibody).
- To diagnose suspected acute HIV infection.
- In case of indeterminate serological test

HIV proviral DNA: This is an integrated form of HIV DNA with the host cell chromosome and is a sensitive marker to establish the infection. This cell associated viral DNA is detected from peripheral blood mononuclear cells (PBMC). The detection is helpful in diagnosis in infants as well as after antiretroviral treatment (ART).

Point of care test (POCT) for HIV PCR: WHO in its updated recommendation in 2021 on "HIV prevention, infant diagnosis, antiretroviral initiation and monitoring", strongly recommends the use of point of care molecular tests (POCT) for diagnosis in infants and children <18 months in order to enable early initiation of ART and to reduce the mortality. Two POCT platforms have so far been approved for HIV diagnosis; Abbott m-PIMA™ and Cepheid GeneXPert®.

Causes of false negative HIV PCR: Though PCR is a highly sensitive and specific test, false negative result can occur: (i) In case of HIV-2, if it is not included in the target, (ii) after antiretroviral therapy because of decrease in viral load below the detection limit, (iii) infection with a new variant HIV, or (iv) in "elite controller" in whom the viral replication may remain suppressed for a long period even without ART.

Quantitative assay: Determination of viral load is an essential tool used for predicting the disease course as well as for monitoring the response to antiretroviral therapy.

HIV RNA load: The plasma HIV RNA load is a marker of disease progression. Determination of plasma HIV-1 RNA load is used a guiding criterion for the clinicians to start, and monitor the antiretroviral therapy and also to decide the therapeutic strategy. The methods employed are either target amplification technique such as; real-time PCR (qPCR), nucleic acid sequence-based amplification

(NASBA) or signal amplification technique like branched-DNA technique (bDNA). These tests can detect up to 20–40 copies of virus/mL plasma.

Cell associated HIV DNA: Recently the total cell associated HIV DNA that includes both integrated and non-integrated viral genomes is considered as an important marker to measure the cell associated HIV reservoir. It can be measured in whole blood, cell or tissues. The total HIV DNA load can act as a predictor of disease kinetics, helps in therapeutic monitoring in terms of predicting the treatment outcome and also has been found to be useful in guiding the therapeutic strategy. It also helps to study the HIV infection status in lymph nodes and tissues and thus can help to study the effect of antivirals on those sites.

Viral load monitoring: This is an inherent component of HIV patient management. It is performed to: (i) Detect the treatment failure at the earliest, (ii) helps in assessing the response to treatment in terms of predicting the risk of future development of treatment failure, and (iii) risk of HIV transmission.

Timing of viral load testing:

- The first viral load determination should be done before initiation of ART to find the baseline. The subsequent viral load monitoring is done at various time interval. The viral load testing algorithm as prescribed by WHO is given in Fig. 21.10.
- In patients with first-line ART, viral load testing needs to be done after 6 months, after 12 months in first year and every year annually.
- In patients with second/third line ART, viral load testing every six months.

WHO's criteria for different point of treatment response:

- *Treatment failure threshold*: ≥1000 copies/mL, with >95% of treatment adherence for each of the last 3 months.
- *Viral suppression and undetectability*: ≤50 copies/mL.
- *Low level viremia:* One or more viral load in one person between 50 and 1000 copies/mL. HIV viral load <1000 copies/mL also has been shown not to be associated with transmission.

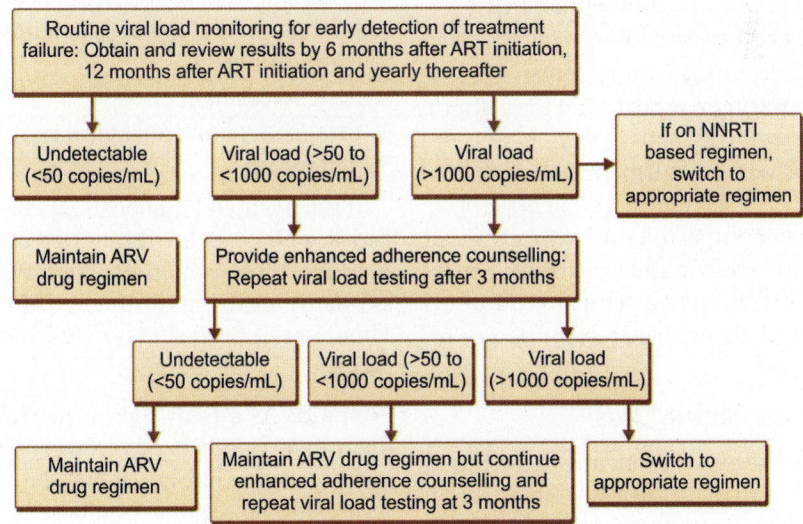

ART: Antiretroviral therapy; NNRTI: Non-nucleoside reverse transcriptase inhibitors

Fig. 21.10: WHO recommended post-ART viral load testing algorithm (*Source*: Updated recommendations on HIV prevention, infant diagnosis, antiretroviral initiation and monitoring (who.int) WHO. March 2021)

The relationship between different ranges of viral load among low level viremia is:

- 50–200 copies/mL shows a trend of predicting towards a future development of virological failure.
- 200–500 copies/mL strongly predicts towards a future development of virological failure.
- 500–1000 copies/mL typically predict a future development of virological failure.

Role of POCT in viral load monitoring: Several POCTs have also been approved by WHO for viral load determination; Cepheid Gene-XPert® and Abbott m-PIMA™. The pooled sensitivity and pooled specificity of Cepheid GeneXPert® has been found to be 96.5% and 96.6% respectively (>96% both) for a treatment failure threshold of 1000 copies/mL. The sensitivity of Abbott m-PIMA™ in two studies carried out in Kenya and Brazil has been reported to be 95.4% and 97.1% (>95% sensitivity), whereas the specificity in these studies was 96% and 76% for a treatment failure threshold of 1000 copies/mL.

In a program-based scenario, viral load testing by POCT is given priority in following scenario (as per WHO's 2021 recommendation):

- Pregnant and breastfeeding women
- Infants, children and adolescents
- People requiring a repeat viral load after a first elevated viral load
- People for whom treatment failure is suspected
- People presenting sick, living with advanced HIV disease or having a known opportunistic infection (TB, cryptococcal infection, etc.)
- First scheduled viral load test for people re-entering care.

POST-EXPOSURE PROPHYLAXIS

This refers to the exposure to infectious blood or body fluids. This can occur either as an occupational exposure to the infected sample or due to sexual exposure with an infected individual which is referred as non-occupational exposure (Table 21.3).

Non-occupational exposure: Exposure to blood-borne pathogens which occurs out side the work place and mostly due to sexual assault or unsafe sex or intravenous drug user. CDC recommends a **28-day course** of non-occupational post-exposure prophylaxis (nPEP) for HIV-uninfected persons who seek care ≤72 hours after a non-occupational exposure to blood, genital secretions, or other potentially infected body fluids of persons known to be HIV infected or of unknown HIV status when that exposure represents a substantial risk for HIV acquisition (men who has sex with men, intravenous drug user, or any other high-risk behavior).

Occupational exposure: Exposure to blood-borne pathogens which occurs during the working period. This pertains to exposure to percutaneous injury, mucous membrane, non-intact skin or intact skin for a prolong period. This occurs mostly in a hospital set up with health care workers. NACO gives a guideline for a comprehensive management in order to minimize the risk of infection after exposure to blood-borne pathogens (Table 21.4). This encompasses:

1. First aid of the site of exposure
2. Counselling
3. Risk assessment
4. Provision of antiretroviral prophylaxis
5. Follow-up

It is to be noted that post-exposure prophylaxis for HIV needs to be started at the earliest possible time, preferably within 2 hours of exposure and maximum within 72 hours. Therefore, it necessitates an early risk assessment.

First aid: As a first step in the management, the area exposed needs to washed or rinsed with water or normal saline. No soap, antiseptic or disinfectant should be used.

Assessment of exposed individual's HIV status: If the exposed individual is not a known HIV positive, his/her sample also

Table 21.3: Risk status according to type of body fluids

Body fluids that are **"at risk"**	Body fluids that are **'not at risk,'**
Blood, semen, vaginal secretions, cerebrospinal fluid, synovial, pleural, peritoneal, pericardial fluid, amniotic fluid, and other body fluids contaminated with visible blood	Tears, sweat, urine and feces, saliva, sputum, vomitus (*unless contain visible blood*)

Table 21.4: NACO guideline for post-exposure prophylaxis (occupational), 2018

Exposure code (EC)	Source HIV status code (SC)	Recommendation for PEP	PEP regimen
1/2/3	Negative	Not warranted	
1	1	Not warranted	
1	2	Recommended × 28 days	Tenofovir (TDF) 300 mg + Lamivudine (3TC) 300 mg one Tab OD
2/3	Unknown	Consider PEP (as above) if HIV prevalence is high	(Next day one tab once OD, continue for 4 weeks) Lopinavir (200 mg) + Ritonavir (50 mg) Two Tab BD. (Next day two-tab BD, continue for 4 weeks)

should be tested for HIV as per the national guideline with counselling and informed consent to know their pre-existing HIV status. If found positive, then person be referred for assessment for ART and should not receive post-exposure prophylaxis.

Assessment of source's risk status: Source should be tested for HIV testing following the national guideline and with informed consent. However, prophylaxis initiation should not be delayed due to pending test result.

Assessment of risk exposure: By establishing source's risk status and level of exposure:

Source's HIV status code (SC)
- SC-1: HIV positive, asymptomatic, low viral load, high CD4 count.
- SC-2: HIV positive, symptomatic, high viral load, low CD4 count.
- SC unknown: Source's status not known and sample also not available for testing.
- HIV negative: HIV test result negative.

Exposure code (EC)
EC-1: Mild exposure: Exposure to mucous membrane or non-intact skin with small volume and less duration.

EC-2: Moderate exposure:
- Mucous membrane or non-intact skin exposure with large volume/long duration, or
- Percutaneous superficial exposure with solid needle.

EC-3: Severe exposure: Percutaneous exposure with large volume.
- Injury with wide bore hollow needle visible contaminated with blood
- Transfer of large volume of blood
- Deep wound
- Injury with material used intravenous or intra-arterial procedure

Conditions when PEP is **NOT** recommended/warranted:
- If source is HIV negative
- Exposed person is HIV positive

- Exposure on intact skin
- When both EC and SC are 1.

Conditions when PEP is Recommended: When any of the EC or SC is 2 or above, PEP is recommended (unless source is known HIV negative).

Bibliography

1. Updated recommendations on HIV prevention, infant diagnosis, antiretroviral initiation and monitoring (who.int), WHO. March 2021.
2. National technical guidelines On Anti retroviral treatment, October 2018. National aids control organization, Ministry of health and family welfare Government of India. https://lms.naco. gov.in/frontend/content/NACO%20-%20National%20Technical%20Guidelines%20on%20ART_October%202018%20(1).pdf
3. Gullett JC, Nolte FS. Quantitative nucleic acid amplification methods for viral infections. Clin Chem. 2015 Jan;61(1):72–8.
4. Clinical guidelines: hiv diagnosis. http://www. who.int/hiv/pub/arv/chapter2.pdf?ua=1
5. Laboratory Testing for the Diagnosis of HIV Infection: Updated Recommendations. 2014. Centers for Disease Control and Prevention. http://www.cdc.gov/hiv/testing/index.html
6. National Guidelines for HIV Testing. National AIDS Control Organisaiton Ministry of Health and Family Welfare, Government of India. 2015. http://www.ilo.org/wcmsp5/groups/public/—ed_protect/—protrav/—ilo_aids/documents/legaldocument/wcms_532685.pdf
7. HIV 2015/16. www.hivbook.com. Hoffmann C and Rockstroh JK eds. 2015 by Medizin Fokus Verlag, Hamburg.
8. Avettand-Fènoël V, Hocqueloux L, Ghosn J, et al. Total HIV-1 DNA, a marker of viral reservoir dynamics with clinical implications. Clin Microbiol Rev 2016;29(4):859–80.

C. MOLECULAR EPIDEMIOLOGY OF HIV*

The two human retroviruses, human immunodeficiency viruses 1 and human immunodeficiency viruses 2 (HIV-1 and HIV-2) are the causative agent of acquired immunodeficiency syndrome (AIDS). Human immunodeficiency virus 1 was discovered in 1983 as the causative agent of AIDS, which was then known as lymphadenopathy associated virus/Human T-lymphotropic virus III. Three years after in 1986, another retrovirus was discovered which was morphologically similar but antigenically distinct than the former HIV which was named HIV-1 and the latter was named HIV-2.

MOLECULAR TAXONOMY AND EPIDEMIOLOGY OF HIV TYPE 1

Phylogenetic analysis of HIV1 isolates in the early 1980 and 90s showed that HIV1 could be divided into at least six distinct *env* types which were equidistant from each other and formed a **"star phylogeny"**. These were called **clades** and named alphabetically from **A to F**. Subsequently, new HIV-1 isolates were discovered which were approximately equidistant to each other and their mutual divergence was comparable to that found between the older clades of HIV-1. These recent strains had an amino acid sequence identity of only about 50% in the *env* gene and formed a distinct phylogenetic group were named group **O for 'outlier'**. While, the older and larger group of HIV-1 strains was called the group **M for 'major'**. Another virus which was distinct from group M and group O was described in 1998 and named **N group or 'non-M, non-O 'group**. Recently, a **P group** of virus was described in two patients from Cameroon.

Currently, the M group has successfully spread globally and causes majority (>95%) of the infections worldwide and has been studied extensively. It is further differentiated into **9 subtypes, namely A, B, C, D, F, G, H, J and K**. Sub-subtypes have been described in subtypes A (A1 to A6) and in subtype F (F1 and F2). An amino acid sequence similarity between 80 and 85% within a subtype is seen, whereas between subtypes it ranges between 65 and 75%.

---

* This chapter has been combinedly contributed by Dr Vinaykumar Hallur and Dr Baijayantimala Mishra.

Recombinant forms: Whole genome sequencing/near complete genome sequencing has led to the discovery of recombinant forms. A recombinant virus is formed when a cell is infected by two or more HIV lineages and there is an exchange of genetic material amongst each other. When there is an evidence that a recombinant virus is transmitted to three or more unrelated persons it is termed a **circulating recombinant form (CRF)**, and in the absence of such an evidence it is called **unique recombinant form (URF)**. To date **79 CRFs and multiple URFs** have been described in HIV-1 group M. HIV-1 group M isolates which are not related to any of the currently defined subtypes or recombinant forms are called unique or unlabeled.

O group: The second largest group of HIV type 1 is the O group, which causes 1% of all HIV infections in Cameroon. The HIV type 1 group O strains exhibit immense intragroup diversity which is higher than that in the M group. Two studies have analyzed the phylogeny of group O HIV viruses. Roques *et al* identified three clusters, namely A, B, and C among the isolates studied and concluded that there was no M group like subtype structure. Whereas Yamaguchi *et al* reported five clusters of isolates along with some non-clustering isolates and concluded that the five clusters had characters similar to that of the group M subtypes. However, these clades and clusters overlap and final classification agreeable to all is yet to be finalized.

N & P groups: Less than 20 infections due to group N and two cases of P group infections have been reported which mainly occur in or originate from Cameroon.

Molecular epidemiological studies to find the origin of HIV type 1 have shown that each of these four groups (M, N, O and P) were established in humans after independent cross species transmission of Simian immunodeficiency virus from chimpanzees (*Pan troglodytes troglodytes* in M, N and O groups) and Cameroonian gorillas (*Gorilla gorilla gorilla* in P groups).

MOLECULAR TAXONOMY AND EPIDEMIOLOGY OF HIV TYPE 2

HIV-2 is believed to have originated from Simian immunodeficiency virus infecting **sooty mangabey monkeys** (SIVsmm). **Nine groups, A to I,** have been described to date in HIV type 2, of which only **A and B are endemic**. The remaining have been found in one patients each except for the F group which has been found in 2 patients. It is mainly found in West Africa, Europe (Portugal and France), India, and United States of America. Only one recombinant form of HIV-2, i.e. **CRF HIV2_CRF01_AB** has been described from patients in Japan.

METHODS FOR STUDYING MOLECULAR TAXONOMY AND EPIDEMIOLOGY

The gold standard for determining the taxonomy and genotype of HIV is sequencing followed by phylogenetic analysis, which identifies the genotype as well as its relationship to reference sequences. In past, it relied mainly on partial genetic sequences of gene targets like *env* and *gag*. Partial gene sequencing and phylogenetic analysis had identified subtypes A to K, but a near complete genome sequencing followed by phylogenetic analysis has led to the conclusion that subtype E and subtype I were recombinants of subtype A and E and subtypes A, G, and unknown subtypes. Hence, more recently near complete genomes and whole genomes are being used to infer the molecular epidemiology and taxonomy of HIV.

Other methods that are used for molecular epidemiology include heteroduplex mobility assay, V-3 serotyping for HIV1 and subtype classification by PCR all of which are useful only in situations where a single subtype predominates in the community. Heteroduplex mobility assay exploits the difference in electrophoretic motility of heteroduplex formed between a PCR amplicon from reference strain and sample and is reliable for subtyping of HIV-1. V-3 serotyping is serological typing using patient's sera which contain antibodies that bind to the peptides from a V-3 loop of envelop from different subtypes. It is a simple assay that depends on immune response to the small antigenic domain and even a single substitution can affect serotyping. V-3 serotyping has not been found to be useful in cases of HIV-2.

After partial or near complete or whole genome sequencing of the HIV isolates, phylogenetic software tools like the PhyloPlace (https://www.hiv.lanl.gov/content/sequence/phyloplace/PhyloPlace.html) and subtyping distance tool (SUDI, https://www.hiv.lanl.gov/content/sequence/SUDI/sudi.html) may be used for phylogenetic/taxonomical analysis. The above softwares have specially been created by the Los Alamos National Laboratory in collaboration with the National Institute of Health and Department of Health and Human Services, USA to help in deciding whether a given sequence or a set sequences belong to a known subtype, or is a new subtype, a sub-subtype or a variant of an existing subtype. While analyzing phylogenetic data generated from software tools, there are no cut off rules on branching index values to decide whether a given sequence is a subtype or a variant. However, a thumb rule subtype specific branching index of 0.66 has been used following its evaluation by Hraber *et al* in 2008. Other softwares that are frequently used in phylogenetics/taxonomy studies include ClustaW, MEGA, and REGA HIV-1 Subtyping Tool: Version 3.0 (http://dbpartners. stanford.edu:8080/RegaSubtyping/stanford-hiv/typingtool/).

Importance of Studying HIV Molecular Epidemiology

HIV type 1 isolates worldwide exhibit huge genetic diversity as the virus possesses an error prone reverse transcription process without proofreading activity, recombination events between strains, immunological pressure, and a rapid turnover. The different types, subtypes, circulating recombinant forms differ in their rate of disease progression, susceptibility to anti-retroviral drugs, transmission dynamics, etc.

It is known that HIV type 2 has a longer asymptomatic period, lower viral copy numbers, and a lower mortality as compared to HIV type 1. Also, HIV-1 group O and HIV-2 isolates are intrinsically resistant to non-nucleoside reverse transcriptase inhibitors. Among HIV-1 group M subtypes, subtype C strains have been observed to frequently use CCR5 co-receptor (more commonly) for entry into the host cells even in the late stages of HIV and switching to CXCR4 receptor use is extremely rare among the Indian subtype C strains. Hence, maraviroc a CCR5 inhibitor can be used for treatment of HIV in India. Although differences in antiretroviral drug susceptibility is observed among different subtypes of HIV-1, large studies have not found increased failure rates among the subtypes.

A study from Thailand has shown an increased probability of transmission of CRF01_AE among injection drug users as compared to subtype B, and another study from Tanzania demonstrated that subtype C showed a high *in utero* transmission compared to other subtypes. Further, the observation that the heterosexual route of transmission drives the AIDS pandemic in Africa and India it is speculated that subtype C especially suited for heterosexual route of transmission. However, since multiple factors play a role in transmission of the infection, it is difficult to pinpoint that the differences in rates of transmission are solely due to the subtype.

Knowledge of the circulating genotypes is essential for the development of a vaccine against HIV. The said vaccine should contain all the circulating forms or conserved epitopes of HIV to be effective prevention of the disease. Thus, study of molecular epidemiology of HIV is important to understand transmission dynamics, devising diagnostic tests, vaccines and public health control of HIV. Continued molecular epidemiological surveillance will help in tracking new infections, susceptible populations and effective delivery of preventive and/or therapeutic measures to contain the HIV pandemic.

GLOBAL SCENARIO

Globally, the **subtype C of the M group is the most predominant** subtype accounting to ~50% followed by subtype A which accounted for 12% and subtype B which accounted for 10% of all HIV type 1 infection. Subtype C is the predominant subtype in India, South Africa, and Ethiopia (Fig. 21.11). Following an observation that heterosexual route of transmission drives the AIDS epidemic in the above countries it is speculated that subtype C may specially be suited for heterosexual route of transmission.

Subtype B is seen mainly in the North America, certain countries in South America, Western Europe, Australia, Northern Africa and Middle East. Subtype A is found in countries constituting the former Soviet Union, Central and Eastern Africa like Kenya, Uganda, Tanzania, and Rwanda. Figure 21.11 depicting the most predominant HIV-1 subtypes around the world.

Circulating recombinant forms and URFs contribute to about 20% of HIV-1 infections and the most common are CRF_01AG and CRF_01AE. All subtypes, CRFs, URFs, can be found in the sub-Saharan Africa. CRF_01AE was previously misclassified as subtype E and accounts for most of HIV infections in south and southeast Asia. CRF_01AG are found predominantly in West and Central Africa and contribute to 8% of HIV infections.

HIV-2 virus is mainly found in West Africa, Europe (Portugal and France), India, and United States of America. The highest prevalence of the infection occurs in the West Africa where an estimated 1–2 million people had HIV-2 infection in 2007. However, a study from Guinea Bissau which had the highest prevalence showed a declining trend of HIV-2 infections. The remainder of the cases usually occur in countries that have historical and/or socioeconomic ties with countries in West

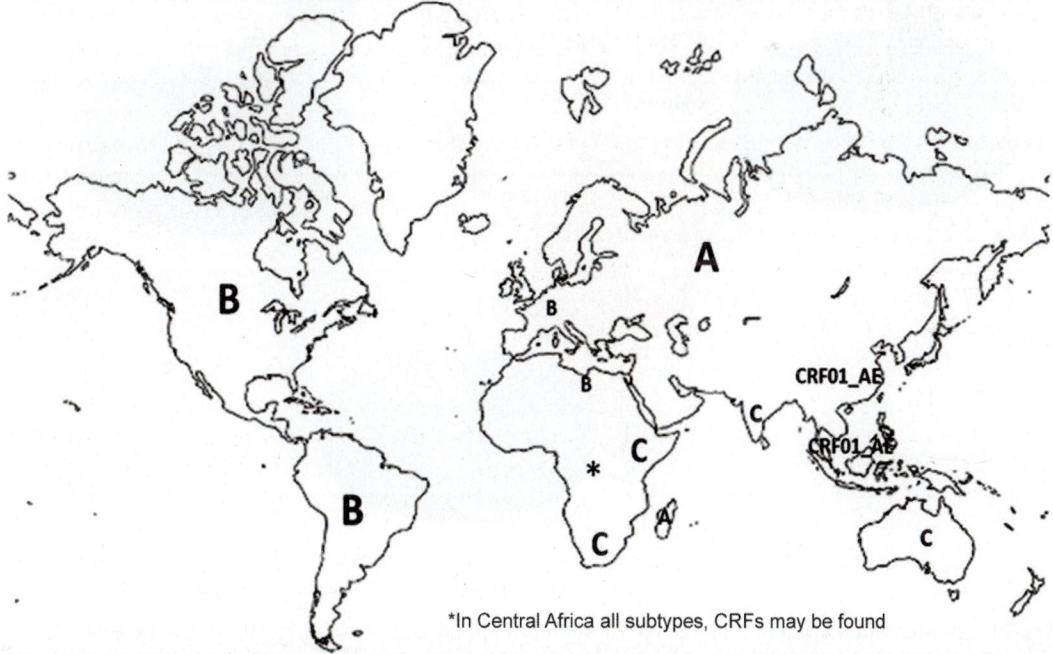

*In Central Africa all subtypes, CRFs may be found

Fig. 21.11: Global distribution of predominant HIV1 subtypes

Africa, for example, Portugal which had colonized these countries in past.

MOLECULAR EPIDEMIOLOGY OF HIV IN INDIA

A large number of studies have concluded that **HIV1 subtype C is the predominant circulating form in India**. The prevalence of subtype C ranges from 78 to 100% in some studies. In the recent years, recombinant strains are increasingly being reported from all over India. The most common recombinant strains from North India and Northeastern India include **URF_BC, URF_A1C, CRF01_AE and CRF02_AG**. While the recombinant forms reported from South India include URF_BC and URF_A1C, and CRF01_AE.

India contributes to 95% of all HIV-2 infections in Asia. In India, HIV-2 has mainly

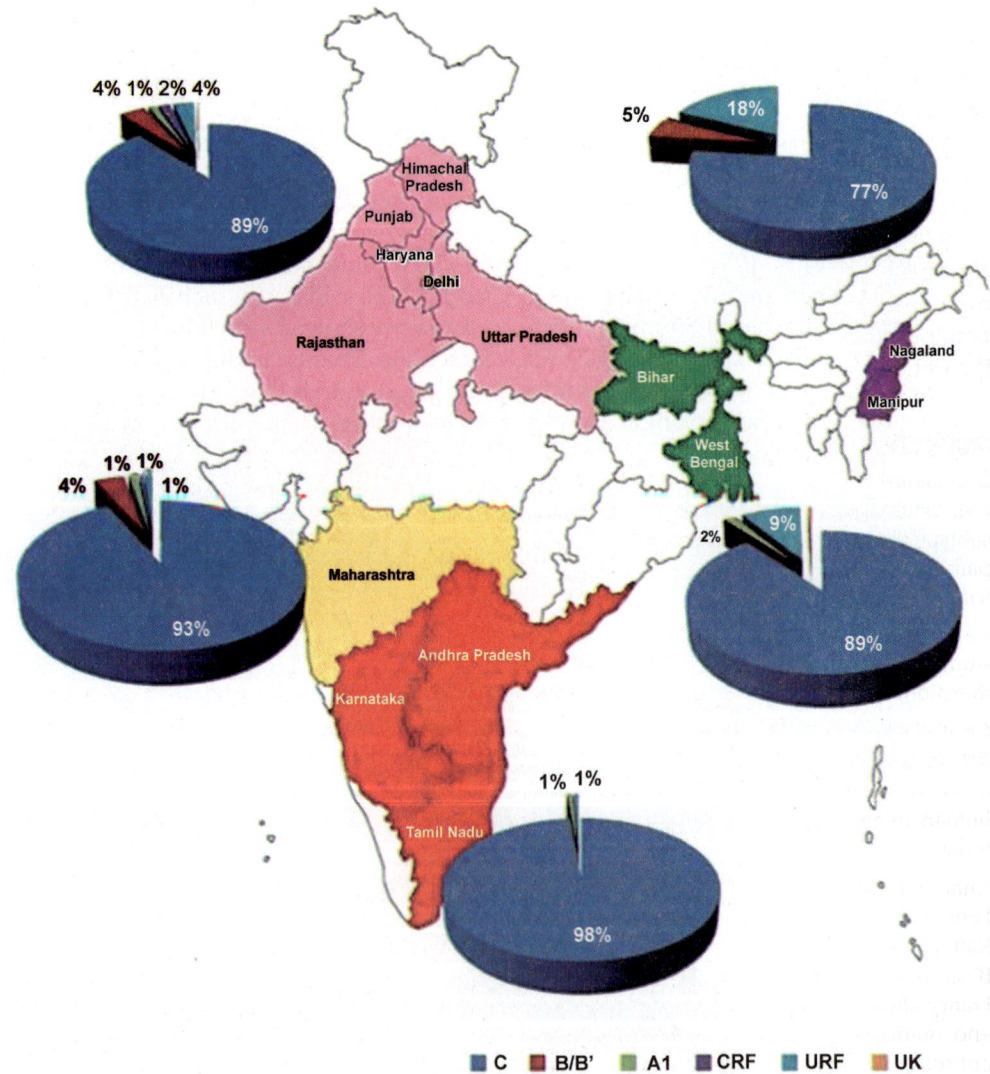

Fig. 21.12: Molecular epidemiology of HIV-1 subtypes circulating in India. 1991–2009. Neogi U (2013). Translational Genomics of HIV-1 subtype C in India: Molecular Phylogeny and Drug Resistance. PhD, Karolinska Institute, Stockholm, Sweden. (Printed with permission from Dr U Neogi)

been reported from southern states (Karnataka and Tamil Nadu) and western India (Maharashtra and Goa), and the prevalence ranges from 0.3 to 2.1%. Dual infections with HIV-1 and HIV-2 may be more frequent than infections by HIV-2 alone. Majority of the HIV-2 strains from India belong to subtype A. A map showing the current understanding of HIV molecular epidemiology in India is given in Fig. 21.12.

Studies on molecular epidemiology of HIV in India are plagued by epidemiological problems like lack of complete geographical representation, selection bias, small sample size, etc. and technical problems like use of single gene for typing which pose a serious subtyping bias. A properly designed study with adequate sample size, representing of all the geographical parts of the countries with multiple gene targets or whole/near complete genome sequencing will help us in better understanding of the molecular epidemiology of HIV in India.

Bibliography

1. Buonaguro L, Tornesello ML and Buonaguro FM. "Human immunodeficiency virus type 1 subtype distribution in the worldwide epidemic: pathogenetic and therapeutic implications." J Virol 2007;81(19):10209–19.

2. Campbell-Yesufu OT, Gandhi RT. Update on human immunodeficiency virus (HIV)-2 infection. Clin Infect Dis 2011;52(6):780–87.

3. Cecilia D, Kulkarni SS, Tripathy SP, Gangakhedkar RR, Paranjape RS, Gadkari DA. Absence of coreceptor switch with disease progression in human immunodeficiency virus infections in India. Virology 2000;271(2):253–58.

4. Chaix ML, Seng R, Frange P, Tran L, Avettand-Fenoel V, Ghosn J, Reynes J, Yazdanpanah Y, Raffi F, Goujard C, et al. Increasing HIV-1 non-B subtype primary infections in patients in France and effect of HIV subtypes on virological and immunological responses to combined antiretroviral therapy. Clin Infect Dis 2013; 56(6): 880–87.

5. Charneau P, Borman AM, Quillent C, Guetard D, Chamaret S, Cohen J, Remy G, Montagnier L and Clavel F. Isolation and envelop sequence of a highly divergent HIV-1 isolate: definition of a new HIV-1 group." Virology 1994;205(1): 247–53.

6. D'Arc M, Ayouba A, Esteban A, Learn GH, Boue V, Liegeois F, Etienne L, Tagg N, Leendertz FH, Boesch C, et al. Origin of the HIV-1 group O epidemic in western lowland gorillas. Proc Natl Acad Sci USA 2015;112(11): E1343–52.

7. Delwart EL, Herring B, Rodrigo AG, Mullins JI. Genetic subtyping of human immunodeficiency virus using a heteroduplex mobility assay. PCR Methods Appl 1995;4(5):S202–16.

8. Foley, BT, Leitner T, Paraskevis D, Peeters M. Primate immunodeficiency virus classification and nomenclature: Review. Infect Genet Evol 2016;46:150–58.

9. Fryer HR, Van Tienen C, Van Der Loeff MS, Aaby P, Da Silva ZJ, Whittle H, Rowland-Jones SL and de Silva TI. Predicting the extinction of HIV2 in rural Guinea-Bissau. AIDS 2015; 29(18):2479–86.

10. Gao F, Robertson DL, Carruthers CD, Li Y, Bailes E, Kostrikis LG, Salminen MO, Bibollet-Ruche F, Peeters M, Ho DD, et al. An isolate of human immunodeficiency virus type 1 originally classified as subtype I represents a complex mosaic comprising three different group M subtypes (A, G, and I). J Virol 1998;72(12): 10234–41.

11. Gaywee J, Artenstein AW, VanCott TC, Trichavaroj R, Sukchamnong A, Amlee P, de Souza M, McCutchan FE, Carr JK, Markowitz LE, et al. Correlation of genetic and serologic approaches to HIV-1 subtyping in Thailand. J Acquir Immune Defic Syndr Hum Retrovirol 1996;13(4):392–96.

12. Grossman Z, Schapiro JM, Levy I, Elbirt D, Chowers M, Riesenberg K, Olstein-Pops K, Shahar E, Istomin V, Asher I, et al. Comparable long-term efficacy of Lopinavir/Ritonavir and similar drug-resistance profiles in different HIV-1 subtypes. PLoS One 2014;9(1):e86239.

13. Gurjar RS, Ravi V, Desai A. Molecular epidemiology of HIV type 2 infections in South India. AIDS Res Hum Retroviruses 2009;25(3): 363–72.

14. Hemelaar J, Gouws E, Ghys PD, Osmanov S, W-U. N. f. H. Isolation and characterisation. Global

trends in molecular epidemiology of HIV1 during 2000–2007. AIDS 2011;25(5):679–89.

15. "HIV Circulating Recombinant Forms." Retrieved 10/12/2016, from https://www.hiv.lanl.gov/content/sequence/HIV/CRFs/CRFs.html.

16. HIV type 1 variation in World Health Organization-sponsored vaccine evaluation sites: genetic screening, sequence analysis, and preliminary biological characterization of selected viral strains. WHO Network for HIV Isolation and Characterization. AIDS Res Hum Retroviruses 1994;10(11):1327–1343.

17. Hraber P, Kuiken C, Waugh M, Geer S, Bruno WJ, T Leitner. Classification of hepatitis C virus and human immunodeficiency virus-1 sequences with the branching index. J Gen Virol 2008; 89(Pt 9):2098–2107.

18. Hudgens MG, Longini IM, Jr, Vanichseni S, Hu DJ, Kitayaporn D, Mock PA, Halloran ME, Satten GA, Choopanya K, Mastro TD. Subtype-specific transmission probabilities for human immunodeficiency virus type 1 among injecting drug users in Bangkok, Thailand. Am J Epidemiol 2002;155(2):159–68.

19. Ibe S, Yokomaku Y, Shiino T, Tanaka R, Hattori J, Fujisaki S, Iwatani Y, Mamiya N, Utsumi M, Kato S, et al. HIV-2 CRF01_AB: First circulating recombinant form of HIV-2. J Acquir Immune Defic Syndr 2010;54(3):241–47.

20. Ingole NA, Sarkate PP, Paranjpe SM, Shinde SD, Lall SS and Mehta PR. HIV-2 Infection: Where Are We Today? J Glob Infect Dis 2013;5(3): 110–13.

21. Kandathil AJ, Ramalingam S, Kannangai R, David S and Sridharan G. Molecular epidemiology of HIV. Indian J Med Res 2005;121(4): 333–44.

22. Kannangai R, Ramalingam S, Prakash KJ, Abraham OC, George R, Castillo RC, Schwartz DH, Jesudason MV, Sridharan G. Molecular confirmation of human immunodeficiency virus (HIV) type 2 in HIV-seropositive subjects in south India. Clin Diagn Lab Immunol 2000; 7(6):987–89.

23. Keele BF, Van Heuverswyn F, Li Y, Bailes E, Takehisa J, Santiago ML, Bibollet-Ruche F, Chen Y, Wain LV, Liegeois F, et al. Chimpanzee reservoirs of pandemic and nonpandemic HIV-1. Science 2006;313(5786):523–26.

24. Langs-Barlow A, Paintsil E. Impact of human immunodeficiency virus type-1 sequence diversity on antiretroviral therapy outcomes. Viruses 2014;6(10):3855–72.

25. Mourez, T, Simon F, Plantier JC. Non-M variants of human immunodeficiency virus type 1. Clin Microbiol Rev 2013;26(3):448–61.

26. Nam JG, Kim GJ, Baek JY, Suh SD, Kee MK, Lee JS, Kim SS. Molecular investigation of human immunodeficiency virus type 2 subtype a cases in South Korea. J Clin Microbiol 2006; 44(4): 1543–46.

27. Nasioulas G, Paraskevis D, Magiorkinis E, Theodoridou M, Hatzakis A. Molecular analysis of the full-length genome of HIV type 1 subtype I: Evidence of A/G/I recombination. AIDS Res Hum Retroviruses 1999;15(8):745–58.

28. Neogi U. Translational Genomics of HIV-1 Subtype C in India: Molecular Phylogeny and Drug Resistance. PhD, Karolinska Institutet, Stockholm, Sweden.2013.

29. Neogi U, Palchaudhuri R, Shet A. High viremia in HIV1 subtype C infection and spread of the epidemic. J Infect Dis 2013;208(5):866–67.

30. Neogi U, Prarthana SB, D'Souza G, Decosta A, Kuttiatt VS, Ranga U, Shet A. Co-receptor tropism prediction among 1045 Indian HIV-1 subtype C sequences: Therapeutic implications for India. AIDS Res Ther 2010;7:24.

31. Neogi U, Sood V, Chowdhury A, Das S, Ramachandran VG, Sreedhar VK, Wanchu A, Ghosh N, Banerjea AC. Genetic analysis of HIV-1 Circulating Recombinant Form 02_AG, B and C subtype-specific envelop sequences from Northern India and their predicted co-receptor usage. AIDS Res Ther 2009;6:28.

32. Paraskevis D, Nikolopoulos GK, Magiorkinis G, Hodges-Mameletzis I, Hatzakis A. The application of HIV molecular epidemiology to public health. Infect Genet Evol 2016;46:159–68.

33. Peters P, Marston B, De Cock K (2014). HIV Epidemiology in the Tropics. Manson's Tropical Diseases. Farrar J, Hotez P, Junghansset T al. China, Elsevier: 71.

34. Pfutzner A, Dietrich U, von Eichel U, von Briesen H, Brede HD, Maniar JK, Rubsamen-Waigmann H. HIV1 and HIV2 infections in a high-risk population in Bombay, India: Evidence for the spread of HIV2 and presence of a

divergent HIV1 subtype. J Acquir Immune Defic Syndr 1992;5(10):972–77.

35. Plantier JC, Leoz M, Dickerson JE, De Oliveira F, Cordonnier F, Lemee V, Damond F, Robertson DL, Simon F. A new human immunodeficiency virus derived from gorillas. Nat Med 2009; 15(8):871–72.

36. Ragupathy V, Casado C, Jacob SM, Samuel NM, Lopez-Galindez C. Circulation of HIV1 subtype A within the subtype C HIV-1 epidemic in Tamil Nadu, India. J Med Virol 2012;84(10): 1507–13.

37. Renjifo B, Gilbert P, Chaplin B, Msamanga G, Mwakagile D, Fawzi W, Essex M, Tanzanian V, HIV S. Group. Preferential in-utero transmission of HIV1 subtype C as compared to HIV-1 subtype A or D. AIDS 2004;18(12): 1629–36.

38. Requejo HI. Worldwide molecular epidemiology of HIV. Rev Saude Publica 2006;40(2):331–45.

39. Robertson DL, Anderson JP, Bradac JA, Carr JK, Foley B, Funkhouser RK, Gao F, Hahn BH, Kalish ML, Kuiken C, et al. HIV1 nomenclature proposal. Science 2000;288(5463):55–56.

40. Roques P, Robertson DL, Souquiere S, Damond F, Ayouba A, Farfara I, Depienne C, Nerrienet E, Dormont D, Brun-Vezinet F. et al. Phylogenetic analysis of 49 newly derived HIV-1 group O strains: high viral diversity but no group M-like subtype structure. Virology 2002;302(2): 259–73.

41. Sarkar R, Pal R, Bal B, Mullick R, Sengupta S, Sarkar K, Chakrabarti S. Genetic Characterization of HIV-1 Strains Among the Injecting Drug Users in Nagaland, India. Open Virol J 2011; 5:96–102.

42. Sarkar R, Sarkar K, Brajachand Singh N, Manihar Singh Y, Mitra D, Chakrabarti S. Emergence of a unique recombinant form of HIV1 from Manipur (India). J Clin Virol 2012;55(3):274–77.

43. Siddappa NB, Dash PK, Mahadevan A, Desai A, Jayasuryan N, Ravi V, Satishchandra P, Shankar SK, Ranga U. Identification of unique B/C recombinant strains of HIV-1 in the southern state of Karnataka, India. AIDS. 2005;19(13): 1426–29.

44. Simon F, Mauclere P, Roques P, Loussert-Ajaka I, Muller-Trutwin MC, Saragosti S, Georges-Courbot MC, Barre-Sinoussi F and Brun-Vezinet F. Identification of a new human immunodeficiency virus type 1 distinct from group M and group O. Nat Med 1998;4(9):1032–37.

45. Tebit DM, Patel H, Ratcliff A, Alessandri E, Liu J, Carpenter C, Plantier JC, Arts EJ. HIV1 Group O Genotypes and Phenotypes: Relationship to Fitness and Susceptibility to Antiretroviral Drugs. AIDS Res Hum Retroviruses 2016;32(7): 676–88.

46. Thomson MM, Perez-Alvarez L, Najera R. Molecular epidemiology of HIV-1 genetic forms and its significance for vaccine development and therapy. Lancet Infect Dis 2002;2(8):461–71.

47. Touloumi G, Pantazis N, Chaix ML, Bucher HC, Zangerle R, Kran AM, Thiebaut R, Masquelier B, Kucherer C, Monforte A, et al. Virologic and immunologic response to cART by HIV-1 subtype in the CASCADE collaboration. PLoS One 2013;8(7):e71174.

48. Tripathy SP, Kulkarni SS, Jadhav SD, Agnihotri KD, Jere AJ, Kurle SN, Bhattacharya SK, Singh K, Tripathy SP, Paranjape RS. Subtype B and subtype C HIV type 1 recombinants in the northeastern state of Manipur, India. AIDS Res Hum Retroviruses 2005;21(2):152–57.

49. Vallari A, Holzmayer V, Harris B, Yamaguchi J, Ngansop C, Makamche F, Mbanya D, Kaptue L, Ndembi N, Gurtler L, et al. Confirmation of putative HIV1 group P in Cameroon. J Virol 2011;85(3):1403–07.

50. Van Heuverswyn F, Li Y, Bailes E, Neel C, Lafay B, Keele BF, Shaw KS, Takehisa J, Kraus MH, Loul S, et al. Genetic diversity and phylogeographic clustering of SIVcpzPtt in wild chimpanzees in Cameroon. Virology 2007; 368(1):155–71.

51. Vessiere A, Rousset D, Kfutwah A, Leoz M, Depatureaux A, Simon F, Plantier JC. Diagnosis and monitoring of HIV1 group O-infected patients in Cameroun. J Acquir Immune Defic Syndr 2010;53(1):107–10.

52. Visseaux B, Damond F, Matheron S, Descamps D, Charpentier C. HIV2 molecular epidemiology. Infect Genet Evol 2016;46:233–40.

53. Walker PR, Pybus OG, Rambaut A, Holmes EC. Comparative population dynamics of HIV1 subtypes B and C: Subtype-specific differences

in patterns of epidemic growth. Infect Genet Evol 2005;5(3):199–208.

54. Woodman Z, Williamson C. HIV molecular epidemiology: Transmission and adaptation to human populations. Curr Opin HIV AIDS 2009; 4(4):247–52.

55. Yamaguchi J, Vallari AS, Swanson P, Bodelle P, Kaptue L, Ngansop C, Zekeng L, Gurtler LG, Devare SG, Brennan CA. Evaluation of HIV type 1 group O isolates: Identification of five phylogenetic clusters. AIDS Res Hum Retroviruses 2002;18(4):269–82.

56. (14/06/2016). "Panel on Antiretroviral Guidelines for Adults and Adolescents. Guidelines for the use of antiretroviral agents in HIV-1-infected adults and adolescents." Retrieved 23/12/2016, 2016, from http://www.aidsinfo.nih.gov/ContentFiles/AdultandAdolescentGL.pdf.

D. HIV VACCINE*

In the absence of curative therapy, preventive measures like sex education, promotion of condoms, prevention of parent to child transmission by using antiretroviral therapy (ART), pre- and post-exposure prophylaxis, topical microbicides, use of highly active ART, screening of blood donors and circumcision have been instituted to control the HIV pandemic. However, despite these measures, it has been found that the estimated number of people with HIV has quadrupled since 1990s. This is because the virus spreads predominantly through sexual mode, the prevention of which requires sustained changes in human sexual behavior, which is difficult to achieve. Further, ART which is therapeutic as well as preventive is not universally available worldwide due to high costs and the drugs themselves are highly toxic. This makes it important to develop vaccine for prevention and therapy of HIV-AIDS.

Despite continued efforts for more than three decades, an effective vaccine against HIV has not been developed. Reasons for this include immunological, experimental, and viral factors like lack of an ideal animal model, incomplete understanding of immune correlates of the disease, multiple portals of entry, ability to cause persistent infection, and huge diversity of HIV-1 strains in the infected individual as well as worldwide.

Since, the majority of the infections around the world are caused by HIV-1 group M and HIV-2 is responsible mainly for localised epidemics in West Africa, efforts are underway to develop a vaccine against HIV-1. The different types of HIV-1 vaccine candidates are discussed below.

1. **Live attenuated virus vaccines:** Animal studies in monkeys using attenuated SIV (obtained by deleting the *nef* gene with/without deletion of other genes) or by using the nonpathogenic humanized SIV (SHIV89.6) have been conducted. These studies initially found that vaccine virus established a lifelong persistent infection in adult monkeys and prevented clinical disease. However, Δ*nef* caused a full blown disease in infant monkeys and reverted to full virulence in some animals. Further attempts to improve the safety profile by deleting more genes resulted in a compromised protective efficacy. Because of concerns like lifelong infection and revert to virulence, these vaccine candidates could not enter human trials.

2. **Whole killed virus vaccine:** Such vaccines have the advantage of presenting all the antigens to the immune system for processing. With older vaccine production techniques, there was a possibility of residual infectious nucleic acid with killed virus vaccines, however, these have been overcome with better purification processes. Whole killed SIV (Simian immunodeficiency virus) was unable to protect monkeys against SIV infection. Because of the above safety concerns, these vaccines have not been investigated in humans.

---

*This chapter has been combinedly contributed by Dr Vinaykumar Hallur and Dr Baijayantimala Mishra.

3. **Subunit vaccines:** Unlike killed vaccines, there is no risk of contamination with residual nucleic acids in subunit vaccines.

   a. *Env subunit vaccines:* Recombinant gp160 subunit vaccine derived from laboratory adapted strains (LAI, MN, SF-2, and IIIB strains) as well as from primary isolates (GNE-8 and CM244 strains) produced in different expression systems like insect cell lines, cancer cell lines, yeasts, etc. have undergone human trials. The major problem with subunit vaccines is that they afford protection only against homologous strains, i.e. they have failed to induce broadly neutralizing antibodies in preclinical or clinical studies. Examples include: AIDSVAX B/B (rgp120: MN, GNE-8), AIDSVAX B/E (rgp120: MN, A244) which have undergone phase III trials showed no efficacy.

   b. *Synthetic peptide:* Synthetic peptide vaccines contain epitopes of interest only and completely omit minor or epitopes with deleterious effects. However, they may not be able to elicit immune response against conformational or non-contiguous epitopes and are usually less immunogenic than native proteins. These shortcomings are overcome by use of carriers or potent adjuvants. Phase I human trials of synthetic peptides have found that they have low immunogenicity. For example, synthetic peptides from V3 loop of multiple strains linked to a lipid carrier.

4. **DNA vaccines:** In this approach, codon optimized synthetic genes encoding HIV proteins of interest are engineered into purified DNA plasmids which are injected into the host cells. Proteins expressed subsequently stimulate both the cellular and humoral immunity. Immunogenicity is further improved by co-expression of cytokines, such as IL-12, IL-15, or M-CSF, which serve as vaccine adjuvants, or by adsorbing DNA onto micro-/nanoparticles, and novel delivery methods like electroporation, e.g. PENNVAX®-B (PV), expressing HIV consensus clade B Gag, Pol, and *Env* with IL-12 or IL-15 plasmid cytokine adjuvants. Phase 1 clinical trials of DNA vaccines have concluded that DNA vaccines are safe and non-reactogenic.

5. **Live vectors:** HIV viral protein gene sequences like e*nv, gag, pol,* and regulatory genes are engineered into live viral or bacterial vectors. Only those vectors which have been rendered replication defective by deletion of essential genes, like New York vaccinia virus strain or those which cannot multiply in humans are used, e.g. canary-pox (ALVAC), fowl pox or attenuated modified vaccinia Ankara are used which adds to their safety. Such vectors stimulate cell-mediated immunity and induce CD8+ T cells. A major disadvantage of using live vectors is the immune response against the vector dampens vaccine efficacy. Two phase 2b clinical trials Step study and Phambili study evaluating replication of a vaccine containing 3 defective adenovirus type 5 with one expressing *gag* gene from CAM-1 HIV strain belonging to clade B, another expressing *pol* gene from HIV-1 strain IIIB and the last expressing *nef* gene from HIV-1 strain JR-FL in a 1:1:1 mixture conducted in men who have sex with men and heterosexual individuals were stopped because there was a higher rate of acquisition of HIV among vaccinees.

6. **Prime boost strategy:** Prime boost strategy uses two different vaccine types wherein priming is done with one vaccine and boosting with another. This strategy has been found to be more effective than repeat doses of the same vaccine in preclinical studies. These findings have been applied to HIV vaccine. Several approaches have been studied. These include DNA prime + subunit protein or vector boost, viral vector prime + protein or heterologous viral vector boost, and vector prime + vector boost. Examples include RV144 trial (consisting of recombinant canarypox vector vCP1521

prime and AIDSVAX B/E rgp120 boost) which has undergone phase 3 trials and HVTN 505 (priming dose using DNA *gag*, *pol*, and *nef* from HIV-1 subtype B and *env* from subtypes A, B, and C, and booster using live vector rAd5 subtype B containing *gag-pol* and *env* from subtypes A, B, and C) which has completed phase 2 clinical trials. While RV 144 showed moderate efficacy, HVTN 505 trial failed to demonstrate any efficacy.

7. **Preventive or therapeutic vaccine:** A preventive vaccine would be used before a person is exposed to HIV and may eliminate the virus at the point of entry, or reduce the initial peak viremia, lower the viral set point and eliminate virus, and/or stop the progression to AIDS. The vaccine candidates described above are all preventive vaccines. A therapeutic vaccine will be used in a person who is already infected by HIV and will augment immune responses against HIV helping it to control HIV replication and stop the progression to AIDS. Several strategies like killed virus, subunit, viral vectors, etc. have been tried but they have shown limited efficacy. The different types of vaccine preparations against HIV are elaborated in Table 21.6.

Of the several HIV vaccine candidates, only six HIV-1 candidates have reached efficiency trials to date. A list explaining the salient features of these trials is given in Table 21.7. Among the 6, only one vaccine trial ALVAC-HIV (vCP1521) + AIDSVAX B/E (RV144) which was started in 2003 and published the results in 2009 has shown moderate efficacy. The trial employed a prime boost strategy which was proven to be effective from previous trials and involved priming with ALVAC-HIV (canarypox vector expressing gp120 from clade E , *gag*, and protease from

clade B, from Sanofi Pasteur) followed by a boost using the AIDSVAX B/E (containing *rgp*120 from MN strains from Vaxgen).

It was based on the concept that the prime boost strategy would induce CD4+ and CD8+ cells as well as binding and neutralizing antibodies which would then recognize and destroy any HIV strains before the infection becomes established. Priming using ALVAC-HIV was done using 4 doses at 0, 4, 12 and 24 weeks followed by boosting with 2 doses of AIDSVAX B/E at weeks 12 and 24 respectively. It was the biggest HIV vaccine trial conducted involving 16402 participants from Thailand and studied the efficacy of the vaccine in preventing infection from sexual exposure. Apart from being the largest study, it had a robust study design (randomized control trial, double blinded and multicentric). It reported an efficacy of 31%.

A study was done by the same group later to identify the immune correlates of RV144 vaccine trial. It found that IgG antibody binding to scaffolded V1V2 *Env* antigen correlated inversely with acquisition of infection, and IgA antibody binding to *Env* antigen correlated directly with acquisition of infection.

Future Prospective

The findings from the RV144 trial have provided fresh impetus to HIV vaccine research, which is now targeting to understand the immune correlates in the early part of the infection as well as to improve on the efficiency of HIV vaccines by use of immunogens that would stimulate B cells in manner which will increase the spectrum of antibodies, (e.g. *env* trimmers, epitope scaffolds), immunogens that will increase the breadth, depth and coverage of cellular immunity [e.g. next generation vectors (NYAC), subunits (gp120), and DNA vaccine, etc.].

Table 21.5: Different types of vaccines for HIV-1

Type	Advantages	Disadvantages and current status
Live attenuated virus, e.g. *nef* deleted SIV virus	*Mimics natural infection*	1. Persistence of the vaccine strains and subsequent reactivation was worrisome. 2. A possibility of the vaccine virus becoming virulent as in polio vaccines. 3. Animal trial showed induction of an immune response but there was increased mortality in the vaccinees.
Whole killed vaccine	Contains all epitopes	1. Infectious nucleic acid, if present, may cause infection. 2. Do not induce neutralizing antibodies or virus specific CD8 T cells 3. Not currently being pursued.
Subunit Vaccines		
Synthetic peptides, e.g. V2 region synthetic peptides, synthetic peptides from V3 loop of multiple strains	Include only the epitopes of interest, omitting the deleterious epitopes	1. Do not stimulate non-contiguous conformational epitopes and less immunogenic. 2. Attempts to improve immunogenicity by use of protein or carriers in conjunction with potent adjuvants.
Envelop proteins: Recombinant gp120 from laboratory adapted strains produced in yeasts, insect or mammalian cell lines.	No contamination with other HIV-1 components	1. Recombinant proteins may differ from wild virus in their tertiary structure, oligomerization and glycosylation which may reduce their efficiency in generating broadly neutralizing antibodies 2. Novel ways to activate germline B cell receptors like *env* trimmers, epitope scaffolds and immunogens which increase spectrum of activity of antibodies generated are being currently in developmental stage.

(Contd.)

Table 21.5: Different types of vaccines for HIV-1 (*Contd.*)

Type	Advantages	Disadvantages and current status
	Live vectors	
Live vectors: For example, poxviruses like vaccinia and its attenuated counterparts likes modified vaccinia Ankara, NYAC, canarypox, adenovirus vectors like Ad5 and CMV vectors	Live vectors induce CD8+ T cells and provide cell mediated immunity	1. Pre-existing immunity against the vector annulus the immunogenicity of the vaccine. 2. Non-attenuated vectors like vaccinia can cause serious illness in the immunocompromised. 3. Non-replicating vectors are less immunogenic than replicating vector systems. 4. Studies currently focus on use of non-replicating vectors. 5. They have been explored in human trials
1. Non-replicating vectors 2. Replicating live vectors		
DNA vaccines	Elicit both cellular and humoral immunity	1. DNA vaccines have been less immunogenic in human trials due to reduced gene expression. Methods to optimize gene expression and improved delivery like electroporation are being explored. 2. DNA vaccines are currently undergoing clinical trials.

Table 21.6: Candidate HIV-1 vaccines that have entered efficacy trials

Trial name (vaccine name, company)	Components	Immunogen	Delivery vehicle	Strategy	Stimulates	Method of administration and dose	Participants recruited (subgroups)	HIV route of transmission	Trial site
VAX003 (AIDSVAX B/B, Vaxgen)	rgp120 produced by expression of the fusion proteins in genetically engineered Chinese hamster ovary cell lines	Clade B/B (isolates MN/ GNE-8) and alum	Protein in alum adjuvant	Recombinant envelop protein	B cells	Intramuscular in deltoid at 0, 1, 6, 12, 18, 24, and 36 months	5095 (MSM) 308 (women)	Sexual	Thailand
VAX004 (AIDSVAX B/E, Vaxgen)	rgp120 produced by expression of the fusion proteins in genetically engineered Chinese hamster ovary cell lines	Clade B/E (MN/CM244)	Protein	Recombinant envelop protein	B cells	Intramuscular in deltoid at 0, 1, 6, 12, 18, 24, and 36 months	2500 (men and women) IDU*	Parenteral	USA/ Netherlands
HVTN 502/Step Trial	1:1:1 mixture of three separate replication— defective Ad5 vectors, one each expressing the gag gene from the HIV-1 strain CAM-1, the pol gene from HIV-1 strain IIIB, and the nef gene from HIV-1 strain JR-FL	Clade B-gag/ pol/nef	Adenovirus serotype 5	Live virus vector	Cell mediated immunity both CD4 and CD8 T cells	1.0 mL Intramuscular injection of 1.5 × 10^10 adenovirus genomes on day 1, week 4, week 26	3000 (MSM; sexual exposure)	Sexual	North America, Caribbean, Australia
HVTN 503/ Phambili Study	1:1:1 mixture of three separate replication— defective Ad5 vectors, one each expressing the gag gene from the HIV1 strain	Clade B-gag/ pol/nef	Adenovirus serotype 5	Live virus vector	Cell mediated immunity both CD4 and CD8 T cells	1.0 mL Intramuscular injection of 1·5 × 10^10 adenovirus genomes on 0, 1, 6 months schedule	801** (heterosexual adults)	Sexual	South Africa

(Contd.)

*IDU: Intravenous drug abuser, **: Trial terminated early. MSM: Men who have sex with men.

Table 21.6: Candidate HIV-1 vaccines that have entered efficacy trials (Contd.)

Trial name (vaccine name, company)	Components	Immunogen	Delivery vehicle	Strategy	Stimulates	Method of administration and dose	Participants recruited (subgroups)	HIV route of transmission	Trial site
	CAM-1, the *pol* gene from HIV1 strain IIIB, and the *nef* gene from HIV1 strain JR-FL								
RV144 [(ALVAC-HIV, Sanofi Pasteur + AIDSVAX B/E, Vaxgen)]	Canarypox vector expressing gp120 from clade E, *gag* and protease from clade B + Clade B/E (MN/CM244)	Clade B-*gag/pol* Clade E-*env* Clade B/E-*env*	Canarypox and protein	Prime with ALVAC-HIV (CP1521) and Boost with AIDSVAX B/E	Both B and T cells	4 doses of ALVAC-HIV (at 0, 1, 3, 6 months) and AIDSVAX B/E vaccine (at 3 and 6 months)	16402 (men and women)	Sexual	Thailand
HVTN 505 DNA/Ad5	**DNA vaccine:** 6 circular plasmids (in a 1:1:1:1:1:1 ratio) which express HIV-1 clade B *gag, pol,* and *nef* and *env* proteins from clades A, B, and C **Ad5 component:** Four rAd5 vectors (in a 3:1:1:1 ratio) expressing an HIV1 clade B *gag-pol* fusion protein and *env* glycoproteins from clades A, B, and C	Clade B-*gag/pol/nef* Clade A/B/C-*env* Clad B-*gag/pol* Clades A/B/C-*env*	DNA plasmid and adenovirus serotype 5	Prime with DNA plasmid and boost with Ad5 vector	Both B and T cells	DNA vaccine: 4 mg dose IM in the deltoid muscle using bioinjector at 0, 4, 8 weeks Ad5 vector: single dose of 10^{10} particle units of vaccine intramuscularly at 24 weeks	2504 men/ transgender men, circumcised without Ad5 infection in past (MSM)	Sexual	USA

Bibliography

1. Belshe RB, Gorse GJ, Mulligan MJ, Evans TG, Keefer MC, Excler JL, Duliege AM, Tartaglia J, Cox WI, McNamara J, et al. Induction of immune responses to HIV1 by canarypox virus (ALVAC) HIV1 and gp120 SF-2 recombinant vaccines in uninfected volunteers. NIAID AIDS Vaccine Evaluation Group. AIDS 1998;12(18): 2407–15.

2. Belyakov IM, Ahlers JD, Nabel GJ, Moss B, Berzofsky JA. Generation of functionally active HIV1 specific CD8+ CTL in intestinal mucosa following mucosal, systemic or mixed prime-boost immunization. Virology 2008;381(1): 106–15.

3. Buchbinder SP, Mehrotra DV, Duerr A, Fitzgerald DW, Mogg R, Li D, Gilbert PB, Lama JR, Marmor M, Del Rio C, et al. Efficacy assessment of a cell-mediated immunity HIV1 vaccine (the Step Study): A double-blind, randomised, placebo-controlled, test-of-concept trial. Lancet 2008;372(9653):1881–93.

4. Cooney EL, McElrath MJ, Corey L, Hu SL, Collier AC, Arditti D, Hoffman M, Coombs RW, Smith GE, Greenberg P. Enhanced immunity to human immunodeficiency virus (HIV) envelop elicited by a combined vaccine regimen consisting of priming with a vaccinia recombinant expressing HIV envelop and boosting with gp160 protein. Proc Natl Acad Sci USA 1993;90(5):1882–86.

5. Corey L, McElrath MJ, Kublin JG. Post-step modifications for research on HIV vaccines. AIDS 2009;23(1):3–8.

6. Corey L, Nabel GJ, Dieffenbach C, Gilbert P, Haynes BF, Johnston M, Kublin J, Lane HC, Pantaleo G, Picker LJ, et al. HIV1 vaccines and adaptive trial designs. Sci Transl Med 2011;3(79):79ps13.

7. Excler JL, Plotkin S. The prime-boost concept applied to HIV preventive vaccines. AIDS 11 Suppl A, 1997;S127–137.

8. Flynn NM, Forthal DN, Harro CD, Judson FN, Mayer KH, Para MF, HIV VSG rgp. Placebo-controlled phase 3 trial of a recombinant glycoprotein 120 vaccine to prevent HIV1 infection. J Infect Dis 2005;191(5):654–65.

9. Gray GE, Allen M, Moodie Z, Churchyard G, Bekker LG, Nchabeleng M, Mlisana K, Metch B, de Bruyn G, Latka MH, et al. Safety and efficacy of the HVTN 503/Phambili study of a clade-B-based HIV1 vaccine in South Africa: a double-blind, randomised, placebo-controlled test-of-concept phase 2b study. Lancet Infect Dis 2011;11(7):507–15.

10. Hammer SM, Sobieszczyk ME, Janes H, Karuna ST, Mulligan MJ, Grove D, Koblin BA, Buchbinder SP, Keefer MC, Tomaras GD, et al. Efficacy trial of a DNA/rAd5 HIV1 preventive vaccine. N Engl J Med 2013;369(22):2083–92.

11. Haynes BF, Gilbert PB, McElrath MJ, Zolla-Pazner S, Tomaras GD, Alam SM, Evans DT, DC Montefiori DT, Karnasuta C, Sutthent R, et al. Immune-correlates analysis of an HIV1 vaccine efficacy trial. N Engl J Med 2012; 366(14):1275–86.

12. Kalams SA, Parker SD, Elizaga M, Metch B, Edupuganti S, Hural J, De Rosa S, Carter DK, Rybczyk K, Frank I, et al. Safety and comparative immunogenicity of an HIV-1 DNA vaccine in combination with plasmid interleukin 12 and impact of intramuscular electroporation for delivery. J Infect Dis 2013;208(5):818–29.

13. Jameson, Fauci AS, Kasper, D. L. *Harrison's Principles of Internal Medicine*, McGraw-Hill Education.

14. Klinman DM, Klaschik S, Tross D, Shirota H, Steinhagen F. FDA guidance on prophylactic DNA vaccines: analysis and recommendations. Vaccine 2010;28(16):2801–05.

15. Lifson JD, Rossio JL, Piatak M, Bess Jr J, Chertova Jr E, Schneider DK, Coalter VJ, Poore B, RF Kiser B, Imming RJ, et al. Evaluation of the safety, immunogenicity, and protective efficacy of whole inactivated simian immunodeficiency virus (SIV) vaccines with conformationally and functionally intact envelop glycoproteins. AIDS Res Hum Retroviruses 2004;20(7):772–87.

16. Paris RM, Kim JH, Robb ML, Michael NL. Prime-boost immunization with poxvirus or adenovirus vectors as a strategy to develop a protective vaccine for HIV1. Expert Rev Vaccines 2010; 9(9):1055–69.

17. Pitisuttithum, P. HIV vaccine research in Thailand: lessons learned. Expert Rev Vaccines 2008;7(3):311–17.

18. Pitisuttithum P, Gilbert P, Gurwith M, Heyward W, Martin M, van Griensven F, Hu D, Tappero JW, Choopanya K, Bangkok G Vaccine Evaluation. Randomized, double-blind,

placebo-controlled efficacy trial of a bivalent recombinant glycoprotein 120 HIV1 vaccine among injection drug users in Bangkok, Thailand. J Infect Dis. 2006;194(12):1661–71.

19. Plotkin S, Orenstein W, Offit P, Edwards KM. Plotkin's vaccine. 7th Edition. Elsevier Publication.

20. Rerks-Ngarm S, Pitisuttithum P, Nitayaphan S, Kaewkungwal J, Chiu J, Paris R, Premsri N, Namwat C, de Souza M, Adams E, et al. Vaccination with ALVAC and AIDSVAX to prevent HIV1 infection in Thailand. N Engl J Med 2009;361(23):2209–20.

21. Russell ND, Graham BS, Keefer MC, McElrath MJ, Self SG, Weinhold KJ, Montefiori DC, Ferrari G, Horton H, Tomaras GD, et al. Phase 2 study of an HIV-1 canarypox vaccine (vCP1452) alone and in combination with rgp120: negative results fail to trigger a phase 3 correlates trial. J Acquir Immune Defic Syndr 2007;44(2):203–12.

22. Safrit JT, Fast PE, Gieber L, Kuipers H, Dean HJ, Koff WC. Status of vaccine research and development of vaccines for HIV1. Vaccine 2016;34(26):2921–25.

23. Sutjipto S, Pedersen NC, Miller CJ, Gardner MB, Hanson CV, GettieA , Jennings M, Higgins J, Marx PA. Inactivated simian immunodeficiency virus vaccine failed to protect rhesus macaques from intravenous or genital mucosal infection but delayed disease in intravenously exposed animals. J Virol 1990;64(5):2290–97.

24. Team AVEGP. Cellular and humoral immune responses to a canarypox vaccine containing human immunodeficiency virus type 1 Env, Gag, and Pro in combination with rgp120. J Infect Dis 2001;183(4):563–70.

25. Whitney JB, Ruprecht RM. Live attenuated HIV vaccines: Pitfalls and prospects. Curr Opin Infect Dis 2004;17(1):17–26.

Prions

The term Prion is derived from "**Proteinaceous Infectious**" particle. These are pathogens of a group of neurodegenerative disorders seen both in human and animals. These are characterized by spongiform lesion affecting the gray matter of the brain with hypertrophy and proliferation of the astroglial cells, and are transmissible in nature. Thus, prion diseases are also known as **"transmissible spongiform encephalopathies" (TSE)**.

In 1967, Griffth first proposed the role of protein particle as pathogenic agent of scrapie, a fatal transmissible spongiform encephalopathy in sheep and goat. In 1982, Stanley Prusiner isolated the infectious material from scrapie-infected hamster brain, confirmed it to be a misfolded protein which he named protein infectious particle or "prions". **Prusiner was awarded Nobel Prize for his discovery and mechanism of prion pathogenesis.** The current working definition as given by Prusiner is, **"proteinaceous infectious particle that lacks nucleic acids"**.

Prion Proteins

Prion proteins are of two types. One is normal or cellular form which is denoted as PrP^c for prion protein cellular form. The other form is the pathogenic form denoted as PrP^{Sc} (Prion protein scrapie).

PrP^c, which is the **physiological form**, is composed of 208 amino acids and is encoded by prion protein gene PRNP located at the short arm of chromosome 20 of human genome. It is also encoded in the gene of most mammals. The protein is located mainly on the cell membrane and expressed predominantly in CNS and lymphoreticular tissues and also in many other tissues such as kidney, skin, myocardium, endothelium. In the central nervous system, the protein is present in neurons (both pre- and post-synaptic), extraneural tissues and glial cells. Expression of PrP^c is more in olfactory bulb, hippocampus, prefrontal cortex and striatum.

The possible physiological functions of PrP^c are:

i. Cell receptor, cell adhesion and signal transduction
ii. Copper transporter
iii. Antioxidant by reducing oxidative stress
iv. Antiapoptotic activity
v. Zinc transporter which may associate it in neurodegeneration
vi. Synaptic formation and maintenance.

These proteins are protease sensitive in contrast to PrP^{Sc} which are protease resistant.

PrP^{Sc}: This is the **pathological form** of prion protein. It is devoid of any nucleic acid. Structurally, it is the abnormal isoform of PrP^c. The amino acid sequence of PrP^c and PrP^{Sc} is same in a single host. PrP^c consists of predominantly alpha helices. Refolding of these alpha helices to β pleated sheets converts PrP^c to PrP^{Sc}. The predominant β pleated sheet structure of PrP^{Sc} makes it more stable and confers it the ability to form aggregate.

Prion replication: The pathological form PrP^{Sc} acts as the template for prion replication. When it comes in contact with the cellular form PrP^c, it catalyses the conversion of PrP^c to PrP^{Sc} resulting in formation of two prions PrP^{Sc} via formation of a heterodimer of PrP^{Sc} and PrP^c. The two PrP^{Sc} now turns two more PrP^c to PrP^{Sc} which in turn, turns four more to PrP^{Sc}. This continues further leading to an exponential transformation and accumulation of prions. More and more accumulation of prions PrP^{Sc} leads to polymerization and formation of fibrillar masses that are visible as plaques (Fig. 22.1).

Differences between PrP^c and PrP^{Sc} are noted in Table 22.1.

Properties of prion proteins: Prion protein (PrP^{Sc}) is highly resistant to protease, UV and γ irradiation. These properties have major implications in sterilization of prion contaminated materials.

Prion proteins do not induce any immunological response or inflammation.

HUMAN PRION DISEASES

Prion diseases can occur both in humans and animals. The uniqueness of prion protein among all other infective agents are: (i) Sporadic or spontaneous, (ii) heritable or genetic, and (iii) acquired or infectious. Accordingly, human prion diseases have been

Table 22.1: Difference between normal (*PrP^C*) and pathological prion proteins (*PrP^{Sc}*)

	PrP^C	*PrP^{Sc}*
Form of prion protein	Normal or cellular	Pathological form
Source	Present naturally in humans and animals	Arises spontaneously, inherited or acquired
Structure	Predominantly α helices	Predominantly β pleated sheets
Property	Not polymerizable	Polymerises, forms aggregates
Protease sensitivity	Sensitive	Resistant

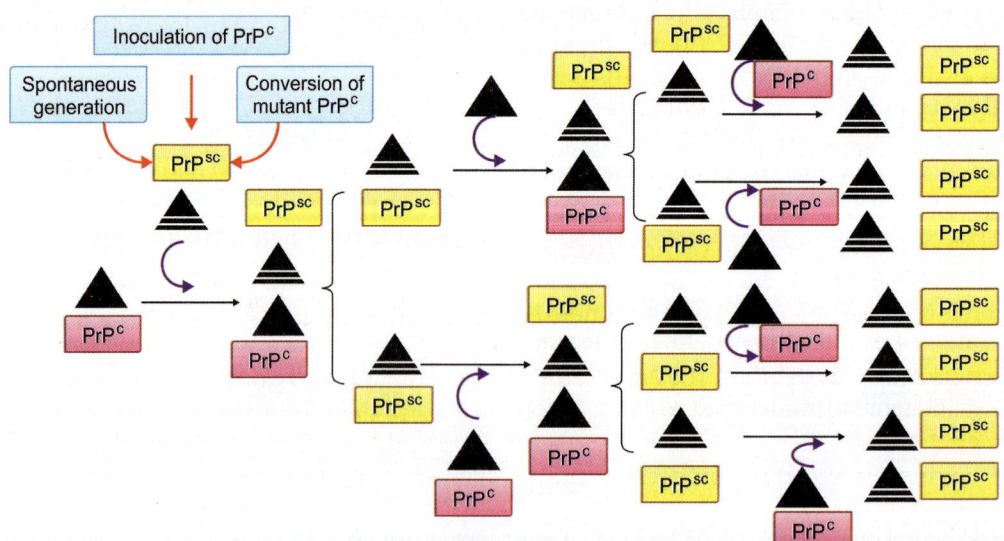

Fig. 22.1: Conversion of PrP^c to PrP^{Sc}

classified into three different types. Examples of different types of human and animal prion diseases are given in Table 22.2.

Sporadic prion disease: Conversion of PrP^c to PrP^{Sc} occurs spontaneously or due to mutations in PRNP gene without prior presence of PrP^{Sc}, e.g. sporadic CJD.

Heritable prion disease: Mutation in PRNP gene occurs which makes the cellular PrP^c more vulnerable for conversion to PrP^{Sc}.

Infectious or acquired prion disease: PrP^{Sc} is acquired by the host from some other tissues carrying the pathological prion protein PrP^{Sc} which transforms the normal PrP^c to pathological PrP^{Sc} and causes the disease.

Sporadic Human Prion Disease

Creutzfeldt-Jakob disease (CJD): This is the most common form of human prion disease. It can present in either of the three types of prion disease; sporadic, hereditary, iatrogenic or infectious. Of these, sporadic CJD is the commonest form, worldwide in distribution and constitute 80–95% of all CJD cases. Heritable or genetic CJD is seen in 10–15% and infectious type is seen in <1%.

The mean age of onset of sporadic CJD is 50–70 years. Majority die within one year of disease onset. The typical clinical manifestations of sporadic CJD are rapid progressive dementia, behavioral abnormality, ataxia and myoclonus. Marked neuronal loss, vacuolation, astrogliosis along with PrP^{Sc} deposits are the characteristic histopathological changes.

Heritable Human Prion Disease

Familial Creutzfeldt-Jakob disease (CJD), Gerstmann-Sträussler-Scheinker (GSS) syndrome and fatal familial insomnia (FFI) are inheritable prion diseases caused by autosomal dominant mutations in the PRNP gene. Familial CJD is the commonest amongst three, whereas the other two are rare.

Acquired Human Prion Disease

Kuru: Kuru was the first identified acquired human prion disease. Kuru is restricted to the Fore ethnic group in Papua New Guinea of Eastern Highlands. The term Kuru in the Fore language of Papua New Guinea means "to tremble from fever". The disease was acquired due to the practice of ritualistic endocannibalism which involved eating of brain and other viscera of a dead relative to pay respect. Infection occurs when the tissues of dead person are contaminated with prion proteins.

Tremor and ataxia are the first symptoms from which the name of disease was derived. Later manifestations are chorea, athetosis and

Table 22.2: Human and animal prion diseases		
Type	*Host*	*Disease*
Sporadic	Human	Sporadic Creutzfeldt-Jakob disease (CJD)
	Human	Variably protease sensitive prionopathy
Heritable	Human	Gerstmann-Straussler-Scheinker syndrome
	Human	Familial CJD
	Human	Fatal familial insomnia
Infectious	Human	Variant CJD
	Human	Kuru
	Human	Iatrogenic CJD
	Sheep, goat	Scrapie
	Cattle	Bovine spongiform encephalopathy
	Mink	Transmissible mink encephalopathy
	Mule deer, elk	Chronic wasting disease
	Cats	Feline spongiform encephalopathy

myoclonus. Death usually occurs within 1–2 years from disease onset. Disease is more common among women and children. Disease has virtually disappeared after the ritual practice of cannibalism was prohibited. Heterozygosity in codon 127 and 129 in PRNP gene, a marker of prion resistance, has been found among the elderly survivors of kuru epidemic.

Variant Creutzfeldt-Jakob disease: Variant CJD is the only zoonotic human prion disease. First case of vCJD was seen during 1994–95 in United Kingdom (UK). The disease was transmitted from cattle with bovine spongiform encephalopathy (also called "mad cow disease").

Origin of vCJD: Cattle in UK were given meat-bone meal prepared from sheep carcasses and offal. When this meal contained prion protein contaminated products from scrapie infected sheep the scrapie prion crossed the species barrier and got established in cattle. This led to development of bovine spongiform encephalopathy (BSE) in cattle which was also called mad cow disease. More and more diseased cattle were slaughtered and provided to produce meat-bone meal. When food products prepared from infected cattle were taken by humans it led to development of prion disease in them which was named variant CJD.

Peak epidemic of BSE occurred in 1992–93 when 37,000 BSE cases were diagnosed in cattle only in a single year. Maximum number of vCJD were detected during 1999–2000 in UK, 6–7 years after the peak BSE outbreak in cattle. Export of cattle feed from UK spread the disease to several other European countries. More than 200 vCJD cases were reported by 2015. Majority of cases were from UK and France. Strict measures were taken to prohibit the use of meat, offal and cattle products in cattle feed, restrictions on animal movement and isolation and euthanasia of cases led to decline in cases. Apart from European countries, vCJD has also been reported from US, Canada, Saudi Arabia, Japan and Taiwan. All these cases possibly had acquired the infection during their stay in UK or golf countries. A total of three cases of vCJD associated with transfusion of blood products also have been documented.

Variant CJD is so named because it differs from classical CJD in many respects. Mean age in vCJD is 28 years. Clinical presentation starts with prodrome of psychiatric and behavioral symptoms which continues for about 6 months before onset of typical neurologic symptoms. vCJD has a longer median period of disease duration of about 14.5 months. It also differs from other human prion diseases in having PrP^{SC} in lymphoreticular tissue in addition to central nervous system. The differences between classical and variant CJD are shown in Table 22.3.

Iatrogenic prion diseases: Iatrogenic CJD can be caused by transplantation or use of prion contaminated neurological tissues for various purpose, e.g. dura mater, human growth hormone, gonadotropin hormone. Three cases of blood transfusion associated vCJD have been reported.

DIAGNOSIS OF PRION DISEASE

Definitive diagnosis: Demonstration of typical histopathology in brain tissue and also in tonsil in vCJD: Spongiform vacuolation, proliferation of astrocytes and microglial cells and neuronal loss with lack of inflammatory response or detection of pathological form of prion proteins in brain tissue by immunohistochemistry or immunoblot assay are the definitive diagnosis of prion disease. However, the disadvantage is collection of brain biopsy which is rarely done antemortem. Thus, the definitive diagnosis is almost limited as post-mortem diagnosis.

Suggestive diagnosis: The following findings are highly sensitive but can be seen in other neuronal disorders.

Magnetic resonance imaging (MRI): High signal in caudate, putamen and cortex is 80%

Table 22.3: Differences between sporadic and variant Creutzfeldt-Jakob disease

Parameters	Sporadic CJD	Variant CJD
Source of infection	Spontaneous	Zoonotic: Consumption of beef products contaminated with BSE
Mean age of onset	Seventh decade	Third decade (28 years)
Disease duration	Rapid disease course Mean age of survival: 6 months	Mean duration: 14 months
Molecular mechanism	Seen in MM and VV homozygous, and MV heterozygous	Only in methionine homozygosity (MM) at codon 129 (with rare exceptions)
Clinical features	Rapid development of neurological symptoms: Cognitive, ataxia, myoclonous	Psychiatry and behavioral prodrome before development of neurologic symptoms
Pathology	Variable features	Florid plaques with surrounding spongiform halo
Location of proteinase resistant proteins	Brain	Brain and lymphoreticular tissue (tonsil)
CSF 14-3-3 protein assay	80–90% positive	50%
EEG	Periodic sharp waves	Not seen
MRI	DWI: High signal in caudate, putamen and cortex	**Pulvinar sign:** Symmetrical hyper-intensity in posterior thalamus compared to anterior putamen

CJD: Creutzfeldt-Jakob disease; BSE: Bovine spongiform encephalopathies; M: Methionine; V: Valine; MRI: Magnetic resonance imaging; DWI: Diffusion weighted image

sensitive and highly specific in sporadic CJD. **Pulvinar sign** (symmetric hyperintensity of posterior thalamus in comparison to anterior putamen) is characteristic of vCJD.

Electroencephalogram (EEG): Characteristic periodic triphasic sharp wave complex is characteristic of sCJD but also seen in other conditions. However, not a feature of vCJD where it may be seen only rarely in terminal cases.

CSF: Increased level of 14-3-3 protein band in CSF is more commonly found in sCJD whereas it is positive only in 50% vCJD cases.

Newer non-invasive methods

Ultrasensitive ELISA: This is used for detection of prion protein in blood sample. It can detect very low concentration of protein (10^{-10} dilution). However, helpful only in vCJD and not sCJD or other prion diseases.

Protein misfolding cyclic amplification (PMCA): The test is *in vitro* amplification of prion protein based on the principle of prion protein replication mechanism.

Sample containing normal prion protein (PrP^c) is mixed with test sample. Amplification will occur in test sample containing pathological form of prion protein and protein aggregates will be formed. These protease resistant prion proteins will be detected by immunoblotting after proteinase K treatment.

Conventionally the detection methods of these protein aggregates are immunoblot method or ELISA which are less sensitive. Detection by surround optical fiber immunoassay (SOFIA) has been shown to be much more sensitive. However, the major disadvantage is use of brain homogenate as source of normal prion protein.

Real-time quaking-induced conversion (RT-QuIC): Principle of the test is same as PMCA.

However, the test is easier than PMCA. It is carried out in 96 well plate, detection of prion aggregates is done by fluorescent plate reader and uses recombinant PrP^c instead of brain homogenate.

Sensitivity, and specificity of CSF sample are 77–97% and 99–100%, respectively.

Nasal/olfactory mucosa: 97% sensitivity and 100% specificity.

Sterilization of prion contaminated instruments: Figure 22.2 describes the WHO recommended sterilization method for prion contaminated instruments.

Surfaces and reusable heat sensitive contaminated items are decontaminated by soaking in 2N NaOH for 1 hour.

All prion contaminate disposable materials, instruments and wastes should be disposed by incineration.

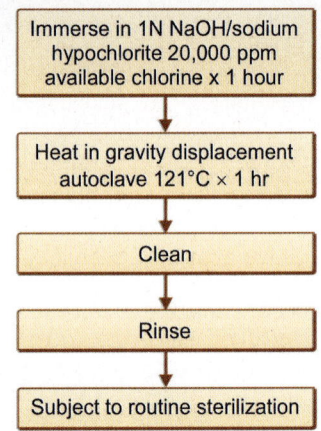

Fig. 22.2: WHO recommended sterilization method for prion contaminated instruments

Bibliography

1. Das AS, Zou WQ. Prions: Beyond a single protein. Clin Microbiol Rev 2016 Jul;29(3):633–58.
2. Diack AB, Head MW, McCutcheon S, et al. Variant CJD. 18 years of research and surveillance. Prion 2014;8(4):286–95.
3. Geschwind MD. Prion diseases. Continuum (Minneap Minn). 2015 Dec; 21 (6 Neuro-infectious Disease):1612–38.
4. https://www.cdc.gov/prions/cjd/infection-control.html
5. Mackenzie G, Will R. Creutzfeldt-Jakob disease: Recent developments. F1000 Res 2017 Nov 27;6:2053.
6. Prusiner SB. Prions. Proc Natl Acad Sci USA. 1998 Nov 10;95(23):13363–83.
7. Tee BL, Longoria Ibarrola EM, Geschwind MD. Prion Diseases. Neurol Clin 2018 Nov;36(4):865–97.
8. Zanusso G, Monaco S, Pocchiari M, Caughey B. Advanced tests for early and accurate diagnosis of Creutzfeldt-Jakob disease. Nat Rev Neurol. 2016 Jun;12(6):325–33.

Chapter

23

Yellow Fever

Yellow fever (YF), a mosquito-borne potentially fatal acute viral hemorrhagic disease, is caused by the agent yellow fever virus (YFV). The disease is endemic in tropical parts of Africa and South America and is mostly limited to this geographical part of the world. It is believed that the virus originated in Africa around 3000 years ago and was introduced into America during the slave trade which then persisted there due to favorable ecological conditions.

Yellow fever is associated with the history of America both politically and developmentally. Repeated epidemics of severe YF with high mortality have ravaged Europeans in New World including the troop of Napoleon Bonaparte. Resurgence of YF during 2016–2018 in various places of America and Africa with epidemics of high mortality and exportation to non-endemic but potentially favorable countries have renewed interest in YF control.

VIRUS

The agent yellow fever virus (YFV) is the prototype virus of the family Flaviviridae and genus Flavivirus. The name of the virus family, Flaviviridae, has been derived from the term flavus which means yellow in Latin. The name "Yellow fever" is due to the manifestation of jaundice which causes yellow discoloration of the infected individuals.

Yellow fever virus (YFV) is an enveloped, positive sense, single-stranded RNA virus. The genome is near 11 kb in length and codes for 10 proteins which include 3 structural and 7 non-structural proteins. The genes coding for proteins are flanked by untranslated regions at 5' and 3'end. The 3 structural proteins are capsid (C), pre-membrane (prM), and envelop (E) and 7 non-structural (NS) proteins are NS1, NS2A, NS2B, NS3, NS4A, NS4B and NS5.

Currently, YFV has single serotype and seven genotypes—two East African, two West African, one central/South African and two South American. Genotypes are based on nucleotide sequence variation of >9% in prM, E and 3'UTR.

EPIDEMIOLOGY

Transmission cycle: YFV infects human and non-human primates. Transmission of YFV occurs through three types of transmission cycle (Fig. 23.1).

I. Sylvatic or jungle cycle: In nature, the virus is maintained by sylvatic cycle. In this, YFV is transmitted **between non-human primates** (monkeys) which are the primary reservoir of YF and sylvatic mosquito species that are found in the forest canopy. The mosquito responsible is predominantly *Aedes* species (*Aedes africanus*) in Africa and Haemagogus species in the Americas. This particularly occurs in the tropical rain forests. Occasionally, humans intruding into the forest acquire the infection when bitten by the infected mosquitoes.

II. Savannah or intermediate cycle: This cycle exists in Africa and is responsible for outbreaks

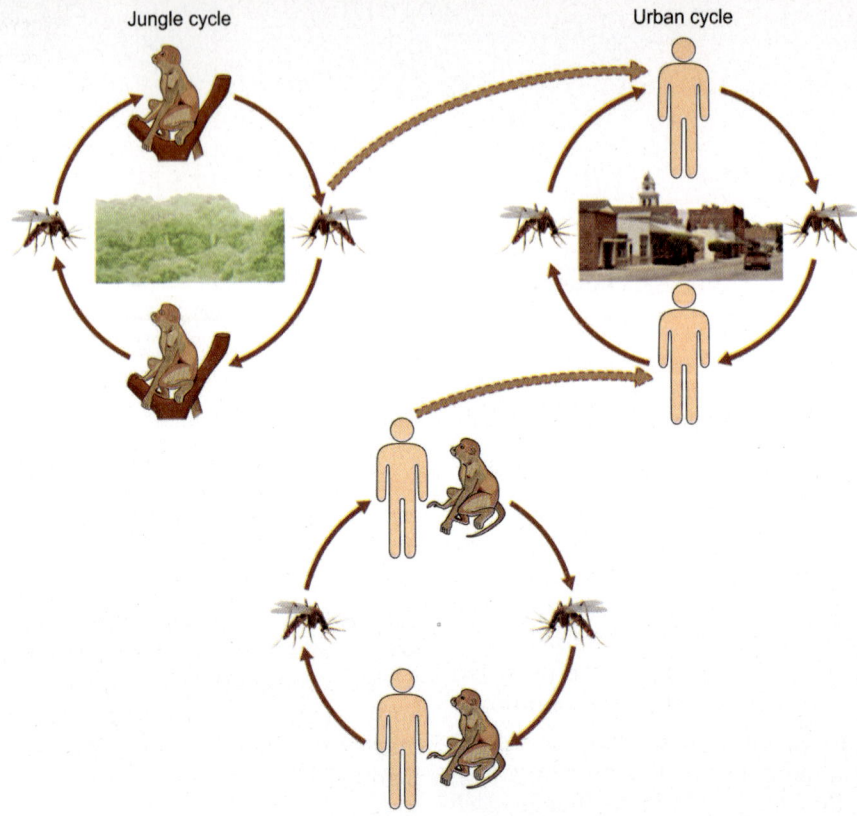

Jungle cycle

Urban cycle

Savannah or intermediate cycle

Fig. 23.1: Transmission cycle of yellow fever virus

in Africa. It involves transmission of virus from infected **mosquitoes to humans living or working in the border areas of jungle**. The virus can be transmitted from monkeys to human or human-to-human by infected mosquitoes. Several different Aedes mosquito species including some peri- or semi-domestic *Aedes* species are involved in this. These mosquitoes breed both in domestic and wild areas and can infect both monkeys and humans. Villages or small towns are commonly affected which are called **emergence zone**. When infected persons from these small towns/villages travel to urban areas it can lead to large outbreaks.

III. Urban cycle: It involves **human-to-human transmission of YFV by urban or domestic mosquito (*Aedes aegypti*)** as the major vector.

The virus is brought to the human dwelling by a viremic man who has got the infection from jungle or savannah. This can lead to large outbreaks when the infected individual enters to a heavily populated locality who have no prior immunity to YFV and having high density of *Aedes aegypti*.

Disease burden: As per WHO, currently 47 countries are endemic to yellow fever. This consists of 34 African and 13 South and Central American countries. As per the estimation carried out by the Yellow Fever Disease Burden Expert Committee under the leadership of World Health Organization, the burden of severe yellow fever cases in Africa during 2013 was 840,000–170,000 and 29,000 to 60,000 deaths each year.

Changing trend in YF transmission: The basic pattern of transmission remains the same. However, the shift is more towards human-to-human transmission. This is mainly due to direct channel from sylvatic to urban cycle, worldwide resurgence of Aedes species that is the primary vector of urban cycle, large population movement, urbanization and deforestation. The concern is more due to vector Aedes particularly *Aedes aegypti* which is also the major vector of many other diseases like dengue, chikungunya, and recently emerged zika, the diseases that are prevalent in Asia, world's most populated region. The recent introduction of YFV into China during 2016 YF outbreak in Angola has raised the concern of YF as a global threat.

Yellow fever outbreaks: In history, the outbreaks of YF in Europe and America due to imported cases are more famous than YF outbreaks in endemic countries. Several major outbreaks of YF have occurred during seventeenth and eighteenth century in East coast of United States including New York, Boston and Charleston. In 1700, YF spread to Europe due to massive slave trade. In 1730, the first described epidemic of YF had occurred in Spain resulting in more than 2000 deaths. Yellow fever has been known to shape the history of early America. During 1793–1798, British troop suffered from near 100,000 deaths due to YF with a case fatality near 70% during their West Indies and Haiti possession. Then the largest ever documented yellow fever outbreak occurred during 1802 when 20,000 to 50,000 French soldiers sent by Napoleon to suppress slave revolt in Haiti got ravaged by the YF virus and was reduced to a few thousands. In both the situations, massive epidemic of yellow fever with high fatality led the Britishers and Napoleon to withdraw their troops from America.

During 2015–2018, several large urban YF outbreaks have occurred in Africa and South America including Angola, Democratic Republic of Congo (DRC), Uganda in Africa and Brazil in South America. Thousands of people were affected by YF in these outbreaks with high death rates.

The unprecedented outbreak in Angola led to (i) spread of outbreak to neighboring countries (DRC), (ii) generation of local transmission, (iii) exportations of YF cases to non-endemic countries including China. This created the cause of global concern as Asian countries are naïve to YFV (possibly due to cross-protection by dengue virus), thickly populated and the vector mosquito (Aedes) is prevalent, (iv) need of YF vaccine stock pile.

It was then realized that YF outbreaks could lead to public health emergencies of international concern. The increased threats of YF outbreaks with international spread led to the development of **"Eliminate Yellow Fever Epidemics" (EYE) strategy** by WHO involving more than 50 partners. The global EYE strategy is based on three strategic objectives:

1. Protect at risk population
2. Prevent international spread
3. Contain outbreaks rapidly.

CLINICAL FEATURES

Yellow fever causes a wide range of symptoms varying from no symptoms to mild febrile illness to severe disease with multiorgan involvement with a high fatality rate.

Incubation period of YF is 3–6 days from the day of mosquito inoculation to development of symptoms with a median of 4 days. Clinical manifestations of YF has been classically divided into three different phases, namely; infection, remission and intoxication (Fig. 23.2).

Period of infection: This includes the onset of symptoms which is typically abrupt and presents with fever, headache, malaise, vomiting. This period corresponds with peak viremia during which YFV can be detected in blood by virus isolation or molecular methods. During this period, the person is infectious to mosquitoes. It lasts for about 3–6 days. Majority

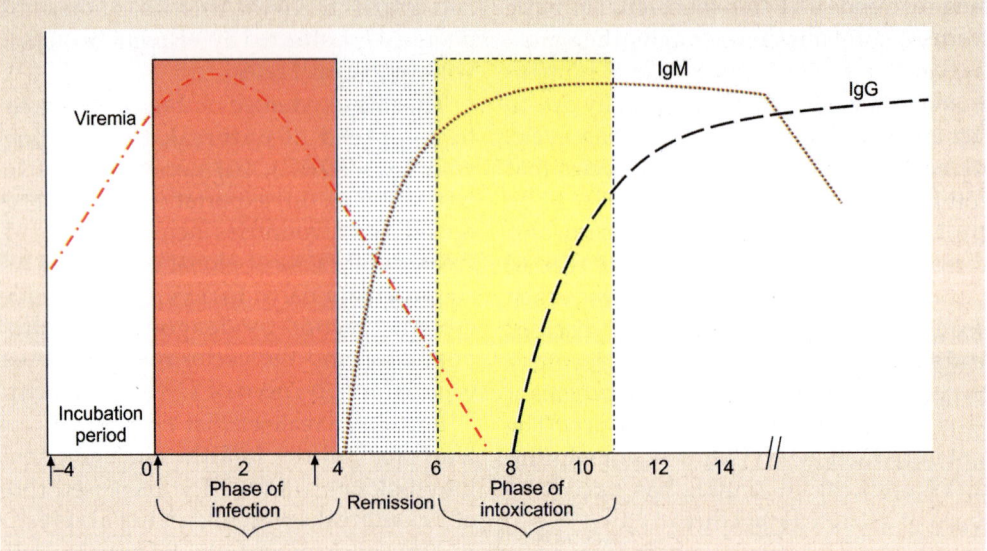

Fig. 23.2: Time course of yellow fever virus infection

of the infected persons recover within one week of initial illness.

Period of remission: Some patients may experience a short period of remission following the infection phase with resolution of fever and improvement in symptoms and signs. Most of the patients recover during this phase.

Period of intoxication: A subset of 15% YFV infected patients develop severe form of disease called phase of intoxication. This phase is marked by return of fever with severe epigastric pain, nausea, vomiting, jaundice (causing yellow skin from which the name yellow fever is given). Hepatic dysfunction, renal failure and bleeding diathesis leading to the manifestations of hemorrhagic fever are the prominent clinical features of severe YF disease. Hemorrhagic manifestations occur due to consumptive coagulopathy and reduce production of clotting factors. Hepatorenal failure along with circulatory collapse leads to metabolic disturbances, coma and death. About 20–50% of patients who enters the period of intoxication die within 7–10 days.

This phase is characterized by absence of viremia and detectable anti-YF antibodies.

PATHOLOGY

Pathological changes in various organs have been described during the period of intoxication. Liver, kidney, and heart are enlarged and edematous. In addition to this, liver is icteric.

The classical histopathology of liver in YF is:
- Eosinophilic degeneration of hepatocytes and Kupffer cells resulting in the formation of **Councilman bodies** and intranuclear eosinophilic granulations due to death of hepatocytes.
- Lytic necrosis of hepatocytes in the midzone of the liver lobule.
- Sparing of cells surrounding the central vein and portal triad.
- No disruption of reticular architecture.

The histological changes in kidney and spleen are mostly eosinophilic degeneration, apoptotic changes and necrosis. Viral antigen can be demonstrated in these tissues by immunohistochemistry.

IMMUNE RESPONSE

Following natural infection, IgM antibody appears during the first week of infection, peak titer reaches during second week after which the titer decreases. However, the titer remains detectable for years. Hence, in YFV infection, positive IgM does not necessarily indicate recent infection.

YF specific IgG is neutralizing in nature and, therefore, indicates protection. It appears towards the end of the first week and persists almost lifelong.

LAB DIAGNOSIS

Laboratory confirmation is required in all suspected YF cases. The detection of YF infection can be made from blood and tissue samples by isolation of virus, viral RNA detection and antigen detection and serology from blood.

Yellow fever virus being a **level 3 pathogen**, the specimens should be handled in biosafety level-3 (BSL-3). As an additional precaution, it is recommended to handle the samples from patients with hemorrhagic manifestations in BSL-4 lab. Laboratory staff should be prior vaccinated for YF. Because of these limitations, the laboratory testing is performed in national or regional reference laboratory and as the disease is more common in rural areas, laboratory confirmation facility is limited in both epidemic and endemic settings.

Isolation of virus: Like many other arboviruses, YFV isolation can be done in cell lines, mosquito inoculation and suckling mice intracerebral inoculation. Isolation of virus can be done from blood or infected tissue samples. Blood should be collected within first few days of illness.

Various kidney cell lines (Vero, BHK, LLC-Mk2) and mosquito cell lines (C6/36 *Aedes albopictus* API-61) are used. Infected cell lines with cytopathic effect (CPE) or plaque formation can be confirmed by antigen detection or RT-PCR.

Virus isolation, however, is time consuming, labour intensive and also less sensitive than RNA detection by molecular techniques. In addition, it may require additional biosafety level. However, this is useful for characterization of the virus strain.

Serology: Detection of YF virus specific IgM and IgG antibodies can help in confirming the infection.

Confirmation of suspected YF case is made by: (i) Detection of YFV specific IgM antibody in a suspected case without vaccination in previous 30 days, (ii) demonstration of four-fold rise in YFV specific IgG antibody in acute and convalescent sera, (iii) detection of neutralizing antibodies by plaque reduction neutralization test (PRNT).

ELISA and indirect immunofluorescence methods are commonly used methods for detection of IgM and IgG antibodies in serum. However, unlike other flaviviruses, reagents are not widely available commercially. CDC has developed and validated an MAC ELISA test for YF IgM antibody detection.

YF IgM antibody mostly appears during the period of remission and peaks during the period of intoxication and persists for a long time. In YF naïve vaccinated individuals, IgM persists for longer period (years) as compared to vaccinated individual from YF endemic regions.

Other serological methods like complement fixation test (CFT), hemagglutination inhibition test (HAI) and plaque reduction neutralization test (PRNT) can be used to detect fourfold rise in antibody titer. However, due to the technical difficulties, these tests are no more in use for patient diagnosis.

Cross-reactivity with other flaviviruses may lead to false positive reaction. Hence, the sample positive for YF serology should be negative for other flaviviruses in order to be considered as specific.

Yellow fever virus RNA detection: Various molecular techniques have been developed for detection of YFV RNA in clinical samples. This includes RT-PCR, real-time PCR, multiplex PCR and isothermal amplification methods. Several commercial PCR assays are available: Single detection or multiplex PCR.

NS1 antigen detection: ELISA for detection of NS1 antigen has been developed and has been shown to be 80% sensitive and 100% specific.

VACCINE

In 1930, two YF vaccines—17D and French neurotropic vaccines were developed. The later one was produced in mouse brain and was discontinued since 1982 due to its association with high incidence of post-vaccinal encephalitis. Whereas 17D is considered as one of the safest and effective vaccines ever developed for which Max Theiler got Nobel Prize in 1951.

All currently available YF vaccines are live attenuated vaccines derived from 17D lineage. The vaccine was derived from a wild type virus isolated in 1927 in Ghana from a man with febrile illness. The strain is called **"Asibi"**strain which was passaged 176 times in mice and chicken tissues to produce the attenuated 17D vaccine virus. This vaccine virus has two substrains—17DD and 17D-204. Both the substrains share 99.9% of sequence homology. The vaccine virus 17D differs from the original Asibi strain by 20 amino acids. It has substantially reduced viscerotropic and neurotropic properties and non-transmissible by mosquitoes.

WHO recommendations for YF 17D vaccine:

- The recommended immunizing dose should not contain less than 3.0 log 10 international unit.
- Storage temperature is 2–8°C, present in lyophilised form, should be reconstituted using sterile diluent and discarded after 1–6 hours.
- Given subcutaneously or intramuscularly.
- Age group: >9 months.
- Should be given to all children at 9–12 months of age in endemic countries and to all unvaccinated travellers going to or coming from the endemic or at-risk countries.
- Single dose is recommended. (The re-commendation of booster dose after 10 years has been stopped since June 2016 except in high-risk groups such as pregnant women and immunocompromised individuals, etc. as the vaccine provides lifelong immunity.) However, the policy differs in different countries.
- India has agreed to WHO recommendation of single dose vaccination policy for international travellers from endemic countries since August 2017.

Immunogenicity of vaccine: Yellow fever 17D vaccine induces rapid and effective immune response. Protective level of neutralizing antibodies develops in 80% of vaccine within 10 days, in 90% by 2 weeks and in 100% by 4 weeks post-vaccination.

Adverse events: Mild symptoms including headache, myalgia with local site side effects are seen in around 25%. Serious adverse events can be:

   i. Anaphylactic reaction: Reported in about 0.8 in 100,000 vaccinations.

   ii. YF associated neurologic diseases leading to symptoms like Guillain-Barre syndrome (GBS), reported in 0.25–0.8 in 100,000 vaccinations.

   iii. YF associated viscerotropic diseases leading to multi-organ dysfunction, reported in 0.25–0.4 in 100,000 vaccinations.

Bibliography

1. Collins ND, Barrett AD. Live attenuated yellow fever 17D vaccine: A legacy vaccine still controlling outbreaks in modern day. Curr Infect Dis Rep 2017 Mar;19(3):14.

2. Domingo C, Charrel RN, Schmidt-Chanasit J, Zeller H, Reusken C. Yellow fever in the diagnostics laboratory. Emerg Microbes Infect 2018 Jul 12;7(1):129.

3. https://www.cdc.gov/yellowfever/transmission/index.html

4. https://www.who.int/csr/disease/yellowfev/YellowFeverBurdenEstimation_Summary2013.pdf

5. https://www.who.int/news-room/fact-sheets/detail/yellow-fever.

6. Vaccines and vaccination against yellow fever. WHO position paper.June 2013. Wkly Epidemiol Rec. 2013 Jul 5;88(27):269–83.

7. Waggoner JJ, Rojas A, Pinsky BA. Yellow fever virus: Diagnostics for a persistent arboviral threat. J Clin Microbiol 2018 Sep 25;56(10). pii:e00827–18.

24

Severe Acute Respiratory Syndrome Coronavirus 2 and COVID-19

Severe acute respiratory syndrome coronavirus 2, the causative agent of COVID-19 is the worst ever pandemic the mankind has ever faced with an unprecedented number of deaths of more than 40 lakh since beginning of the pandemic which is still continuing. COVID-19 pandemic brought the world to a halt. The unprecedentedly high number of deaths across several countries led the entire world under lockdown. For the first time in the history of mankind, across the continents, educational institutes, offices, shopping complexes, everything got closed excepting the health care system. World came to a standstill.

In December 2019, a cluster of cases of severe acute pneumonia due to unknown etiology was reported from the Wuhan city of China's Hubei province. The cases started spreading to the near by areas. As the causative agent was not known, the cases were labelled as "pneumonia of unknown etiology". The causative agent was later found to be a novel Betacoronavirus which was named by World Health Organization (WHO) as 2019-novel coronavirus (2019nCoV). As per the International Health Regulations, the outbreak was declared as a Public Health Emergency of International Concern (PHEIC) by WHO on 30th January 2020. The virus was renamed by Coronavirus Study Group of International Committee on Taxonomy of Viruses (ICTV) on 11 February 2020 as "Severe acute respiratory syndrome coronavirus 2" or SARS-CoV-2 due to its phylogenetic similarity with severe acute respiratory syndrome coronavirus and

the disease caused by SARS-CoV-2 was named as Coronavirus Disease 2019 or COVID-19. The previously called SARS-CoV is now called SARS-CoV-1. Within two months, the disease spread to all over China and then to several other countries affecting thousands of people. On **11 March 2020, WHO declared the COVID-19 as pandemic**. Since then, the COVID-19 pandemic is ongoing throughout the world in an unprecedented manner affecting more than 19 crore people leading to more than 40 lakh deaths within last one and a half year.

TAXONOMY

SARS-CoV-2 belong to the order Nidovirales, family Coronaviridae and subfamily Orthocoronavirinae. The subfamily Orthocoronavirinae has four genera: Alphacoronavirus, Betacoronavirus, Gammacoronavirus and Deltacoronavirus. Of these, SARS-CoV-2 belong to the genus Betacoronavirus along with 4 other human coronaviruses such as SARS-CoV and MERS CoV and HKU1, HKU OC43. The genus Betacoronavirus has five subgenera: Sarbecovirus, Merbecovirus, Hibecovirus, Nibecovirus, and Embecovirus. SARS-CoV-2 belongs to the subgenus Sarbecovirus along with SARS-CoV. The taxonomical position of SARS-CoV-2 is shown in Table 24.1.

ORIGIN OF SARS-CoV-2

SARS-CoV-2 shares less than 90% identity with other members of Betacoronavirus genus unlike other members of the genus and belong

Table 24.1: Taxonomy of SARS-CoV-2	
Order	Nidovirales
Family	Coronaviridae
Subfamily	Orthocoronavirinae
Genus	Betacoronavirus
Subgenus	Sarbecovirus

to a new evolutionary branch. The nucleotide identity between SARS-CoV-2 and SARS-CoV has been found as 80% and only 51.8% with MERS CoV. Phylogenetic analysis revealed SARS-CoV-2 is most closely related to bat SARS-CoV RaTG13 with over all genome sequence homology of 96.2% and spike protein sequence homology of about 98%. The evolutionary analysis indicates that RaTG13 bat related SARS-CoV and SARS-CoV-2 might have evolved from common ancestor. The origin of SARS-CoV-2 is thought to be evolved from bat coronavirus because of the following evidences:

- High sequence homology (96%) between bat coronavirus RaTG13 and SARS-CoV-2.
- Both have come from the same evolutionary branch.
- Both the viruses recognize the same receptor angiotensin-converting enzyme 2 (ACE-2).
- Both viruses have the same ability to infect cells.
- Both the viruses possess the same four residues in their receptor binding motif (RBM) which binds to the ACE receptor.

Besides the high sequence homology with bat coronavirus, SARS-CoV-2 also shows high similarity with Malayan pangolin beta-coronavirus. The epidemiological link of the Wuhan pneumonia cluster with the Huanan city's local live animal and seafood market had raised the possibility of a zoonotic source of this virus. Extensive search for a similar SARS-CoV in all possible animals that were sold in the market during the outbreak revealed that, Malayan pangolin harbor a very similar betacoronavirus with high sequence homology with SARS-CoV-2. This led to a possibility of pangolin as the inter-mediate host for transmission of SARS-CoV-2

from bat to human. However, after extensive evolutionary and epidemiological analysis, it was concluded that Malayan pangolin is a natural host of beta CoV but the role of pangolin as intermediate host could not be substantiated.

EPIDEMIOLOGY

Since the beginning of the COVID-19 pandemic till 25 July 2021, the SARS-CoV-2 virus has infected near 200 million population with an inordinate number of deaths of more than 4 million across the world with an average case fatality ratio of 4%. The pandemic has affected more than 200 countries all over the world. USA, India and Brazil are the worst three affected countries with the highest number of cases and more than 4 lakh deaths in each of these countries. Several European and South American countries have also reported high number of cases with more than 4 million cases and >1 lakh deaths in each of these countries.

India is one of the worst affected countries in the world and has reported the second highest cumulative number of COVID positive cases after USA. More than 30 million cases have so far been reported amounting to >4 lakh deaths by 25 July 2021. Since March 2020, India has experienced two waves of pandemic. The first one lasted from the month of June to November 2020 with a peak during September. The second wave started in March 2021 and took an abrupt surge in 1st week of April which took peak during the month of May with more than 4 lakh cases everyday. The cases of second wave have been decreased to less than 1 lakh daily cases but still continuing. The second wave occurred in a much higher proportion than the first wave. The average number of daily cases during the peak of second wave was more than 4 lakhs as compared to 1 lakh of first wave.

VIRUS

SARS coronavirus 2 is a single-stranded, positive sense, enveloped RNA virus with helical nucleocapsid. Virus particle is spherical

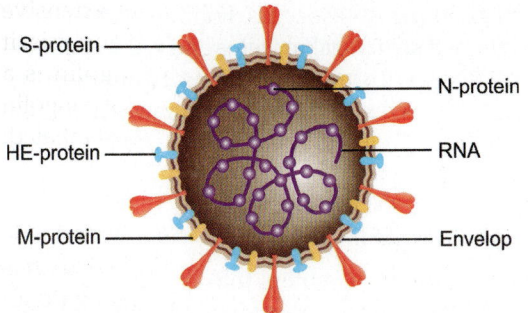

Fig. 24.1: Schematic diagram showing structure of SARS-CoV-2

in shape with approximately 100 nm in diameter. The envelop contains several surface projection proteins which give the corona-like appearance to the virus (Fig. 24.1).

The size of the RNA genome is around 30 kb. The genomic mRNA has a 5′-cap and a 3′-poly (A) tail. The genome contains 14 open reading frames (ORF) of which ORF1a and ORF1ab are the two most important ones, comprise two-thirds of the genome and produce two large polyproteins pp1a and pp1ab, respectively. These polyproteins are cleaved by virally encoded proteases into 16 non-structural proteins (nsp1–nsp16) which combinedly form the replication transcription complex (RTC) and are responsible for virus transcription and replication. Non-structural proteins also play role in the pathogenesis mainly by cytokine expression and suppression of host's innate immune response.

The structural protein's spike (S), envelop (E), membrane (M) and nucleocapsid (N) proteins are encoded by ORF located in 3′ one-third of the genome. They mainly take part in binding and entry of the virus particle into the host cell (Fig. 24.2).

Spike protein: Spike protein (S) is a type I transmembrane glycoprotein. This is located on the outer surface of the virus. Its size is 180–200 kDA and it is consists of an extra-terminal N terminus, transmembrane and intraterminal C terminus. S protein is responsible for the virus-host interaction and binds to the angiotensin-converting enzyme 2 (ACE-2) receptor present on the host cell and facilitates the entry of virus into the host cell. As the protein is present on the surface of virus, it is the major immunogenic protein and induces neutralizing antibody and the candidate protein for therapeutic target and vaccine.

S protein is a clove-shaped trimeric structure. Each monomer consists of two subunits—S1 and S2. S1 is the globular head and the ectodomain, whereas S2 is the stalk, transmembrane and the intracellular component. Receptor binding domain (RBD) is present in S1. RBD constitutes of core domain and receptor binding motif (RBM). It is the RBM which makes the direct contact with the host cell receptor ACE-2. Upon contact of virus ligand with the host cell receptor, cellular protease-transmembrane protease, serine 2 (TMPRSS2) prime the inactivate S protein which gets cleaved to S1 and S2. The fusion domain in S2 gets exposed and mediates the fusion of virus with the host cell membrane and facilitates the entry of virus into the host cell.

Nucleoprotein (N): N protein is the capsid protein. It forms complex with the RNA genome and forms the helical ribonucleoprotein and thereby gives stability to the viral genome. It takes part in virus replication cycle, deregulates the host cell cycle by inhibiting

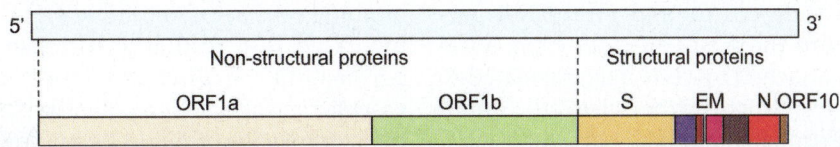

Fig. 24.2: Genome of SARS-CoV-2

S-phase process. It is shown to inhibit the production of interferon and upregulate the cyclooxygenase 2 (COX-2) one of the important proinflammatory mediators.

Envelop (E) protein: E protein plays role in virus morphogenesis and interaction with the host cell.

Membrane (M) protein: It is the most abundant protein and plays important role in virus fusion, morphogenesis and replication. It also helps in the formation of virus envelop, release of virus through budding and transport of nutrients across the cellular membrane.

Host cell receptor: Angiotensin-converting enzyme 2 is a homolog of angiotensin-converting enzyme. It converts ACE-1 to ACE1–9. ACE-2 also acts as the receptor for SARS-CoV and human coronavirus NL63. Cluster of differentiation 147 (CD147) or extracellular matrix metalloproteinase inducer (EMMPRIN) which is also known as Basigin is thought to be another receptor for SARS-CoV-2.

SARS-CoV-2 LIFE CYCLE

The entry of SARS-CoV-2 virus into the host cell occurs through the interaction between the S protein of the virus and the receptor ACE-2 present on the host cell. ACE-2 is distributed mainly in lungs, heart, kidney and alveolar epithelial cell II and gastrointestinal tract. The receptor binding motif (RBM) present on RBD of S1 subunit of S protein recognizes the receptor and is responsible for host and tissue tropism. After attachment to the host cell receptor, S protein gets clipped to S1 and S2 subunits where the latter acts as the viral fusion peptide and mediates the fusion with the host cell membrane resulting in entry of the virus into the host cell. Following this, uncoating occurs and viral RNA gets released into the host cell cytoplasm. Viral RNA gets attached to ribosome, translated to polyproteins which further cleaved to nonstructural proteins that take part in the viral replication. Following translation, transcription of genomic RNA occurs. Following the translation of structural proteins and encapsidation of RNA with nucleoprotein, it buds out by exocytosis through the membrane as a mature virion article which further infects other cells.

PATHOGENESIS

When the virus enters into the host cell, the viral antigens are recognized by the various antigen presenting cells through various pattern recognition receptors. Antigen peptides are presented to CD8+ T cells along with major histocompatibility complex I (MHC). CD+ T cells get activated and develop into effector and memory cells. Effector CD8+ T cells kills the virally infected cells, whereas CD4+ T cells help the B cells to produce neutralizing antibody. Other mechanisms such as activation of natural killer cells and production of proinflammatory cytokines also adapted to restrict the viral infection. As long as the immune response is well-coordinated, it acts to clear the virus. Dysregulated and uncontrolled systemic immune response leads to exaggerated cytokine production and cytokine storm causes collateral immune damage to the host body. Cytokine storm is characterized by the increased production of proinflammatory cytokines, such as IL-1b, IL-2, IL-6, IL-7, IL-8, IL-9, IL-10, and IL-17; granulocyte-macrophage colony stimulating factor (GM-CSF); TNF-α, IFN-γ and chemokines; CCL2, CCL3, CCL5, CXCL8, CXCL9, CXCL10, etc. by immune effector cells. It induces the immune-mediated damage to lungs leading to ARDS, respiratory failure, shock, tissue damage and multiorgan failure and potentially death.

Amongst the cytokines, IL-6, TNF-α, IL-17 play crucial role in immunopathogenesis. IL-6 is produced by various immune and non-immune cells across multiple organ systems. Increased IL-6 results in increase vascular permeability leading to interstitial edema, vascular leakage, and hypotension. It weakens the cytotoxicity of NK cells by downregulating the perforin and granzyme β expression.

TNF-α is another potent multifunctional inflammatory cytokine produced by macrophage/monocyte. Besides its role in inducing fever, augmenting systemic inflammation and activates antimicrobial response by IL-6. However, in high level rather than resisting infection it is known to cause necrosis and accentuates T cell apoptosis. Being a potent inducer of NF-kβ, it also induces the expression of multiple proinflammatory cytokine genes. TNF-α has been implicated in severe immune-mediated pulmonary injury.

IL-17 cytokine is another important cytokine that is produced by Th17 cell. IL-17 plays its role by recruiting monocyte/macrophage and neutrophil to the site of infection and also stimulate the cytokine cascade like IL-6 and IL1-β.

The increase in the level of chemokines stimulates the leukocyte infiltration and aggravates the disease severity and leads to the development of pulmonary centric disease.

SARS-CoV-2 infection induces a robust humoral immune response. Like other viral infection, the antibody profile has a typical pattern of IgM and IgG antibody production. Antibodies produced are mainly against N and S proteins. Seroconversion occurs around 7–14 days of symptom onset and persists for a few weeks after clearance of the virus. IgM antibody persists for about 3 months or 12 weeks post onset of symptoms whereas IgG persists longer and may be protective in nature. The production of antibody helps the host by binding to the virus and thereby inhibits the binding and invading of virus into the host cell. However, once the virus has entered into the cell, they become inaccessible to the antibodies and cannot be neutralized by them. In SARS-CoV-2, antibody to S protein has the ability of neutralizing the virus. The level and duration of protectivity of antibody produced after natural infection is not certain which gets more complicated with appearance of virus variants.

Immunopathology: Diffuse alveolar damage is the hallmark of the ARDS. It has two phases—acute phase and organic phase. Acute phase is characterized by formation of hyaline membrane, interstitial edema, alveolar collapse and pneumocyte desquamation and organic phase is characterized with loosely organized fibrosis and hyperplasia of type II pneumocyte. Organization of fibrosis begins at around 10 days of illness. The main histopathologic features of diffuse alveolar damage are infiltration of mononuclear cell, alveolar fibrin deposit, hyperplasia of type II pneumocytes and thickened alveolar septa.

MODES OF TRANSMISSION

Respiratory transmission: SARS-CoV-2 is transmitted from person-to-person while in close contact that is within 1 meter of distance. Infection occurs through aerosol or droplets carrying the virus either by: (i) Inhalation, or (ii) inoculation to mucosal surface of nose, mouth or conjunctiva; or (iii) by touching the mucosa by hand contaminated with virus containing respiratory droplet or fomites. Transmission mostly occurs from a symptomatic person through sneezing, coughing and talking when the infected droplets are produced and get dispersed in the air.

Asymptomatic person transmission: Transmission from asymptomatic COVID-19 positive individuals is also common as the virus shedding occurs through nasal and pharyngeal mucosa. The infected particles from nose and oral cavity can infect the nearby person. This is also common as it may take 2–14 days to develop symptoms after acquiring the virus (incubation period) and also could be due to many infected individuals have mild symptoms which go unnoticed. During this period, virus replication and shedding occurs which makes the person a potential source of infection but as the person appears otherwise normal, they are not usually considered as an overt source of infection. So, essentially asymptomatic persons are silent transmitter of the infection.

The average size of the droplet particle is 5–10 μm which usually gets deposited within 1 meter area, therefore, transmission usually occurs due to close contact with the infected person within this distance.

Aerosol transmission: Lower size of respiratory particle carrying the virus may remain suspended in the air for a long-time and can get dispersed to a longer distance (>1 meter) possessing the risk of airborne transmission. In general, the risk through air-borne transmission is less excepting crowded places where the possibility cannot be ruled out. This is based on the finding of less concentration of viral RNA from aerosols of various isolation wards, ventilated patients and public places. The chance of aerosol transmission is more in crowded places and closed and poorly ventilated rooms. The risk is higher when the process of aerosolization is involved such as endotracheal intubation, bronchoscopic procedure, nebulization or cardiopulmonary resuscitation, etc.

Fomite transmission: Indirect transmission through fomites is also a possible source of transmission. Viral RNA has been detected from environmental surfaces and fomites from rooms occupied by SARS-CoV-2 infected symptomatic or asymptomatic individuals. However, there is no evidence regarding the infectivity and transmissibility through fomites.

Feco-oral transmission: The potential of feco-oral route of transmission is considered due to the expression of ACE-2 and TMPRESS2 in the gastrointestinal tract epithelium and thereby infection and viral replication in GI tract. SARS-CoV-2 RNA has been detected in stool sample for up to 42 days and usually positive in stool for prolonged period than in the nasopharyngeal swab. Several studies have shown stool RT-PCR positivity in about 30% of COVID-19 positive cases. Viability of virus has been shown by successful culture of

virus from stool sample in Vero cells and detection of viral RNA from toilet surface and door handle of room of a patient whose stool sample was RT-PCR positive. Sewage aerosol had caused an outbreak of SARS-CoV in an apartment in Hong Kong in 2003. Considering the evidences regarding presence and viability of SARS-CoV-2 virus in stool sample, possibility of spread of SARS-CoV-2 virus through similar mechanism also exist.

The potentiality of vertical and sexual route of transmission of SARS-CoV-2 has also been shown, however, so far no convincing evidence have come up to consider these routes as the mode of transmission.

Vertical transmission: COVID-19 positivity has been demonstrated in neonate born to COVID-19 positive mothers. Babies have been found to be positive for COVID-19 IgM antibodies in most of the studies. However, there is no consensus regarding the vertical transmission as many other studies have shown lack of such evidence. Preterm delivery and miscarriage have been reported in COVID-19 positive pregnant ladies. However, so far there is no evidence of fetal malformation due to COVID-19 infection.

Sexual transmission: Some studies have shown the SARS-CoV-2 RNA detection in semen of COVID-19 positive males particularly during the acute phase of infection. However, there is no concrete proof of sexual route of transmission though the viral RNA has been detected in semen.

CLINICAL FEATURES

The incubation period to develop symptoms in COVID-19 ranges from 2–14 days. Average incubation period is 5 days and majority develop symptoms by 10–11 days. Median interval from onset of symptoms to hospital admission is about 7 days. Majority of the infected individuals (80%) develop mild symptoms, whereas 15% develop moderate symptoms and 5% develop critical manifestations with respiratory failure and multiorgan dysfunction syndrome and septic shock. Amongst the hospitalized patients, 15–20% require high dependency or intensive care facility.

COVID-19 infection can occur at any age, but symptomatic and severe manifestations are mostly seen in adult age group. The severity and death among male gender is observed more commonly than females. Children are mostly asymptomatic or pauci-symptomatic, and hospitalization and severe manifestations are rare. Death among <18 years though reported but extremely rare.

COVID-19 infected individuals may remain asymptomatic or may develop symptoms. It has been observed that majority of individuals who are asymptomatic at the time of positive detection, develop some or other mild symptom in next a few days.

The clinical severity in symptomatic patients ranges from mild, moderate to severe and critical. The disease course of COVID-19 severity is shown in Fig. 24.3.

Mild: Fever is the most common symptom present in almost all symptomatic patients. Cough, sore throat, malaise and body ache, fatigue are the other common manifestations. Some proportion of patients develop loss of smell and alteration in taste. Headache, diarrhea also seen in 5–8% patients. Respiratory rate is <24 and normal SpO_2 on room air.

Moderate: Persistence of high fever and cough with shortness of breath, features of pneumonia with development of hypoxemia and corroborative findings on high resolution chest CT. Respiratory rate is between 24 and 30/minute and SpO_2 on room air is 90–93%.

Severe: Fever or suspected respiratory infection, plus one of the following: Respiratory rate >30 breaths/minute, severe respiratory distress with respiratory rate >30 and **SpO_2 <90% in room air**.

Critical: New or worsening respiratory symptoms with development of acute respiratory syndrome (ARDS) requiring invasive mechanical ventilation, along with features of shock, coagulation defects, acute kidney injury, cardiac failure and encephalopathy. Chest imaging showing bilateral opacities, not fully explained by effusions, lobar or lung collapse, or nodules.

Most of the moderately severe patients need hospitalization. The common manifestations among the hospitalized patients are fever, shortness of breath, cough, myalgia and fatigue. The presence of comorbidity acts as the risk factor for development of severe disease. Diabetes, hypertension, chronic kidney disease, cardiovascular disease, chronic liver disease and malignancy are the common comorbid conditions.

COVID-19 in children: Majority of the infected children remain asymptomatic or develop

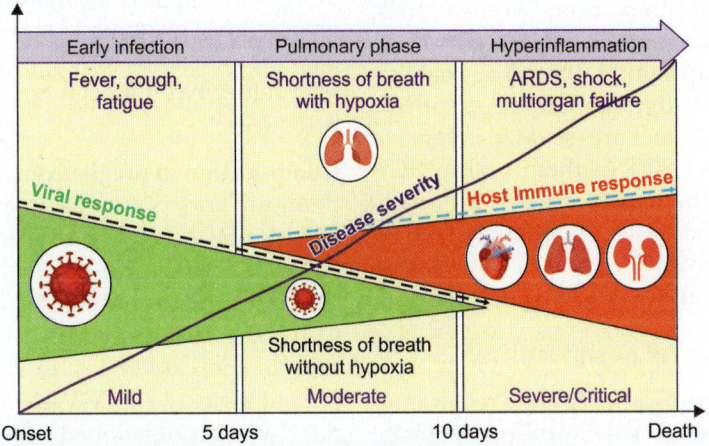

Fig. 24.3: Time course of COVID-19 disease

only mild symptoms. Fever and cough are the two most common symptoms in children and occur in about 50% of cases. COVID-19 positive children account for <1.5% of all COVID-19 related hospitalizations. Among the hospitalized children, about 30–70% have underlying comorbid conditions, such as immunosuppressive conditions, chronic pulmonary or cardiac diseases, or neurologic disease. In children, a newly recognized spectrum of disease known as **multiorgan inflammatory syndrome in children (MIS-C)** has been associated with COVID-19. The syndrome was noticed when many children were admitted to hospital in Europe and North America with symptoms resembling Kawasaki disease. Later it was named multi-organ inflammatory syndrome in children (MIS-C) or pediatric inflammatory multisystem syndrome **(PIMS)**. The estimated incidence is 1 in 100,000. A confirmed case of MIS-C is defined by increased level of two or more inflammatory markers and virologic evidence with SARS-CoV-2 RNA detection or IgM or IgG antibody against SARS-CoV-2 within 10 days of admission. Suspected case of MIS-C is defined when clinical criteria are fulfilled in presence of history of contact with a known COVID-19 case in last 6 weeks but without virologic evidence. Most of the cases present 2–4 weeks after SARS-CoV-2 infection and are positive for SARS-CoV-2 antibodies. Fever, chill, pain abdomen, rash, myocarditis, vomiting, and diarrhea are the common presenting symptoms. Level of C-reactive protein, lactate dehydrogenase, proBNP, troponin and D-dimer are elevated in most of the cases. About 80% of them require ICU admission of them 10% need mechanical ventilation. Majority of the patients recover with treatment with IVIG and glucocorticoid. Mortality of about 2% has been reported.

COMPLICATIONS OF COVID-19

Respiratory complication: Acute respiratory distress syndrome or ARDS is the predominant manifestation and most common complication of SARS-CoV-2 as lungs is the main target organ. Histopathology of ARDS lungs shows the diffuse alveolar damage, development of hyaline membrane, edema and interstitial infiltration. Damage to lungs occurs due to direct attack of virus to pneumocytes. The development of ARDS increases the mortality rate by 45%.

Cardiovascular involvement: COVID-19 patients have been reported to develop myocarditis, cardiomyopathy and arrhythmia during the disease process. High mortality rate has been found to be in patients with heart failure.

Kidney: Acute kidney injury is one of the significant factors for COVID-19 associated deaths. The epithelial cells of proximal tubular cells and podocytes are rich in ACE-2 receptor and thus are the direct target of SARS-CoV-2. In a meta-analysis study, the incidence of AKI has been found to be >30% in severe and critical cases and >50% in COVID-19 associated deaths.

Pancreas: Both islet cells and exocrine cells of pancreas express the ACE-2 receptor, hence are the direct target of SARS-CoV-2. Pancreatic injury may lead to hyperglycemia and may be responsible for precipitating the ketoacidosis.

Liver: Abnormal level of aminotransferase (AST and ALT) is often observed in up to 50% of COVID-19 patients. Significant increase in level of liver enzymes is more seen in severe patients than mild COVID-19 cases. However, liver injury has not been associated with bad prognosis.

Complication in pregnancy: Increased oxygen demand and physiologic anemia during pregnancy are the potential factors that could exacerbate the severity of COVID-19. The information on clinical implications of COVID-19 during pregnancy is limited to case reports and case series. In studies based on patient series, 14% COVID-19 infected pregnant women developed pre-eclampsia and about 6% required ICU admission.

Secondary Infections

Bacterial infections are the most common secondary infections associated with COVID-19. However, systemic fungal infections in COVID-19 patients such as COVID associated mucormycosis (CAM) and COVID associated pulmonary aspergillosis (CAPA), have raised concern due to high mortality. The mortality rate is about 56% in COVID patients with secondary infection as compared to 10.6% in total COVID admitted patients as reported in a multicentric ICMR study.

COVID *associated mucormycosis*: Mucormycosis is an angioinvasive disease caused by mold fungi of class—Zygomycetes and order—Mucorales. The genus Rhizopus, Mucor, Rhizomucor, Cunninghamella and Absidia are the common causes. This has been associated mostly with moderate-to-severe COVID-19, however, also been reported in mild COVID patients. The overall prevalence of CAM is <3%. The prevalence in ICU patient is higher as compared to patients of general ward. CAM is reported more in male than females.

Multiple factors play role in facilitating the germination of mucorale spores in COVID-19 patients. The environment of low oxygen, hyperglycemia, acidic medium, immunosuppression and high iron level play the major risk factor. Hyperglycemia (high blood glucose) has been attributed as the prime predisposing factor of CAM. This can be present in patients with known diabetic, newly onset hyperglycemia or due to steroid-induced hyperglycemia. SARS-CoV-2 itself has been shown to be diabetogenic by causing

damage to the β cell of pancreas. High blood glucose level leads to the increased expression of endothelial receptor GRP78 resulting in polymorphonuclear dysfunction, decreased chemotaxis and defective intracellular killing. Steroid act as another major risk factor by inducing immunosuppression (particularly by impairing the neutrophil migration and inhibiting the phagolysosome fusion) and inducing hyperglycemia (steroid-induced hyperglycemia). COVID-19 disease process itself causes endothelial damage, thrombosis and lymphopenia which predisposes to secondary infections. Iron is one of the essential growth factors for Mucorales. Mucorales has the ability to acquire iron from the host. During the acidic medium in the host due to metabolic acidosis or diabetic ketoacidosis, free irons are available in the serum which are efficiently taken up by them through siderophores or iron permease which increases the virulence of these molds. The summary of the factors responsible for COVID associated mucormycosis is given in Table 24.2.

Rhinocerebro-orbital is the most common type of mucormycosis in COVID patients followed by pulmonary variety. The spectrum of rhinocerebro-orbital involves sinonasal disease to limited rhino-orbital to rhino-orbital-cerebral disease. The involvement of orbital and cranial nerves are ominous signs. The tissue necrosis leading to black necrotic lesion, eschar or black discharge are the clinical hallmark of disease. The diagnosis is confirmed by demonstration of fungal element **(broad aseptate hyphal structures)** from tissue samples

Table 24.2: Risk factors for COVID-19 associated mucormycosis

	Factors responsible
Hyperglycemia	Pre-existing diabetic
	Steroid induced hyperglycemia
	SARS-CoV-2 causing pancreatic islet cell injury
Immunosuppression	COVID-19 disease: By causing lymphopenia
	Diabetes
	Steroid-induced immune suppression
	IL-6 inhibitor: Tocilizumab

by KOH or calcofluor fluorescent method and histopathology or by culture and molecular methods. **Liposomal amphotericin B is the drug of choice.** Isavuconazole and posaconazole are recommended as salvage therapy.

COVID associated pulmonary aspergillosis (CAPA) is another important superinfection which contributes to increased mortality due to COVID-19. Immunosuppression particularly decrease in T cell population and defective function is the major predisposing factor for CAPA or invasive fungal infection. The confirmation of diagnosis is often difficult as bronchoscopic sample is the most relevant specimen which is not usually recommended in a COVID-19 positive patients due to risk of transmission to health care workers. The diagnosis, therefore depends mostly on radiological features, demonstration of aspergillus-like fungal elements in sputum, tracheal aspirate and serum galactomannan and aspergillus PCR.

Secondary bacterial infections: Secondary bacterial infections in COVID patients have been reported mostly due to gram-negative bacilli. The overall incidence in admitted patients is less than 5% as reported in a multicentric ICMR network study in India. Multi-drug resistant *Klebsiella pneumoniae* and *Acinetobacter baumannii* are the predominant pathogens reported.

SARS-CoV-2 VARIANTS

A variant has one or more mutations that differentiate it from other variants in circulation. They have demonstrably different phenotype difference in terms of antigenicity, transmissibility, and virulence. Center for Disease Prevention and Control, (CDC) Atlanta has classified the SARS-CoV-2 variants into three categories—variants of interest, variants of concern and **variants of high consequences**. World Health Organization (WHO) has classified into variants of interest and variants of concern.

Variants of interest (VOI): Variants of interest of SARS-CoV-2 has been defined by having the following two characters:

1. SARS-CoV-2 with specific genetic markers that are predicted to affect the virus characteristics in terms of transmission, diagnostics, disease severity, immune, diagnostic or therapeutic escape.
2. Identified as a cause of significant community transmission or multiple COVID-19 clusters, in multiple countries with increasing relative prevalence alongside increasing number of cases over time or other apparent epidemiological impacts to suggest an emerging risk to global public health.

Examples of VOI is shown in Table 24.3.

Variants of concern (VOC): Variants of concern of SARS-CoV-2 has been defined by having the following characters in addition to the attributes of VOI:

- Evidence of reduced effectiveness to treatment.
- Evidence of reduced effectiveness to vaccine.
- Evidence of significant decreased neutralization by antibodies generated during previous infection.
- Evidence of increased transmissibility.
- Evidence of increased disease severity.
- Evidence of escaping diagnostic detection.

The example of VOC is shown in Table 24.4.

The properties imparted by different mutations in variants are given in Table 24.5.

Table 24.3: Variants of interest			
WHO label of variant	Pango lineage	Earliest documented sample	Date of designation
Eta	B.1.525	Multiple countries, December 2020	17 March 2021
Iota	B.1.526	USA, November 2020	24 March 2021
Kappa	B.1.617.1	India, October 2020	4 April 2021
Lambda	C.37	Peru, December 2020	14 June 2021

Table 24.4: Variants of concern

WHO label of variant	Pango lineage*	Month of first detection	Country first detected	Important spike mutations
Alpha	B.1.1.7	Sep 2020	UK	D614G,N501Y
Beta	B.1.351	May 2020	South Africa	D614G, N501Y, E484K, K417N
Gamma	B.1.1.28.1	Nov 2020	Brazil (P1)	D614G, N501Y, E484K
Delta	B1.617	Oct 2020	India	D614G, E484Q, L452R, P681R

*Pango lineage: Phylogenetic Assignment of Named Global Outbreak. This is a software tool developed by Laboratory of Andrew Rambaut and Center of Genomic Pathogen Surveillance (COVID-19 Genomic UK Consortium).

Table 24.5: Properties of important mutations

Mutation	Characteristics
D614G: Mutation in spike protein at position 614 the amino acid aspartic acid is replaced with glycine	• Mutation in spike protein • First detected in March 2020 in Italy • Spread all over the globe and became the most dominant strain in two months • More transmissible: Increased infectivity in airway cells, higher replication, higher shedding of virus • Not more severe: No higher replication in lungs is shown • No immune escape: G614 got neutralized by antibody developed due to infection with D614
N501Y: Substitution of asparagine to tyrosine at position 501 of spike protein	• Improved the affinity of the viral spike protein for cellular receptors hACE • Increases replication in upper airway cells and more shedding • Increases transmissibility • It is the major spike determinant of enhanced transmission of the UK variant, also present in the South African and Brazilian variants • As all these variants (UK, Brazil and SA) replace their previous strain, indicates the power of N501Y mutation in making the variant much more infectious
L452R: Substitution of leucine to arginine	• Enhances ACE-2 receptor binding ability • Reduces vaccine-stimulated antibodies from attaching to this altered spike protein • Reduces/abolishes neutralizing activity of 14 out of 35 RBD-specific mAbs • >4-fold decrease in antibody titer in convalescent serum • Increases viral infectivity, potentially promoting viral replication
K417N: Substitution of lysine to asparagine	• Presents in beta and delta variants • Resistance to monoclonal antibody therapy casirivimab and imdevimab • Reduced efficacy to vaccine • Immune escape mutant

(contd.)

Table 24.5: Properties of important mutations (contd.)	
Mutation	*Characteristics*
E484K: Substitution of glutamic acid to lysine	• Presents in beta and gamma variants • Known as immune escape mutant • Reduces neutralizing activity by monoclonal and convalescent serum-derived antibodies: 10 to 60 times
E484Q: Substitution of glutamic acid to glutamine	• Presents in delta variant • Enhances ACE-2 receptor binding ability • Reduces efficacy of vaccine-induced antibodies from attaching to this altered spike protein

Characteristics of variants of concern

Alpha variant: Known for its high transmissibility and higher severity.
- Mutations: N501Y, D614G, P681H.
- Highly infectious: 70% increased transmission.
- Higher severity with higher case fatality.
- No impact on susceptibility to monoclonal antibodies.
- Minimal impact on neutralizing antibody by convalescent serum.
- Minimal impact on neutralizing antibody by post-vaccination serum: Almost by all available vaccine.

Beta variant (B.1.351: South Africa): Known mostly for its immune evasion property.
- Mutations: D614G, N501Y, E484K, K417N.
- Higher infectivity: Shows 50% increased transmission.
- Escape neutralization: Significant decrease in susceptibility to monoclonal antibodies.
- Reduced neutralization by convalescent sera and post-vaccination sera.
- Astra Zenecea (ChAdOx1nCoV-19-AZD1222): No protection against mild-mod infection.
- Sputnik: Mod-marked reduction in neutralization.
- Moderna and Pfizer: 6.5–8.6-fold reduction in neutralization.

Gamma variant (P1: Brazil): Known mostly for its immune evasion property
- Mutations: K417T, E484K, N501Y, D614G, H655Y.
- Significant decrease in susceptibility to monoclonal antibodies.

- Reduced neutralization by convalescent sera and post-vaccination sera.

Delta variant: Known for its high transmissibility, higher severity and immune evasion.

Initially B.617 pango lineage was labelled as "Delta variant". It has 3 sub-lineages—B.1.617.1, B.1.617.2, and B.1.617.3. Of the three, 1 and 2 are more transmissible. Currently, **B.1.617.2 is named as "Delta variant".** It is presently spreading all over the globe and has been detected in more than 90 countries and has been declared as National Variant of Concern in UK.

Initially it was popularly called "Double mutant" due to presence of E484Q and L452R mutations which are responsible for their immune evasion and transmissibility properties. However, evidence showed, whether combinedly present or singly present, these mutations show equal fold of immune evasion. Thus, refuting the labelling of double mutant. The presence of mutation P681R in this variant has shown enhanced capacity of cell-cell fusion and syncytium formation indicating that it is responsible for higher pathogenicity.

Delta-AY.1 (delta with K417N): Delta variant containing the mutation K417N is popularly known as **"Delta Plus"** though such labelling is not made by WHO or CDC. The mutation K417N is also present in beta and gamma variants and responsible for immune escape property of the variant. It has two clades AY.1 and AY.2, of which the prior one is internationally distributed. Delta variant with

417N mutation is already detected in 10 countries including India.

Important properties of delta variant:

- Highly transmissible:
  - Transmissibility: 50% higher than aplha and 97% higher than other VOC/VOI
  - Predominant in India, UK and Moscow, Russia
  - Secondary attack rate: Higher than alpha (11 *vs* 8%)
- Impact on severity: Mostly due to P681R mutation
  - More severe
  - Higher viral load in lungs and severe lung damage in experimental animal
  - Fusion and syncytium formation
- Effect of monoclonal antibody (mAb): Reduced neutralization by approved mAb for therapy (bamlanivimab, etesevimab, casirivimab, and imdevimab).
- Effect of convalescent sera: Sixfold less as compared to alpha variant.

Variants of high consequence (VHC): Variants of high consequence have the following criteria in addition to the features of VOC:

- Significant reduction in vaccine effectiveness
- Significant failure to approved therapeutics
- Failure to be detected by the diagnostics
- More number of severe diseases and hospitalizations

So far, no variants have been labelled as Variants of high consequence.

Effect of vaccine: It has been shown that in fully vaccinated general population or vulnerable population the risk of infection is much lower as compared to those who are partially or not vaccinated.

Several studies have shown 3 to 8-fold reduction of neutralization by mRNA vaccines (Moderna and Pfizer) and about twofold reduction in neutralization as compared to B.1 (D614G) and B.1.1.7 (UK strain) by Covaxin and Covishield.

Vaccine efficacy against symptomatic disease due to delta variant is less as compared to that of alpha variant. This is more so after the first dose of vaccine (33% in delta variant vs 51% in alpha variant). Whereas, the efficacy after the second dose is almost similar against both the variants and more than 80% in both when tested with Pfizer and AstraZeneca.

Vaccine efficacy against hospitalization due to either delta or alpha variant is similar and about 75% after the first dose and more than 90% in both variants after the second dose when tested with Pfizer and AstraZeneca.

Durability of vaccine effect: Johnson and Johnson single-shot COVID-19 vaccine shows neutralizing antibody activity against the delta (B.1.617.2) variant. Both humoral and cellular immune responses lasted through at least eight months.

Effect of herd immunity: More than 90% vaccine-induced herd immunity is required to keep a check on delta variant. In general, higher vaccine coverage is required in case of higher transmissibility of the virus variants and poor vaccine efficacy.

Strategies to reduce the emergence of variants:

a. Vaccine coverage should be >90% with complete dose and with minimum gap between first and second doses and should be done rapidly in order to reduce the population of partially/non-vaccinated individuals.

b. In addition to this, care should be taken to reduce the duration of virus replication particularly in immunocompromised individuals where usually there occurs delay in virus clearance due to poor immunity. Prolong replication gives scope for mutation.

c. Use of antibody based therapy such as monoclonal antibodies, plasma therapy where their efficacy is limited or undemonstrated efficacy should be avoided as partially effective measures may facilitate the evolution of variants.

LABORATORY DIAGNOSIS

The SARS-CoV-2 infection is confirmed by detection of virological markers such as viral RNA detection or viral antigen detection in clinical specimens by various laboratory tests. Antibody detection against SARS-CoV-2 virus is a marker of current/recent or past infection depending on the type of antibody detected. Hence, not used as a diagnostic method to detect active COVID-19 infection.

Real-time PCR

Viral RNA detection by real-time polymerase reaction (RT-PCR) is considered as the gold standard and the most commonly used and the recommended test for SARS-CoV-2 infection. Figure 24.4 depicts the viral and serological kinetics of SARS-CoV-2 infection.

Clinical specimens: RT-PCR for diagnosis is done from respiratory samples, such as nasal/ pharyngeal or nasopharyngeal swab, sputum, or bronchoalveolar lavage (BAL). Viral RNA also has been detected in stool, saliva, blood and urine samples. However, these samples are not the recommended clinical specimen for diagnosis because of their poor positivity and inconsistent result. Of the respiratory samples, BAL has shown highest sensitivity (93%) followed by sputum (72%), nasal swab (63%), and pharyngeal swab (32%). Practically, nasopharyngeal or oropharyngeal samples

are most commonly used. Swabs are collected in viral transport medium. All samples need to be transported to the laboratory in cold chain. Samples should be stored at 4°C till tested.

RT-PCR steps: Samples are subjected to lysis for inactivation of the virus followed by nucleic acid extraction, reverse transcription to generate complementary DNA (cDNA) and then amplification of the cDNA for detection of the virus target genes by using target specific primers and probes.

Target genes: Most of the currently available RT-PCR kits for COVID-19 are multiplex RT-PCR having the detection system for more than one target gene in one run. A variety of target genes are used by different manufacturers, mostly consisting of genes of structural proteins like envelop (E), nucleocapsid (N) or spike (S) or non-structural proteins such as RNA-dependent RNA polymerase (RdRp), and ORF1 genes. E gene target is mostly used as the screening gene indicating the positivity for Sarbecovirus, whereas other genes are used as the specific target genes for SARS-CoV-2. The sensitivity of all the target genes

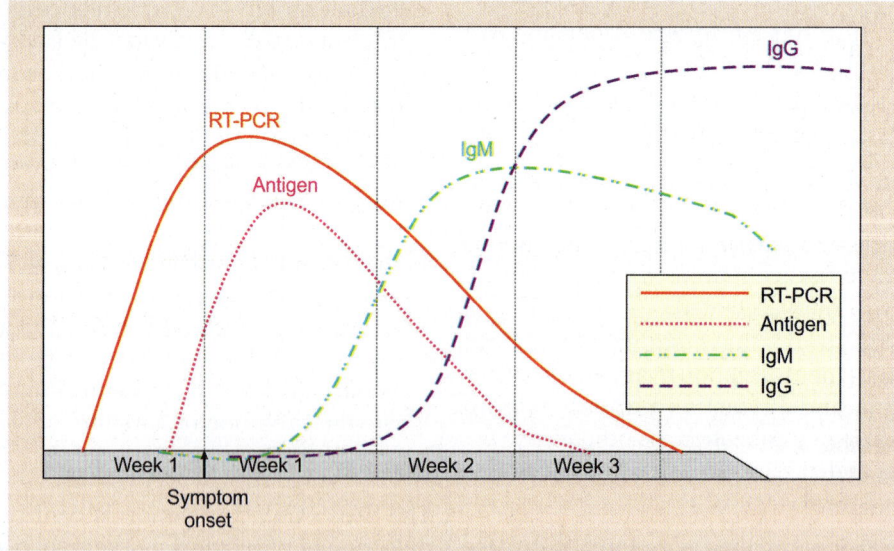

Fig. 24.4: Viral and serological kinetics in SARS-CoV-2

is comparable except the rdRp gene which has shown lower sensitivity. Assay having more than one target gene is more sensitive than the assay having only one target gene detection system.

Cycle threshold: Cycle threshold value is the cycle at which the fluorescence signal crosses the defined background threshold. The cycle threshold (Ct) value less than 40 is clinically reported as PCR positive. However, different commercially available approved RT-PCT kits for COVID-19 have different Ct cut-off for consideration as positive. Indian Council of Medical Research (ICMR) has recommended cut-off of 35 Ct value as positive. This is based on various study results that in samples of Ct>35, virus is non-cultivable indicating the non-viability of the virus and as RT-PCR detects viral RNA which can be positive even in sample containing non-viable virus.

Sensitivity: The clinical sensitivity of various RT-PCR assays is reported from 70–>90%. The analytical sensitivity depicting the limit of detection has been found to be as low as <5 log10 copies/mL. The positivity of RT-PCR is affected by various factors. False negative RT-PCR could be due to low viral load in the patient, too early or too late stage of infection, early viral clearance, viral replication predominantly occurring in lower respiratory tract, RT-PCR and nucleic acid extraction kit variation, type and number of target genes used in RT-PCR assay, type of clinical specimen, quality of specimen, transport and storage system, etc. Variation in any of these factors may lead to false negative RT-PCR result.

Specificity: The specificity of RT-PCR is often 100% as the currently used primer probes have been tested as specific for SARS-CoV-2 virus with no cross-reactivity with other viruses including other human coronaviruses. However, false positive may occur due to reagent contamination or carry over contamination.

Result: The result of RT-PCR can occur as: Positive, negative, inconclusive or invalid.

Positive RT-PCR is often taken as per the cut-off Ct value of the assay used. Both Ct value and the appearance of proper sigmoid amplification curve should be taken into account to label the sample as positive. However, after several studies have shown the non-viability of virus in samples with Ct >35, Ct <35 is presently widely used. The cut-off Ct to consider positive may vary as per the policy of different countries or different health agency.

Negative: Sample showing no amplification or Ct value beyond the defined cut off are considered as negative.

Inconclusive: Samples labelled in this category are basically border line positive samples, where both the genes are not within the defined cut-off of Ct value or the sigmoid curve is not proper. The criteria of inconclusive is different in different RT-PCR kits used.

Invalid: Non-amplification of the in-built internal control in the reagent assay is considered as invalid. This could be due to inadequate sample, presence of PCR inhibitor in clinical specimen, failure of extraction or failure of amplification process.

Clinical implication of COVID-19 RT-PCR results: RT-PCR positivity depends on the viral kinetics. Several studies have shown the lowest Ct value at the symptom onset. However, many studies have shown lower Ct from 5 days before to 10 days after the onset of symptom. Majority become positive by day 3 or day 4. Most of the mild cases become negative by day 8–day 10 of symptom onset. In symptomatic cases, RT-PCR usually become negative during second week of illness. In patients with steroid treatment, viral clearance may take longer duration, so RT-PCR may remain positive for a longer time than average duration. The RT-PCR status in a proportion of COVID-19 positive cases may follow a positive-negative pattern for 3–6 weeks as observed in several studies in hospitalized patients.

In hospital admitted patients after day 10 of illness, the Ct value usually becomes more than 30 which is considered as non-transmissible as virus from those samples have been found to be non-cultivable. Hence, after 10 days of symptom onset and when the patient become afebrile for 3 consecutive days without taking antipyretic, patient is considered as non-infectious irrespective of RT-PCR positivity status as in majority shows high Ct value.

The quality of the conventional RT-PCR test result demands a stringent quality control. Involvement of several steps in this makes the test procedure time consuming and cumbersome. The major drawback of the molecular assay is high turn around time. It needs batch testing. It takes minimum 4–6 hours which may be more depending on the number of samples to be tested in one batch, infrastructure available. Several FDA approved high throughput platforms are available for RT-PCR testing for COVID-19. Cobas 6800 is one such platform which is widely used.

Cobas 6800 and 8800 instruments are fully automated RT-PCR testing platforms. It consists of sample supply module, transfer module, processing module and analytic module. For detection of SARS-CoV-2, the system uses two target genes; ORF1, a non-structural region and E gene for pan Sarbecovirus

detection, a target from a conserved region. The overall agreement with other standard RT-PCR system is reported as 96–98%.

Point of Care Molecular Test

Several other molecular test platforms have come up with rapid turn around time giving the RT-PCR result in 30 minutes to <1 hour. These systems are helpful in following situations:

- Emergency situations such as surgery to rapidly evaluate the patient
- Severe acute respiratory cases (SARI) for rapid evaluation and isolation of patient
- Low throughput laboratories
- Surveillance setting with relative low sample numbers per day.

Several rapid molecular tests have been currently approved for use under Emergency Use Authorization (EUA) by US Food and Drug Administration (FDA): Xpert Xpress SARS-CoV-2 (Cepheid), ID NOW COVID-19 (Abbott), Accula SARS-CoV-2 (Mesa Biotech), and Cue COVID-19 (Cue Health Inc.). Of these, the Xpert Xpress and the ID NOW assays are widely used.

Table 24.6 shows the detail comparison of Xpress COVID, ID NOW and TrueNat POCT.

Table 24.6: Comparison of different POCT based on RNA detection								
		Target	Sample volume	Detection limit	Sensitivity	Specificity	Run time	Biosafety requirement
Xpress COVID (Cepheid)	Cartridge based, all steps integrated	E, N2	300 µL in VTM	250 copies/mL	99% (pooled)	97% (pooled)	45 minutes	Yes
ID NOW (Abbott)	Isothermal, all steps integrated	RdRp	Swab in specified liquid	125 ge/mL	93.4%	98.4%	15 minutes	Yes
TrueNat (MolBio, India)	Chip based. Three steps: Extraction, screening PCR, confirmation PCR	E, RdRp	Swab in lysis buffer	100 copies/µL	100%	100%	60 minutes	Not required

Antigen

Several antigen detection tests from various manufacturers have been approved and are commercially available. Most of these tests detect the nucleoprotein of the SARS-CoV-2. Two types of antigen detection tests are available: (i) Lateral flow based immunochromatographic test, and (ii) fluorescence immunoassay. The major advantages of antigen detection tests are: (i) Result available within 15–30 minutes, (ii) easy to perform as mostly based on lateral flow assay, (iii) require no instrument, (iv) can be done in community, and (v) less expensive as compared to RT-PCR test . Thus fulfils all the criteria to be used as a point of care test (POCT). The test can also be done at home by the patient himself.

Antigen detection test is performed on respiratory specimens, often nasal swab is the recommended sample. The sensitivity of rapid antigen test depends on mainly three parameters:

i. Symptom status: Sensitivity is higher in symptomatic as compared to asymptomatic.

ii. Duration of disease: Sensitivity is higher within 7 days of illness.

iii. Ct value of RT-PCR or viral load: Lower the Ct value higher is the sensitivity.

The overall sensitivity in symptomatic adults is reported as 67–92% and in symptomatic children 45 to 78% as reported in different studies.

The pooled sensitivity in symptomatic adults within 7 days of illness is 84% and after 7 days 62%.

Within 7 days of symptoms, the sensitivity is near ~100% with Ct value <20, sensitivity is ~95% when Ct value is <25 and ~75% when Ct value is <30.

In asymptomatic adults and children, the sensitivity is less than 45%.

The specificity in all studies irrespective of symptom, duration of illness and Ct value and age of patients is reported as 100%.

In clinical setting, rapid antigen test is mostly employed in symptomatic cases particularly in severe cases, emergency setting where rapid decision is required to isolate the patient. In such cases, if rapid antigen is negative, it should be confirmed by RT-PCR.

In a community setting for screening purpose where the pre-test probability is less, the test is not recommended. However, in high-risk contacts, the test may be done, but all negative tests should be repeated by RT-PCR before ruling out SARS-CoV-2 infection.

Antibody

Several antibody detection assays are available commercially for detection of SARS-CoV-2 specific IgM and IgG antibody against nucleoprotein (N) and spike (S) protein. Antibody detection against N protein is more commonly used than S protein as antibody to N protein is a more sensitive marker of infection. Whereas, antibody to S protein more commonly known as marker of protection.

Lateral flow assay, chemiluminescence and ELISA, all types of methods are available for detection of SARS-CoV-2 antibody detection.

IgM antibody has been observed to appear as early as 4th day post onset symptom. In various studies, IgM antibody appears around 5th day of symptom with a sensitivity about 20% which rises to near 60% during the second week of illness reaching the peak positivity. IgM persists thereafter for another 1–2 weeks (till 3rd and 4th week of symptom onset) then starts declining. The overall positivity is higher in symptomatic as compared to asymptomatic individuals (40% *vs* 25%).

IgG antibody appears mostly during second week of illness and persists for a few months.

Positivity of only IgM antibody indicates current infection, both IgM and IgG positivity indicates current or recent infection whereas only IgG positivity indicates past infection.

When considered either IgM or IgG antibody, the average positivity by end of 1st week is 45%, which increases to >80% by end of second week. From third week onwards, the positivity of IgG antibody rises to >90% by end of 4 weeks.

The sensitivity of antibody testing increases when both IgM and IgG antibodies of either N or S proteins are considered where the positivity is 75% during first week, 95% in second week and almost 100% in third week.

Clinical implication of antibody testing: IgM antibody is the marker of acute infection. In SARS-CoV-2, the sensitivity of IgM antibody is variable and mostly present during second week of illness. Due to the absence of IgM antibody at the time of symptom onset in majority cases, this is not a recommended test to diagnose acute phase infection. However, as the presence of either IgM and IgG acts as a marker of infection, in certain special situation antibody detection helps in diagnosis of suspected COVID-19 infection such as:

1. Suspected COVID-19 pneumonia where RT-PCR is repeatedly negative.
2. Multisystem inflammatory syndrome in children (MIS-C).
3. Inconclusive result of molecular test.
4. Late onset of post-infectious complications.
5. Community based seroepidemiological surveillance.

Role of Biomarkers in COVID-19

Several blood markers have been shown to be associated with severe COVID-19 disease. The level of various biomarkers reflect the patho-physiology of disease and help in assessing the disease severity and thereby play impor-tant role in appropriate patient management on right time.

Lymphocyte count: Lymphocyte plays an important role in immune regulation and inflammatory response to protect the body against viral infections. Several meta-analyses have shown the association of lymphopenia with poor disease outcome in hospitalized patients. The possible hypothesis is lympho-cyte expresses the receptor ACE-2 and thus attacked and killed by the virus SARS-CoV-2 or increase level of proinflammatory cytokines

leads to lymphocyte induced apoptosis. Decrease in lymphocyte count decreases the innate immune power of the host facilitating the progression of disease leading to poor outcome.

C-reactive protein (CRP): This is considered as the most sensitive and reliable biomarker in predicting the severity of COVID-19 disease. This is a non-specific acute phase reactant. It is induced by IL-6 in liver. Elevated CRP level has been observed as a unique feature in COVID-19 unlike other viral infections. The level of CRP is directly related to the level of inflammation and disease severity. It is considered as an early marker. The high early CRP level has been shown to be correlated with lung lesion and reflect the disease severity.

Lactate dehydrogenase (LDH): LDH is found in almost all living cells and catalyzes the interconversion of lactate and pyruvate. Tissue injury releases LDH to the bloodstream, thereby increases LDH level is an indicator of tissue damage. The level of LDH acts as a good early marker of lung injury. High LDH level has been associated with risk of development of severe disease and higher risk of mortality. Several studies also have proposed as prog-nostic marker of COVID-19.

D-dimer: It is a fibrin degradation product. It is a small protein fragment produced when a blood clot gets degraded by fibrinolysis. Increased inflammatory response and hypoxia due to severe pneumonia lead to activation of coagulation and fibrinolysis leading to elevated D-dimer level. High level of D-dimer has been associated with poor outcome in COVID-19 patients. This is the basis of prophylactic use of anticoagulant in moderate to severe hospitalized COVID-19 patients.

Procalcitonin (PCT): Procalcitonin is a pre-cursor of hormone calcitonin. It is produced by several types of cells in the body often in bacterial infections and also due to tissue injury. High level of PCT is one of the established biomarkers for systemic bacterial infection and sepsis. Several meta-analyses have shown association of high level of PCT or constantly rising level with bacterial infection and severe pneumonia or ARDS.

Several other markers like AST, ALT, CK and creatinine have also been associated with severe COVID-19 disease.

TREATMENT

Several therapeutic agents including antivirals, monoclonal antibodies, antiparasitic drugs, steroid and plasma therapy have been tried in COVID-19 patients during early part of COVID-19 pandemic without any scientific evidence based on their efficacy on other viral infections or similar clinical conditions. Over the span of one and half year of COVID-19 pandemic, based on the results of various clinical trials, currently only a few therapeutic agents are recommended for use in COVID-19 patients. Based on the line of management guided by WHO and CDC, Ministry of Health and Family Welfare (MOHFW), India has come up with a management guideline. The below mentioned management of COVID-19 is in line with MOHFW, India.

In asymptomatic patients on home isolation: No medication is recommended for this group of patients. Advised to take healthy balanced diet with adequate fluid.

In mild patients: Antipyretic, antitussive and inhalational budesonide are advised for symptomatic relief.

In patients with moderate and severe COVID-19, oxygen support to maintain SpO_2 92–95% and steroid, antiviral, immunomodulator and anticoagulant are to be given as indicated based on the results of relevant parameters.

Remdesivir: This is an adenosine analog of adenosine triphosphate, inhibits the viral RNA dependent RNA polymerase. It has been shown broad antiviral activity against RNA viruses including Ebola virus, SARS-CoV and middle east respiratory syndrome coronavirus (MERS-CoV). In non-human primate infected with MERS-CoV, it has shown prophylactic and therapeutic effects by decreasing the pulmonary lesions. The safety profile of remdesivir in patients who received for Ebola was favorable. Based on the limited scientific evidence globally, remdesivir presently approved for use in hospitalized patients in many countries including USA, European Union, India. It is to be used only in hospitalized patient with moderate to severe category who are on supplemental oxygen within 10 days of disease onset. It is given intravenously, 200 mg/day on day 1 followed by daily maintenance with 100 mg/day up to 10 days.

Tocilizumab: This is a humanized monoclonal antibody against IL-6 receptor. It has been approved for use in severe and critically ill COVID-19 patients who (i) show no signs of improvement in oxygen requirement after 24–48 hours of treatment, (ii) has significantly increased level of inflammatory markers with C-reactive protein ≥75 mg/L. As this is an immunosuppressive agent, it should be ensured that there is no secondary bacterial, fungal or tubercular infection in the patient. It is administered as a single dose of 8 mg/kg body weight in 100 mL normal saline over 1 hour.

Steroid: Steroids are not indicated in asymptomatic or patients with mild COVID-19 infection and it may be harmful in these patients as it may delay the clearance of virus and increase the viral load. It is indicated only in hospitalized patients when the oxygen saturation goes below 92%. Dexamethasone or methylprednisolone or hydrocortisone can be given initially for 10 days or till the time of discharge either orally or intravenous. Monitoring of blood glucose is mandatory to keep a check on hyperglycemia.

Anticoagulants: Prophylactic anticoagulant is indicated in moderate and severe COVID-19 patients with low molecular weight heparin or unfractionated heparin. Therapeutic dose is indicate, if evidence of thromboembolism is there.

Monoclonal antibodies: Before COVID-19, several monoclonal antibodies have been developed for many viral infections such as HIV, Ebola, respiratory syncytial virus, influenza, Zika virus, etc. Monoclonal antibody (mAb) against RSV is the only one which have FDA approval, whereas mAb against Ebola has shown promising results. Several monoclonal antibodies got the emergency use authorization of FDA for use in COVID-19 patients.

Bamlanivimab: Also known as LY-CoV555, or LY3819253 was originally derived from the blood of one of the first US patients who recovered from COVID-19. It is a recombinant neutralizing monoclonal antibody directed against the receptor binding domain (RBD) of SARS-CoV-2 spike protein.

Etesivumab: Also known as LY-CoV016 or JS016, or LY3832479 is a monoclonal antibody directed against the overlapping but different epitopes of RBD of SARS-CoV-2 spike protein.

Casirivimab (previously REGN10933) and *imdevimab* (previously REGN10987) are recombinant human monoclonal antibodies that bind to non-overlapping epitopes of the S protein RBD of SARS-CoV-2.

The combination products of bamlanivimab plus etesevimab and casirivimab plus imdevimab have got the FDA, EUA approval.

Recommendation of NIH (National Institute of Health) guideline as of July 2021: Use of bamlanivimab 700 mg plus etesevimab 1,400 mg OR casirivimab 1,200 mg plus imdevimab 1,200 mg are recommended for non-hospitalized COVID-19 patients with mild to moderate severity who are at high-risk of clinical progression to severe disease. Treatment should be started as soon as possible of symptom within 10 days of symptom onset.

NIH recommends against the use of monoclonal antibodies in patients who are hospitalized because of COVID-19, who are on oxygen therapy due to COVID-19 or non-COVID-19

underlying comorbidity or who require increase oxygen flow than baseline. However, the combination may be used for persons with mild to moderate COVID-19 who are hospitalized for a reason other than COVID-19 but who otherwise meet the EUA criteria.

SARS-CoV-2 variant having mutation E484K or L452R or K417N have shown decreased activity to bamlanivimab, combination of bamlanivimab plus etesevimab and also to casirivimab.

VACCINE

COVID-19 pandemic is the deadliest ever pandemic the world has ever faced in the modern era. Having no effective antiviral in place, vaccine is the only hope to get over the menace of SARS-CoV-2 virus. High mortality and emergence of variants made the entire world to put all its effort to prepare vaccine for this virus. For the first time in the history of medical science, vaccine was prepared in less than a year for any infectious disease. Currently, there are more than 20 vaccines for COVID-19 are already authorized by various national authorities and several more in pipeline (Table 24.7). Different vaccine technologies have been used for the development of COVID-19 vaccines such as nucleic acid (RNA or DNA vaccine), inactivated virus, live attenuated virus, non-replicating viral vector, recombinant protein, subunit or peptide, virus-like particle. Of these vaccine technologies, RNA vaccine came to use for human use for the first time for COVID-19 vaccine.

RNA vaccine: This works with the principle that the RNA segment containing the information to produce a specific antigen is inserted to plasmid. Plasmid containing this RNA is injected to host which is taken up by the host cell which then uses the RNA which then translated the protein or antigen. This in turn mounts the adaptive immune response and produces antibody.

The advantage of RNA vaccine is, it is safe as unlike DNA vaccine RNA does not possess the risk of integrating into the host cell genome. The production of RNA vaccine is simpler and rapid. However, the disadvantages are the risk of disintegration of RNA which is taken care by coformulation with lipid nanoparticle which protects the RNA strand and helps its absorption into host cell. Storage of RNA vaccine requires –70°C or –20°C depending on various modifications adapted to stabilize the RNA. Storage at this temperature is a major drawback for use of RNA vaccine in developing countries because of lack of availability of –20°C or –70°C deep freezer, stable electricity supply across the country.

Type of vaccine	Name	Country of origin	Dosage	Storage temp.	Pre-marketing study result
RNA vaccine	Moderna COVID vaccine	USA	2 doses 4 weeks interval	–20°C	Phase III (n = 30000), randomized, placebo controlled efficacy 94%
	Pfizer-BioNTech: RNA in lipid nanoparticle	German, USA	2 doses 3–4 weeks interval	–70°C	Phase III (n = 43998), randomized, placebo controlled efficacy 95%
Adeno-virus vector based vaccine	Oxford-AstraZeneca (Vaxzevria, Covishield): ChAdOx1	UK, Sweden	2 doses, 4–12 weeks interval	2–8°C	Phase III (n = 30000), randomized, placebo controlled efficacy: After 1st dose: 76%; After 2nd dose: 81%
	Sputnik V: Recombinant Ad5, Ad26	Russia	2 doses 3 weeks interval	≤18°C	Phase III (n = 40000), randomized, placebo controlled efficacy: 91.6%
	Janssen COVID-19: Recombinant Ad26	USA, Netherland (Janssen vaccine, Johnson and Johnson)	Single dose	2–8°C	Phase III (n = 40000), randomized, placebo controlled efficacy: Mild–moderate 62% Severe disease: 85% South Africa: 64% USA: 72% Persistence humoral and T cell response for 18 months
Inactivated COVID vaccine	Covaxin	India Bharat Biotech and ICMR	2 doses	2–8°C	Phase III, randomised, observer blinded, placebo controlled efficacy: 78%
	Coronavac (Sinovac)	China	2 doses 2–4 weeks interval	2–8°C	Phase III, double blind, randomised, placebo controlled efficacy: Turkey: 83.5% Chile: Against symptomatic: 65% Against hospitalization: 87% Against ICU admission: 90%

Table 24.7: Details of SARS-CoV-2 vaccines

Till date, two messenger RNA vaccine have got WHO approval for COVID-19. Pfizer-BioNTech COVID-19 vaccine and Moderna vaccine.

Recommendation for vaccination: The current recommendation for COVID-19 vaccination is for all citizens above 18 years age unless contraindicated for medical reasons.

Vaccination in pregnancy: For women who are pregnant, the choice of vaccination is to be offered with the condition that the pregnant women may be informed about the risks of exposure to COVID-19 infection along with the risks and benefits associated with the COVID-19 vaccines available in the country. Risk of exposure to COVID-19 disease in pregnancy usually leads to severe infection and may affect the fetus whereas so far there is no evidence of any associated risk of vaccination during pregnancy.

Vaccination in children: The severity and mortality due to COVID-19 is much less in children as compared to adults. However, children do suffer from the infection and play important role in virus transmission and also though less they may suffer from severe COVID-19 and mortality also has been reported. So, it goes undisputed that children need vaccine against COVID-19. Pfizer-BioNTech's mRNA BNT162b2 vaccine has been approved for 12–15 years. Vaccination in this age group is ongoing in many countries. Currently, trials are going on in children for several other vaccines in India; Covaxin, Bharat BioTech's nasal vaccine, Zydus Cadila's ZyCov-D and Novavax/Covavax.

The details of some of the widely studied vaccines are given in Table 24.7.

Other preventive measures: Mask, social distancing and use of sanitizer are the three most effective measures to prevent the acquisition of SARS-CoV-2 infection when followed properly.

Bibliography

1. Ning S, Yu B, Wang Y, Wang F. SARS-CoV-2: Origin, evolution, and targeting inhibition. Front Cell Infect Microbiol 2021 Jun 17;11: 676451.

2. Yang L, Liu S, Liu J, Zhang Z, Wan X, Huang B, Chen Y, Zhang Y. COVID-19: Immunopathogenesis and immunotherapeutics. Signal Transduct Target Ther 2020 Jul 25;5(1):128.

3. Quan C, Li C, Ma H, Li Y, Zhang H. Immunopathogenesis of coronavirus-induced acute respiratory distress syndrome (ARDS): Potential infection-associated hemophagocytic lymphohistiocytosis. Clin Microbiol Rev 2020 Oct 14; 34(1):e00074–20.

4. Fajgenbaum DC, June CH. Cytokine storm. N Engl J Med 2020 Dec 3;383(23):2255–73. doi: 10.1056/NEJMra2026131. PMID: 33264547; PMCID: PMC7727315.

5. Patel KP, Vunnam SR, Patel PA, et al. Transmission of SARS-CoV-2: An update of current literature. Eur J Clin Microbiol Infect Dis 2020 Nov;39(11):2005–11. doi: 10.1007/s10096-020-03961-1. Epub 2020 Jul 7.

6. Onakpoya IJ, Heneghan CJ, Spencer EA, Brassey J, Plüddemann A, Evans DH, Conly JM, Jefferson T. SARS-CoV-2 and the role of fomite transmission: A systematic review. F1000Res. 2021 Mar 24;10:233. doi: 10.12688/f1000 research. 51590.3. PMID: 34136133; PMCID: PMC8176 266.

7. Patel A, Agarwal R, Rudramurthy SM, et al; MucoCovi Network3. Multicenter epidemiologic study of coronavirus disease-associated mucormycosis, India. Emerg Infect Dis 2021 Jun 4;27(9). doi: 10.3201/eid2709.210934. Epub ahead of print.

8. Singh AK, Singh R, Joshi SR, Misra A. Mucormycosis in COVID-19: A systematic review of cases reported worldwide and in India. Diabetes Metab Syndr 2021 May 21;15(4):102146. doi: 10.1016/j.dsx.2021. 05.019. Epub ahead of print.

9. Vijay S, Bansal N, Rao BK, Veeraraghavan B, et al. Secondary infections in hospitalized COVID-19 patients: Indian experience. Infect Drug Resist 2021 May 24;14:1893–1903.

10. Koehler P, Bassetti M, Chakrabarti A, et al; European Confederation of Medical Mycology; International Society for Human Animal Mycology; Asia Fungal Working Group; INFOCUS

LATAM/ISHAM Working Group; ISHAM Pan Africa Mycology Working Group; European Society for Clinical Microbiology; Infectious Diseases Fungal Infection Study Group; ESCMID Study Group for Infections in Critically Ill Patients; Interregional Association of Clinical Microbiology and Antimicrobial Chemotherapy; Medical Mycology Society of Nigeria; Medical Mycology Society of China Medicine Education Association; Infectious Diseases Working Party of the German Society for Haematology and Medical Oncology; Association of Medical Microbiology; Infectious Disease Canada. Defining and managing COVID-19-associated pulmonary aspergillosis: the 2020 ECMM/ISHAM consensus criteria for research and clinical guidance. Lancet Infect Dis 2021 Jun;21(6):e149–62.

11. Parasher A. COVID-19: Current understanding of its Pathophysiology, Clinical presentation and Treatment. Postgrad Med J 2021 May; 97(1147): 312–20. doi: 10.1136/postgrad-medj-2020-138577. Epub 2020 Sep 25. PMID: 32978337.

12. Kordzadeh-Kermani E, Khalili H, Karimzadeh I. Pathogenesis, clinical manifestations and complications of coronavirus disease 2019 (COVID-19). Future Microbiol 2020 Sep;15: 1287–1305. doi: 10.2217/fmb-2020-0110. Epub 2020 Aug 27. PMID: 32851877; PMCID: PMC7493723.

13. Velikova TV, Kotsev SV, Georgiev DS, Batselova HM. Immunological aspects of COVID-19: What do we know? World J Biol Chem 2020 Sep 27;11(2):14–29.

14. Elhamzaoui H, Rebahi H, Hachimi A. Coronavirus disease 2019 (COVID-19) pathogenesis: A concise narrative review. Pan Afr Med J 2021 May 3;39:8.

15. Kashyap H, Kumar RNS, Gautam S, et al. Multisystem inflammatory syndrome in children (MIS-C) associated with COVID-19 infection. Indian J Pediatr 2021 Jul 16:1.

16. Bogunovic D, Merad M. Children and SARS-CoV-2. Cell Host Microbe 2021 Jul 14;29(7): 1040–42.

17. Dufort EM, Koumans EH, Chow EJ, et al. New York State and Centers for Disease Control and Prevention Multisystem Inflammatory Syndrome in Children Investigation Team. Multisystem Inflammatory Syndrome in Children in New York State. N Engl J Med 2020 Jul 23;383(4): 347–58.

18. Tracking SARS-CoV-2 variants (who.int)

19. SARS-CoV-2 variants of concern and variants under investigation in England. Technical briefing 15. 11 June 2021.

20. Plante JA, Liu Y, Liu J, et al. Spike mutation D614G alters SARS-CoV-2 fitness. Nature. 2021 Apr;592(7852):116–21. doi: 10.1038/s41586-020-2895-3. Epub 2020 Oct 26. Erratum in: Nature. 2021 Jul;595(7865):E1.

21. Sadoff J, Gray G, Vandebosch A, et al. ENSEMBLE Study Group. Safety and efficacy of single-dose Ad26.COV2.S vaccine against COVID-19. N Engl J Med 2021 Jun 10;384(23): 2187–2201.

22. Altmann DM, Boyton RJ, Beale R. Immunity to SARS-CoV-2 variants of concern. Science 2021 Mar 12;371(6534):1103–04.

23. Yadav PD, Sapkal GN, Abraham P, et al. Neutralization of variant under investigation B.1.617 with sera of BBV152 vaccinees. Clin Infect Dis 2021 May 7:ciab411.

24. Hodgson D, Flasche S, Jit M, Kucharski AJ; CMMID COVID-19 Working Group; Centre for Mathematical Modelling of Infectious Disease (CMMID) COVID-19 Working Group. The potential for vaccination-induced herd immunity against the SARS-CoV-2 B.1.1.7 variant. Euro Surveill 2021 May;26(20): 2100428.

25. Baric RS. Emergence of a Highly Fit SARS-CoV-2 Variant. N Engl J Med 2020 Dec 31;383(27): 2684–86.

26. Madhi SA, Baillie V, Cutland CL, Voysey M, et al. NGS-SA Group; Wits-VIDA COVID Group. Efficacy of the ChAdOx1 nCoV-19 COVID-19 Vaccine against the B.1.351 Variant. N Engl J Med 2021 May 20;384(20):1885–98.

27. Isabella Ferreira I, Rawlings Datir R, Kemp S et al. The Indian SARS-CoV-2 Genomics Consortium (INSACOG), The CITIID-NIHR BioResource. COVID-19 Collaboration, Sato K, James L, Agrawal A, Gupta RK. SARS-CoV-2 B.1.617 emergence and sensitivity to vaccine-elicited antibodies. bioRxiv preprint doi: https://doi.org/10.1101/2021.05.08.443253.

28. Jang S, Rhee JY, Wi YM, Jung BK. Viral kinetics of SARS-CoV-2 over the preclinical, clinical,

and postclinical period. Int J Infect Dis 2021 Jan;102:561–65.

29. Young BE, Ong SWX, Kalimuddin S, et al; Singapore 2019 Novel Coronavirus Outbreak Research Team. Epidemiologic features and clinical course of patients infected with SARS-CoV-2 in Singapore. JAMA 2020 Apr 21; 323(15):1488–94.

30. Chang D, Mo G, Yuan X, et al. Time kinetics of viral clearance and resolution of symptoms in novel coronavirus infection. Am J Respir Crit Care Med 2020 May 1;201(9):1150–52.

31. Engelmann I, Alidjinou EK, Ogiez J, et al. Preanalytical issues and cycle threshold values in SARS-CoV-2 real-time RT-PCR testing: Should test results include these? ACS Omega. 2021 Mar 6;6(10):6528–36.

32. Nalla AK, Casto AM, Huang MW et al. Comparative performance of SARS-CoV-2 detection assays using seven different primer-probe sets and one assay kit. J Clin Microbiol 2020 May 26;58(6):e00557–20.

33. LeBlanc JJ, Gubbay JB, Li Y, Needle R, et al. Real-time PCR-based SARS-CoV-2 detection in Canadian laboratories, J Clin Virol 128 (2020) 104433.

34. Vogels CBF, Brito AF, Wyllie AL, et al. Analytical sensitivity and efficiency comparisons of SARS-CoV-2 RT-qPCR primer-probe sets. Nat Microbiol 2020 Oct;5(10):1299–1305.

35. Basawarajappa SG, Rangaiah A, Padukone S, Yadav PD, Gupta N, Shankar SM. Performance evaluation of Truenat™ Beta CoV & Truenat™ SARS-CoV-2 point-of-care assays for coronavirus disease 2019. Indian J Med Res 2020 Nov 4.

36. Gupta N, Augustine S, Narayan T, et al. Point-of-Care PCR Assays for COVID-19 Detection. Biosensors (Basel). 2021 May 1;11(5):141.

37. Lee J, Song JU. Diagnostic accuracy of the Cepheid Xpert Xpress and the Abbott ID NOW assay for rapid detection of SARS-CoV-2: A systematic review and meta-analysis. J Med Virol 2021 Jul;93(7):4523–31.

38. Goldenberger D, Leuzinger K, Sogaard KK, et al. Brief validation of the novel GeneXpert Xpress SARS-CoV-2 PCR assay. J Virol Methods 2020 Oct;284:113925.

39. Xiang F, Wang X, He X, Peng Z, Yang B, Zhang J, Zhou Q, Ye H, Ma Y, Li H, Wei X, Cai P, Ma WL. Antibody detection and dynamic characteristics in patients with coronavirus disease 2019. Clin Infect Dis 2020 Nov 5;71(8):1930–34.

40. Boum Y, Fai KN, Nicolay B, et al. Performance and operational feasibility of antigen and antibody rapid diagnostic tests for COVID-19 in symptomatic and asymptomatic patients in Cameroon: A clinical, prospective, diagnostic accuracy study. Lancet Infect Dis 2021 Mar 25:S1473-3099(21)00132–8.

41. Monoclonal Antibodies (idsociety.org)

42. Anti-SARS-CoV-2 Monoclonal Antibodies| COVID-19 Treatment Guidelines (nih.gov)

Index

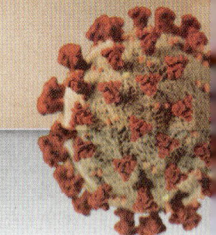